Advanced Structured Materials

Volume 194

Series Editors

Andreas Öchsner, Faculty of Mechanical Engineering, Esslingen University of Applied Sciences, Esslingen, Germany

Lucas F. M. da Silva, Department of Mechanical Engineering, Faculty of Engineering, University of Porto, Porto, Portugal

Holm Altenbach , Faculty of Mechanical Engineering, Otto von Guericke University Magdeburg, Magdeburg, Sachsen-Anhalt, Germany

Common engineering materials are reaching their limits in many applications, and new developments are required to meet the increasing demands on engineering materials. The performance of materials can be improved by combining different materials to achieve better properties than with a single constituent, or by shaping the material or constituents into a specific structure. The interaction between material and structure can occur at different length scales, such as the micro, meso, or macro scale, and offers potential applications in very different fields.

This book series addresses the fundamental relationships between materials and their structure on overall properties (e.g., mechanical, thermal, chemical, electrical, or magnetic properties, etc.). Experimental data and procedures are presented, as well as methods for modeling structures and materials using numerical and analytical approaches. In addition, the series shows how these materials engineering and design processes are implemented and how new technologies can be used to optimize materials and processes.

Advanced Structured Materials is indexed in Google Scholar and Scopus.

Holm Altenbach · Konstantin Naumenko
Editors

Creep in Structures VI

IUTAM Symposium Proceedings

Editors
Holm Altenbach
Chair of Engineering Mechanics, Institute of Mechanics
Otto von Guericke University Magdeburg
Magdeburg, Sachsen-Anhalt, Germany

Konstantin Naumenko
Chair of Engineering Mechanics, Institute of Mechanics
Otto von Guericke University Magdeburg
Magdeburg, Sachsen-Anhalt, Germany

ISSN 1869-8433　　　　　　　　ISSN 1869-8441　(electronic)
Advanced Structured Materials
ISBN 978-3-031-39069-2　　　　ISBN 978-3-031-39070-8　(eBook)
https://doi.org/10.1007/978-3-031-39070-8

© The Editor(s) (if applicable) and The Author(s), under exclusive license to Springer Nature Switzerland AG 2023

This work is subject to copyright. All rights are solely and exclusively licensed by the Publisher, whether the whole or part of the material is concerned, specifically the rights of translation, reprinting, reuse of illustrations, recitation, broadcasting, reproduction on microfilms or in any other physical way, and transmission or information storage and retrieval, electronic adaptation, computer software, or by similar or dissimilar methodology now known or hereafter developed.
The use of general descriptive names, registered names, trademarks, service marks, etc. in this publication does not imply, even in the absence of a specific statement, that such names are exempt from the relevant protective laws and regulations and therefore free for general use.
The publisher, the authors, and the editors are safe to assume that the advice and information in this book are believed to be true and accurate at the date of publication. Neither the publisher nor the authors or the editors give a warranty, expressed or implied, with respect to the material contained herein or for any errors or omissions that may have been made. The publisher remains neutral with regard to jurisdictional claims in published maps and institutional affiliations.

This Springer imprint is published by the registered company Springer Nature Switzerland AG
The registered company address is: Gewerbestrasse 11, 6330 Cham, Switzerland

Preface

Creep mechanics deals with theoretical and experimental approaches to the analysis of time-dependent changes of stress and strain states in engineering components up to the critical stage of rupture. Since 1960 there is a tradition to organize IUTAM symposium *Creep in Structures* every ten years:

1. 1960 - Stanford,
2. 1970 - Gothenburg,
3. 1980 - Leicester,
4. 1990 - Cracow, and
5. 2000 - Nagoya.

The IUTAM symposium Advanced Materials Modelling for Structures, held in Paris in 2012, was continuation and a new version of *Creep in Structures* with the focus on new materials and on generalized and unified models of inelastic deformation.

During the last decade, many advances and new results in the field of Creep Mechanics were established. Examples include: interlinks of mechanics with materials science in multi-scale analysis of deformation and damage mechanisms over a wide range of stresses and temperature; development and analysis of new alloys for (ultra)high-temperature applications; formulation and calibration of advanced constitutive models of inelastic behavior under transient loading and temperature conditions; development of efficient procedures and machine learning techniques for identification of material parameters in advanced constitutive laws; introduction of gradient-enhanced and non-local theories to account for damage and fracture processes; and application of new experimental methods, such as digital image correlation, for the analysis of inelastic deformation under multi-axial stress state.

To discuss these recent developments the well-established IUTAM series is continued. This volume of the *Advanced Structured Materials* Series contains a collection of contributions on advanced approaches of Creep Mechanics. They are presented in the IUTAM Symposium *Creep in Structures VI* in Magdeburg, Germany, 18-22 September 2023.

We would like to acknowledge the series editor Professor Andreas Öchsner for giving us the opportunity to publish this volume. We would like to acknowledge

Dr. Christoph Baumann from Springer Publisher for the assistance and support during the preparing the book.

Magdeburg,
September 2023,

Holm Altenbach,
Konstantin Naumenko

Contents

1 **Phase-Field Damage Modeling in Generalized Mechanics by using a Mixed Finite Element Method (FEM)** 1
  Bilen Emek Abali
  1.1 Introduction to Standard Phase-Field Formulation 1
  1.2 Extension to Generalized Mechanics 4
  1.3 Strain Gradient Parameters 6
  1.4 Numerical Implementation and Results 7
  1.5 Conclusion... 10
  References 11

2 **Creep-Damage Processes in Cyclic Loaded Double Walled Structures** 19
  Holm Altenbach, Dmytro Breslavsky, and Oksana Tatarinova
  2.1 Introduction 19
  2.2 Constitutive Equations 21
    2.2.1 Static Loading 22
    2.2.2 Cyclic Loading. Stresses Lower the Yield Limit 22
    2.2.3 Cyclic Load. Overloading with Transition to Plastic Deformation 26
  2.3 Problem Statement 33
  2.4 Comparison Between Data of Direct Approach and Use of Averaged Function K 37
  2.5 Numerical Simulation of the Cyclic Creep-damage in DWTC System Model 38
    2.5.1 Description of the Calculation Model 39
    2.5.2 Determination of the Temperature and Stress Field in the Blade 40
    2.5.3 Creep Calculations for a Two-dimensional Model of a Blade Made of Nickel Based Alloy 41
    2.5.4 Creep Calculations for a Two-dimensional Model of a Blade Made of an Inconel X Alloy 47
  2.6 Conclusions.. 52

	References		54
3	**Creep Mechanics – Some Historical Remarks and New Trends**		**57**
	Holm Altenbach, Johanna Eisenträger, Katharina Knape, and Konstantin Naumenko		
	3.1	Starting Point - the Early Period of Creep Mechanics	57
	3.2	IUTAM Symposia and Other Events Devoted to Problems in Creep Mechanics	58
	3.3	Research Directions and Magdeburg's Contributions	59
		3.3.1 Kachanov-Rabotnov Approach and Mechanism-Based Models	59
		3.3.2 Non-Classical Creep	60
		3.3.3 Benchmark Tests for Creep Problems	60
		3.3.4 Rheological Models	61
		3.3.5 Thesis	61
	3.4	Outlook	62
	References		62
4	**Various State-of-the-Art Methods for Creep Evaluation of Power Plant Components in a Wide Load and Temperature Range**		**69**
	Eike Blum, Yevgen Kostenko, and Konstantin Naumenko		
	4.1	Introduction	70
	4.2	Applied Creep Models	71
		4.2.1 Norton-Bailey Equation	71
		4.2.2 Modified Garofalo Eequation	71
		4.2.3 Constitutive Model	72
	4.3	Structural Analysis	74
		4.3.1 Verification of the Creep Models Based on Creep Tests	74
		4.3.2 Relaxation Test with Cube-one-Element Model	75
		4.3.3 Pipe Benchmark FE Model	75
		4.3.4 Performance Evaluation of User-Creep Routines	77
		4.3.5 Temperature Interpolation for Norton-Bailey Creep Equation	78
		4.3.6 Isothermal Steam Turbine Valve FE Model with a Constant Loading	79
	4.4	Conclusions	79
	References		81
5	**Creep and Irradiation Effects in Reactor Vessel Internals**		**83**
	Dmytro Breslavsky and Oksana Tatarinova		
	5.1	Introduction	83
	5.2	Problem Statement and Description of Solution Approaches	86
	5.3	Constitutive Equations	88
		5.3.1 Materials with Isotropy of Properties	88
		5.3.2 Materials with Transversal Isotropy of Properties	89
	5.4	Deformation, Damage Accumulation and Fracture in RVI	91

	5.4.1	Creep of T-joint of Tubes	91
	5.4.2	Damage Accumulation and Fracture of Reactor Fuel Element	92
	5.4.3	Transversal-Isotropic Creep-Damage Behaviour of Aluminium Notched Plate	95
5.5	Conclusions		100
References			101

6 Analysis of Damage and Fracture in Anisotropic Sheet Metals Based on Biaxial Experiments 105
Michael Brünig, Sanjeev Koirala, and Steffen Gerke
6.1	Introduction	105
6.2	Constitutive Framework	106
6.3	Numerical Simulations and Results	109
6.4	Conclusions	113
References		113

7 Effect of Physical Aging on the Flexural Creep in 3D Printed Thermoplastic 115
Marcel Fischbach and Kerstin Weinberg
7.1	Introduction	115
7.2	Theoretical Background	117
	7.2.1 Viscoelasticity of Thermoplastics	117
	7.2.2 Physical Aging	119
7.3	Material and Methods	121
	7.3.1 Test Specimens	121
	7.3.2 Sequential Creep Tests	122
	7.3.3 Long Term Creep Test	124
7.4	Experimental Results	124
	7.4.1 Sequential Creep Tests	124
	7.4.2 Long Term Creep Test	127
7.5	Discussion	127
References		129

8 Development of a Microstructure-Based Finite Element Model of Thermomechanical Response of a Fully Metallic Composite Phase Change Material 131
Elisabetta Gariboldi, Matteo Molteni, Diego André Vargas Vargas, and Konstantin Naumenko
8.1	Introduction	132
8.2	Microstructure-Based FE model of a Al-Sn C-PCM with Free Expansion	133
8.3	Results and Discussion	136
8.4	Final Remarks	139
References		140

9	**The Effect of Dynamic Loads on the Creep of Geomaterials** 143	

Andrei M. Golosov, Evgenii P. Riabokon, Mikhail S. Turbakov, Evgenii V. Kozhevnikov, Vladimir V. Poplygin, Mikhail A. Guzev, and Hongwen Jing

- 9.1 Introduction 143
- 9.2 Materials and Methods 144
  - 9.2.1 Materials 144
  - 9.2.2 Methods 145
- 9.3 Results and Discussion 147
- 9.4 Conclusion 149
- References 149

10 A Novel Simulation Method for Phase Transition of Single Crystal Ni based Superalloys in Elevated Temperature Creep Regions via Discrete Cosine Transform and Maximum Entropy Method 151

Hideo Hiraguchi

- 10.1 Introduction 151
- 10.2 Materials and Experiments 152
  - 10.2.1 A Single Crystal Ni Based Superalloy, CMSX-4 152
  - 10.2.2 Creep Tests 152
  - 10.2.3 Two Dimensional Discrete Cosine Transform 154
  - 10.2.4 Maximum Entropy Method 155
- 10.3 Estimation of Phase Transition and Results 156
- 10.4 Discussion 156
- 10.5 Conclusion 156
- References 159

11 Anisotropic Creep Analysis of Fiber Reinforced Load Point Support Structures for Thermoplastic Sandwich Panels 161

Jörg Hohe and Sascha Fliegener

- 11.1 Introduction 161
- 11.2 Material Model 162
  - 11.2.1 Basic One-Dimensional Formulation 163
  - 11.2.2 Generalization to Three Dimensions 164
  - 11.2.3 Unidirectionally Fiber Reinforced Thermoplastics 165
  - 11.2.4 Discontinuously Fiber Reinforced Thermoplastics 165
- 11.3 Experimental Investigation 166
  - 11.3.1 Coupon Experiments 166
  - 11.3.2 Structural Experiments 167
- 11.4 Multiscale Simulation 168
- 11.5 Results 168
  - 11.5.1 Parameter Identification on Coupon Experiments 168
  - 11.5.2 Validation on Structural Level 170
- 11.6 Summary and Conclusion 172
- References 173

12 Time-Swelling Superposition Principle for the Linear Viscoelastic Properties of Polyacrylamide Hydrogels 175
Seishiro Matsubara, Akira Takashima, So Nagashima, Shohei Ida, Hiro Tanaka, Makoto Uchida, and Dai Okumura
- 12.1 Introduction 176
- 12.2 Experiment 177
  - 12.2.1 Materials 177
  - 12.2.2 Mixed Solvents for Transient Equilibrium Swelling 178
  - 12.2.3 Measurement of Swelling 179
  - 12.2.4 Measurement of Dynamic Moduli 180
- 12.3 Experimenta Results 180
  - 12.3.1 Transient Equilibrium Swelling 180
  - 12.3.2 Linear Viscoelastic Behavior 182
- 12.4 Swelling–Dependent Linear Viscoelasticity 186
  - 12.4.1 Model Formulation 187
  - 12.4.2 Time-Swelling Superpostion Principle 190
- 12.5 Discussion 193
  - 12.5.1 Master Curves of Dynamic Moduli 193
  - 12.5.2 Swelling Dependence of Linear Viscoelastic Properties 194
  - 12.5.3 Frequency Dependence of Complex Shear Moduli 195
- 12.6 Conclusion 198
- Appendix A: Validity for Transient Equilibrium Swelling Using Ethanol 200
- Appendix B: Experimental Data 202
- References 202

13 Application of Nonlinear Viscoelastic Material Models for the Shrinkage and Warpage Analysis of Blow Molded Parts 205
Patrick Michels, Christian Dresbach, Esther Ramakers-van Dorp, Holm Altenbach, and Olaf Bruch
- 13.1 Introduction 206
- 13.2 Material Models 208
  - 13.2.1 Linear Viscoelastic Material Model 208
  - 13.2.2 Abaqus Parallel Rheological Framework Model 210
- 13.3 Shrinkage and Warpage Analysis 211
- 13.4 Calibration Strategy 215
  - 13.4.1 Experimental Data 216
  - 13.4.2 Implementation of a One-Dimensional Model to Reduce Computation Time 217
  - 13.4.3 Reduction of Material Parameters 220
  - 13.4.4 Calibration Workflow 223
- 13.5 Results 224
- 13.6 Discussion and Outlook 227
- References 230

14 Modeling Solid Materials in DEM Using the Micropolar Theory.... 233
Przemysław Nosal and Artur Ganczarski
- 14.1 Introduction 233
- 14.2 Formulation of the Thermo-Elasto-Viscoplastic Contact Model ... 234
  - 14.2.1 Short introduction to DEM basics 234
  - 14.2.2 DEM Interaction Force Model Based on Micropolar Theory 236
  - 14.2.3 Visco-Plasticity 240
- 14.3 Model Verification 242
  - 14.3.1 Simulation Set-up 243
  - 14.3.2 Results and Discussion 244
- 14.4 Concluding Remarks 246
- References 247

15 The Development of a Cavitation-Based Model for Creep Lifetime Prediction Using Cu-40Zn-2Pb Material 249
Mbombo Amejima Okpa, Qiang Xu, and Zhongyu Lu
- 15.1 Introduction 250
- 15.2 Stress Breakdown and Creep Lifetime 250
- 15.3 Creep Cavitation and Cavitation Data Concerns 252
  - 15.3.1 How to Use Cavitation Data 253
  - 15.3.2 Current Approach to Cavitation Modelling and Creep Life Prediction 253
- 15.4 Aims 254
- 15.5 Experimental Data and Method 255
  - 15.5.1 Experimental Data 255
  - 15.5.2 Method 255
  - 15.5.3 Determination of Cavitation Constants 256
  - 15.5.4 Cavity Size Distribution Modelling 256
- 15.6 Conclusions and Future Work 260
- References 261

16 Self-heating Analysis with Respect to Holding Times of an Additive Manufactured Aluminium Alloy 265
Lukas Richter, Holger Sparr, Daniela Schob, Philipp Maasch, Robert Roszak, and Matthias Ziegenhorn
- 16.1 Introduction 265
- 16.2 Thermomechanical Experiment 267
  - 16.2.1 Experimental Set-Up 267
  - 16.2.2 Temperature and Deformationfield Measurement by Digital Image Correlation 270
  - 16.2.3 Experimental Results with Respect to Holding Time 271
- 16.3 Theoretical Framework 273
  - 16.3.1 Energy Balance and Heat Conduction 273
  - 16.3.2 Material Model 277
- 16.4 Modelling Methods 279

| | 16.4.1 | Parameter Identification .... 279 |
| | 16.4.2 | Concept for Thermomechanic FE Analysis .... 281 |

- 16.5 Results 281
- 16.6 Discussion 284
- 16.7 Conclusion and Outlook...................................... 284
- References 286

17 Creep Under High Temperature Thermal Cycling and Low Mechanical Loadings 289
Romana Schwing, Stefan Linn, Christian Kontermann, and Matthias Oechsner

- 17.1 Introduction 289
- 17.2 Experimental Methods 290
  - 17.2.1 Materials 290
  - 17.2.2 Creep Test Equipment................................ 291
- 17.3 Observation of Accelerated Creep Under Anisothermal Testing Conditions 293
  - 17.3.1 Anisothermal Creep Tests............................ 295
  - 17.3.2 Influencing Factors 296
  - 17.3.3 Observations on the Creep Behavior Within a Cycle..... 300
- 17.4 Discussion 302
  - 17.4.1 Possible causes of accelerated creep under thermal cycling 302
  - 17.4.2 Microstructural Processes Under Anisothermal Creep Testing 305
- 17.5 Summary 307
- References 307

18 The Development and Application of Optimisation Technique for the Calibrating of Creep Cavitation Model Based on Cavity Histogram . 309
Qiang Xu, Bilal Rafiq, Xuming Zheng, and Zhongyu Lu

- 18.1 Introduction 309
- 18.2 Background Theories and Knowledge 311
  - 18.2.1 Cavitation Model Theory 311
  - 18.2.2 Current Calibration Methods 311
- 18.3 Optimisation with Excel Solver 313
- 18.4 Cavitation Data 313
- 18.5 Results 313
- 18.6 Discussion and Conclusion 317
- References 318

19 A Temperature-Dependent Viscoelastic Approach to the Constitutive Behavior of Semi-Crystalline Thermoplastics at Finite Deformations 321
Le Zhang, Bo Yin, Robert Fleischhauer, and Michael Kaliske

- 19.1 Introduction 321
- 19.2 Preliminaries of the Finite Thermo-Viscoelasticity 323
- 19.3 Constitutive Formulation of Finite Thermo-Viscoelasticity 325

		19.3.1	Helmholtz Energy 325
		19.3.2	Creep Law 327
		19.3.3	Governing Equations............................. 328
	19.4	Numerical Study 328
		19.4.1	Stress-Controlled loading 329
		19.4.2	Strain-Controlled loading 332
	19.5	Summary 333
References 333

List of Contributors

Bilen Emek Abali
Uppsala University, Division of Applied Mechanics, Department of Materials Science and Engineering, Uppsala, Sweden, e-mail: bilenemek@abali.org

Holm Altenbach
Lehrstuhl für Technische Mechanik, Institut für Mechanik, Otto-von-Guericke-Universität Magdeburg, D-39106 Magdeburg, Germany,
e-mail: holm.altenbach@ovgu.de

Eike Blum
Siemens Energy Global GmbH & Co. KG, Müheim/Ruhr, Germany,
e-mail: eike-marcel.blum.ext@siemens-energy.com

Dmytro Breslavsky
Department of Computer Modelling of Processes and Systems, National Technical University "Kharkiv Polytechnic Institute", UKR-61002, Kharkiv, Ukraine,
e-mail: dmytro.breslavsky@khpi.edu.ua

Olaf Bruch
Dr. Reinold Hagen Stiftung, Kautexstr. 53, 53229 Bonn & Bonn-Rhein-Sieg University of Applied Sciences, Grantham Allee 20, 53757 Sankt Augustin, Germany, e-mail: o.bruch@hagen-stiftung.de, olaf.bruch@h-brs.de

Michael Brünig
Institut für Mechanik und Statik, Universität der Bundeswehr München, 85577 Neubiberg, Germany, e-mail: michael.bruenig@unibw.de

Christian Dresbach
Bonn-Rhein-Sieg University of Applied Sciences, Von-Liebig-Str. 20, 53359 Rheinbach, Germany, e-mail: christian.dresbach@h-brs.de

Johanna Eisenträger
Lehrstuhl für Technische Mechanik, Institut für Mechanik, Fakultät für Maschinenbau, Otto-von-Guericke-Universität Magdeburg, Universitätsplatz 2, 39106 Magdeburg, Germany, e-mail: johanna.eisentraeger@ovgu.de

Marcel Fischbach
Chair of Solid Mechanics, University of Siegen, Paul-Bonatz-Straße 9-11, 57076 Siegen, Germany, e-mail: marcel.fischbach@uni-siegen.de

Robert Fleischhauer
Institute for Structural Analysis, Technische Universität Dresden, 01062 Dresden, Germany, e-mail: robert.fleischhauer@tu-dresden.de

Sascha Fliegener
Fraunhofer-Institut für Werkstoffmechanik IWM, Wöhlerstr. 11, 79108 Freiburg, Germany, e-mail: sascha.fliegener@iwm.fraunhofer.de

Artur Ganczarski
Cracow University of Technology, Jana Pawła II Av. 37, 31-864 Cracow, Poland, e-mail: artur.ganczarski@pk.edu.pl

Elisabetta Gariboldi
Politecnico di Milano, Dipartimento di Meccanica, Via La Masa 1, 20156 Milano, Italy, e-mail: elisabetta.gariboldi@polimi.it

Steffen Gerke
Institut für Mechanik und Statik, Universität der Bundeswehr München, 85577 Neubiberg, Germany, e-mail: steffen.gerke@unibw.de

Andrei M. Golosov
Perm National Research Polytechnic University, Perm, Russian Federation
e-mail: a-dune@mail.ru

Mikhail A. Guzev
Perm National Research Polytechnic University, Perm & Institute for Applied Mathematics of the Far Eastern Branch of the Russian Academy of Sciences, Vladivostok, Russian Federation, e-mail: guzev@iam.dvo.ru

Hideo Hiraguchi
The Institution of Professional Engineers, Japan (IPEJ), 3-5-8, Shibakoen, Minato-ku, Tokyo, Japan, e-mail: hideoh@abox2.so-net.ne.jp

Jörg Hohe
Fraunhofer-Institut für Werkstoffmechanik IWM, Wöhlerstr. 11, 79108 Freiburg, Germany, e-mail: joerg.hohe@iwm.fraunhofer.de

Shohei Ida
Faculty of Engineering, The University of Shiga Prefecture, 2500, Hassaka-cho, Hikone-City, Shiga, 522-8533, Japan, e-mail: ida.s@mat.usp.ac.jp

List of Contributors

Hongwen Jing
State Key Laboratory for Geomechanics and Deep Underground Engineering,
China University of Mining and Technology, Xuzhou, P.R. China
e-mail: hwjing@cumt.edu.cn

Michael Kaliske
Institute for Structural Analysis, Technische Universität Dresden, 01062 Dresden,
Germany, e-mail: michael.kaliske@tu-dresden.de

Katharina Knape
Lehrstuhl für Technische Mechanik, Institut für Mechanik, Fakultät für Maschinenbau,
Otto-von-Guericke-Universität Magdeburg, Universitätsplatz 2, 39106 Magdeburg,
Germany, e-mail: katharina.knape@ovgu.de

Sanjeev Koirala
Institut für Mechanik und Statik, Universität der Bundeswehr München, 85577
Neubiberg, Germany, e-mail: sanjeev.koirala@unibw.de

Christian Kontermann
Institut für Werkstoffkunde TU Darmstadt, Grafenstraße 2, 64283 Darmstadt,
Germany, e-mail: christian.kontermann@tu-darmstadt.de

Yevgen Kostenko
Siemens Energy Global GmbH & Co. KG, Mülheim/Ruhr, Germany,
e-mail: yevgen.kostenko@siemens-energy.com

Evgenii V. Kozhevnikov
Perm National Research Polytechnic University, Perm, Russian Federation,
e-mail: kozhevnikov_evg@mail.ru

Stefan Linn
Institut für Werkstoffkunde TU Darmstadt, Grafenstraße 2, 64283 Darmstadt,
Germany, e-mail: stefan.linn@tu-darmstadt.de

Zhongyu Lu
Department of Technology and Engineering, School of Computing and Engineering,
University of Huddersfield, Huddersfield, HD1 3DH, UK, e-mail: j.lu@hud.ac.uk

Philipp Maasch
Brandenburg University of Technology Cottbus-Senftenberg, Universitätsplatz 1,
01968 Senftenberg, Germany, e-mail: philipp.maasch@b-tu.de

Seishiro Matsubara
Department of Mechanical Systems Engineering, Nagoya University, Furo-cho,
Chikusa-ku, Nagoya 464-8603, Japan,
e-mail: seishiro.matsubara@mae.nagoya-u.ac.jp

Patrick Michels
Bonn-Rhein-Sieg University of Applied Sciences, Grantham Allee 20, 53757 Sankt
Augustin, Germany, e-mail: patrick.michels@h-brs.de

Matteo Molteni
Politecnico di Milano, Dipartimento di Meccanica, Via La Masa 1, 20156 Milano, Italy, e-mail: matteo1.molteni@polimi.it

So Nagashima
Department of Mechanical Systems Engineering, Nagoya University, Furo-cho, Chikusa-ku, Nagoya 464-8603, Japan,
e-mail: so.nagashima@mae.nagoya-u.ac.jp

Konstantin Naumenko
Lehrstuhl für Technische Mechanik, Institut für Mechanik, Fakultät für Maschinenbau, Otto-von-Guericke-Universität Magdeburg, Universitätsplatz 2, 39106 Magdeburg, Germany, e-mail: konstantin.naumenko@ovgu.de

Przemysław Nosal
AGH University of Science and Technology, Mickiewicza Av. 30, 30-059 Cracow, Poland , e-mail: pnosal@agh.edu.pl

Matthias Oechsner
Institut für Werkstoffkunde TU Darmstadt, Grafenstraße 2, 64283 Darmstadt, Germany, e-mail: matthias.oechsner@tu-darmstadt.de

Mbombo Amejima Okpa
Department of Technology and Engineering, School of Computing and Engineering, University of Huddersfield, Huddersfield, HD1 3DH, UK,
e-mail: Mbombo.okpa@hud.ac.uk

Dai Okumura
Department of Mechanical Systems Engineering, Nagoya University, Furo-cho, Chikusa-ku, Nagoya 464-8603, Japan,
e-mail: dai.okumura@mae.nagoya-u.ac.jp

Vladimir V. Poplygin
Perm National Research Polytechnic University, Perm, Russian Federation,
e-mail: poplygin@bk.ru

Bilal Rafiq
Department of Technology and Engineering, School of Computing and Engineering, University of Huddersfield, Huddersfield, HD1 3DH, UK,
e-mail: U2181632@unimail.hud.ac.uk

Esther Ramakers-van Dorp
Bonn-Rhein-Sieg University of Applied Sciences, Von-Liebig-Str. 20, 53359 Rheinbach, Germany, e-mail: esther.vandorp@h-brs.de

Evgenii P. Riabokon
Perm National Research Polytechnic University, Perm, Russian Federation,
e-mail: riabokon.evgenii@gmail.com

List of Contributors

Lukas Richter
Brandenburg University of Technology Cottbus-Senftenberg, Universitätsplatz 1,
01968 Senftenberg, Germany, e-mail: lukas.richter@b-tu.de

Robert Roszak
Brandenburg University of Technology Cottbus-Senftenberg, Universitätsplatz 1,
01968 Senftenberg, Germany, e-mail: robert.roszak@b-tu.de

Daniela Schob
Brandenburg University of Technology Cottbus-Senftenberg, Universitätsplatz 1,
01968 Senftenberg, Germany, e-mail: daniela.schob@b-tu.de

Romana Schwing
Institut für Werkstoffkunde TU Darmstadt, Grafenstraße 2, 64283 Darmstadt,
Germany, e-mail: romana.schwing@tu-darmstadt.de

Holger Sparr
Brandenburg University of Technology Cottbus-Senftenberg, Universitätsplatz 1,
01968 Senftenberg, Germany, e-mail: holger.sparr@b-tu.de

Akira Takashima
Department of Mechanical Systems Engineering, Nagoya University, Furo-cho,
Chikusa-ku, Nagoya 464-8603, Japan,
e-mail: takashima.akira@j.mbox.nagoya-u.ac.jp

Hiro Tanaka
Department of Mechanical Engineering, Osaka University, 2-1 Yamadaoka, Suita,
Osaka, 565-0871, Japan, e-mail: htanaka@mech.eng.osaka-u.ac.jp

Oksana Tatarinova
Department of Computer Modelling of Processes and Systems, National Technical
University "Kharkiv Polytechnic Institute", UKR-61002, Kharkiv, Ukraine,
e-mail: ok.tatarinova@gmail.com

Mikhail S. Turbakov
Perm National Research Polytechnic University, Perm, Russian Federation
e-mail: msturbakov@gmail.com

Makoto Uchida
Department of Mechanical Engineering, Osaka Metropolitan University, 3-3-138,
Sugimoto, Sumiyoshi-ku, Osaka, 558-8585, Japan,
e-mail: uchida@osaka-cu.ac.jp

Diego André Vargas Vargas
Politecnico di Milano, Dipartimento di Meccanica, Via La Masa 1, 20156 Milano,
Italy, e-mail: diegoandre.vargas@mail.polimi.it

Kerstin Weinberg
Chair of Solid Mechanics, University of Siegen, Paul-Bonatz-Straße 9-11, 57076
Siegen, Germany, e-mail: kerstin.weinberg@uni-siegen.de

Qiang Xu
Department of Technology and Engineering, School of Computing and Engineering, University of Huddersfield, Huddersfield, HD1 3DH, UK, e-mail: Q.Xu2@hud.ac.uk

Bo Yin
Ansys Germany GmbH, 99423 Weimar, Germany, e-mail: liam.yin@ansys.com

Le Zhang
Institute for Structural Analysis, Technische Universität Dresden, 01062 Dresden, Germany, e-mail: le.zhang@mailbox.tu-dresden.de

Xuming Zheng
Department of Technology and Engineering, School of Computing and Engineering, University of Huddersfield, Huddersfield, HD1 3DH, UK,
e-mail: Xuming.Zheng@hud.ac.uk

Matthias Ziegenhorn
Brandenburg University of Technology Cottbus-Senftenberg, Universitätsplatz 1, 01968 Senftenberg, Germany, e-mail: matthias.ziegenhorn@b-tu.de

Chapter 1
Phase-Field Damage Modeling in Generalized Mechanics by using a Mixed Finite Element Method (FEM)

Bilen Emek Abali

Abstract Material modeling is applied for bulk materials where the length-scale of the geometry is adequately larger than any voids within the material. Indeed, material is composed of a lattice in alloys or chains in polymers, but this structural dependency is negligible since these are multiple order smaller than the geometric dimensions. By using an additive manufacturing, we create so-called metamaterials or architectured materials, where at the same length-scale, a microscale is introduced. The materials response is then predicted accurately by means of the generalized mechanics that uses higher gradients in its formulation. In the case of damage mechanics, this generalization is still lacking. We demonstrate a possible approach for filling this gap because the generalized damage mechanics achieves additional regularization by means of adding higher gradients to the model. Phase-field approach is employed for the damage variable implementation by using a mixed formulation in the Finite Element Method (FEM) in order to solve strain gradient elasticity model with higher gradients in damage formulation.

1.1 Introduction to Standard Phase-Field Formulation

Damage mechanics is developed as a continuum approach of a collection of contiguous particles, where a fracture is a discontinuity between them. The theory is well-known and visualized as existing microcracks and their agglomeration for forming a fracture. Therefore, an additional governing equation is needed that is modeling microcrack evolution at the nanometer length-scale (microscale) in order to estimate a formation of a fracture at the micrometer or millimeter length-scale (macroscale). Hence, for

---

Bilen Emek Abali
Uppsala University, Division of Applied Mechanics, Department of Materials Science and Engineering, Uppsala, Sweden
e-mail: bilenemek@abali.org

damage evolution, we have a multiscale problem modeled by a scalar field, a so-called damage variable [1–4].

The damage variable increases monotonously for engineering materials such as metal alloys or plastics in an irreversible fashion, we may introduce an initial 1 increasing to infinity [5, 6]. For a numerical treatise, from 1 until ∞ may be challenging such that often its inverse is used from 1 to 0, or equivalently, one minus this variable is introduced as ω for modeling $\omega = 0$ as an initial "virgin" material without fracture and $\omega = 1$ denotes fracture. All numbers between $\omega \in [0, 1]$ is allowed and this order parameter is the solution of a differential equation. Indeed, we may understand a particle as a composition of virgin material (one phase) and fracture (another phase), where ω is a phase-field function indicating a volumetric percentage of the fracture phase. In this way, actually, ω connects two length-scales without any mathematical difficulties [7]. This phase-field approach is widely applied as a diffusive fracture modeling in fracture mechanics in staggered [8–11] and in monolithic approaches [12–15].

Instead of a smeared crack formulation, one may try to introduce a loss of bond between molecules. This vanishing stiffness between neighboring particles leads to a jump in the displacement. Balance on singular surfaces is then used to bring the phase-field modeling in consistence with configurational forces [16, 17]. Both viewpoints result in the same parameter, either called a damage or phase-field variable, although historically developed by using different visions [18]. The governing equation for the phase-field may be suggested by using a thermodynamical formulation [19, 20].

The parameters in the formulation are yet lacking to have clearly defined experiments. Formulations are proposed as benchmarks [21, 22] in order to allow suggestions to be verified. Some difficulties in the damage mechanics involve rate dependency [23] and finite speed [24] in crack propagation, as well as the tip zone and localization of stress [25, 26]. Often, these multiscale analyses are justified as a size effect [27–29]. One common way of modeling this size effect [30, 31] relies on strain gradient elasticity or so-called generalized mechanics [32, 33], since additional forces on the fracture tip play a significant role in the damage evolution [34].

The phase-field is mathematically an order parameter. The mechanism is based on a potential [35] developed in [36] as a minimization problem. In [37], it is derived by a Γ-convergence analysis. Technically, it models microcracks as a bump function that is 1 within a finite domain (compact support) and 0 otherwise. This assumption is backed by our physical understanding of a fracture and mathematically a differential equation is formally proposed in form of an evolution equation. For this manner, a sharp crack is regularized by a crack surface density, $\gamma = \gamma(\omega, \omega_{,i})$, where a comma notation denotes a (partial) spatial derivative. We obtain in a continuum body, $\mathcal{B}$, an integral relation with an accumulated (dissipative) energy density, $^{\mathrm{d}}\psi$, and so-called critical energy release rate per surface, G_c, as follows:

$$\int_{\mathcal{B}} {}^{\mathrm{d}}\psi \, \mathrm{d}V = \int_{\mathcal{B}} G_c \gamma \, \mathrm{d}V . \qquad (1.1)$$

1 Phase-Field Damage Modeling in Generalized Mechanics

By using a given shape of the smeared crack like an exponential, it is possible [38] to approximate the type of γ, as follows:

$$\gamma(\omega, \omega_{,i}) = \frac{1}{c_\alpha \ell}\left(\alpha + \ell^2 \omega_{,i}\omega_{,i}\right), \tag{1.2}$$

with a normalized constant $c_\alpha = 4\int_0^1 \alpha^{0.5}\,d\omega$ depending on the choice of crack geometry function $\alpha = \alpha(\omega) \in [0,1]$. Herein we use index notation in Cartesian coordinates and apply EINSTEIN's summation convention over repeated indices. Now by constructing a variational formulation

$$\int_\mathcal{B} {}^d\psi\, dV = R,$$
$$\delta \int_\mathcal{B} {}^d\psi\, dV = \int_\mathcal{B} R\delta\omega\, dV, \tag{1.3}$$

with

$$R = \frac{\partial \mathcal{L}}{\partial \omega} \tag{1.4}$$

and a LAGRANGE function

$$\mathcal{L} = -\mathfrak{f} + \rho g_i u_i, \quad \mathfrak{f} = g\,{}^e\psi, \tag{1.5}$$

in statics given by a free energy density, $\mathfrak{f}$, and a potential energy, where $|g_i| = 9.81\,\text{N/kg}$ is the specific body force due to the gravitational forces. The so-called degradation function, $g = g(\omega)$, and stored energy density, ${}^e\psi$, needs to be chosen according to a model. Let us consider the case of linear elasticity, ${}^e\psi = {}^e\psi(\varepsilon)$, where the strain is given by displacement gradient, for example the simplest measure reads $\varepsilon_{ij} = (u_{i,j} + u_{j,i})/2$. Therefore, we obtain the damage evolution by

$$\delta \int_\mathcal{B} {}^d\psi\, dV = \int_\mathcal{B} \frac{\partial \mathcal{L}}{\partial \omega}\delta\omega\, dV,$$
$$\int_\mathcal{B} \left(\frac{\partial\,{}^d\psi}{\partial \omega}\delta\omega + \frac{\partial\,{}^d\psi}{\partial \omega_{,i}}\delta\omega_{,i}\right) dV = -\int_\mathcal{B} \frac{\partial g}{\partial \omega}\,{}^e\psi\,\delta\omega\, dV. \tag{1.6}$$

One usual choice for the degradation function is a quadratic one, $g = (1-\omega)^2$, and $\alpha = \omega$ as in [39] leading to $c_\alpha = 8/3$ such that we obtain

$${}^d\psi = \frac{3G_c}{8\ell}\left(\omega + \ell^2 \omega_{,i}\omega_{,i}\right). \tag{1.7}$$

Hence the damage is evolving by

$$\int_{\mathcal{B}} \left(\frac{3G_c}{8\ell} \delta\omega + \frac{3G_c\ell}{8} 2\omega_{,i} \delta\omega_{,i} \right) dV = -\int_{\mathcal{B}} \frac{\partial g}{\partial \omega} {}^e\psi \delta\omega \, dV \,,$$
$$\int_{\mathcal{B}} \left(\frac{3G_c}{8\ell} \delta\omega + \frac{3G_c\ell}{4} \omega_{,i} \delta\omega_{,i} + \frac{\partial g}{\partial \omega} {}^e\psi^* \delta\omega \right) dV = 0 \,, \quad (1.8)$$

where ${}^e\psi^* = \max({}^e\psi)$ is used algorithmically in order to ensure a monotonously increasing damage variable.

1.2 Extension to Generalized Mechanics

We aim for a formulation of phase-field with higher gradients in displacement by using a variational approach following [40, 41]. Also for ductile materials [42] and in general [43–45], higher gradient formulation introduces additional regularization adequate for numerical procedure. We refer to [46] for parameter sensitivity, to [47, 48] for including plasticity, to [49, 50] for involving anisotropy, to [51] for viscoelastoplasticity, to [52] for crack branching, to [53] for thermal damage, and to [54] for multiphysics applications.

One possible approach is to extend the crack surface density in Eq. (1.2) and add higher order terms in the damage variable, $\gamma(\omega, \omega_{,i}, \omega_{,ij})$. This approach is valid yet unclear how the interaction with displacement needs to follow. Therefore, we use a variational approach by following [55] and begin with an assertion that the LAGRANGEan depends on first and second time and space derivatives of primitive variables that are displacement and damage variable. By using then an action:

$$\mathcal{A} = \int_{\Omega} \mathcal{L} \, dt \, dV + \int_{\partial\Omega} W_s \, dt \, dA + \int_{\partial\partial\Omega} W_e \, dt \, d\ell \,, \quad (1.9)$$

where the energy is prescribed (given) on surface (first-order) and edges (second-order) of the computational domain Ω. When we model the application with the following LAGRANGEan density (per volume):

$$\mathcal{L} = \frac{1}{2} \rho_0 \dot{u}_i \dot{u}_i - \mathfrak{f} + \rho_0 g_i u_i - c_1 \omega \dot{\xi} - m\xi - c_2 \omega_{,i} \xi_{,i} - c_3 \omega_{,ij} \xi_{,ij} \,, \quad (1.10)$$

with one term for damage, m, and for displacement, the so-called free energy density, $\mathfrak{f}$, yet to be defined. Kinetic energy density, $\rho_0 \dot{u}_i \dot{u}_i/2$, and potential energy density, $\rho_0 g_i u_i$, are known terms for obtaining the balance of momentum. Use of a quadratic kinetic energy introduces second rate of displacement called inertial terms. We use an auxiliary variable, ξ, in order to obtain first rate of damage variable in the formulation. For simplicity, we use $W_s = \hat{t}_i u_i$ and $W_e = 0$. In this way, we enforce a zero damage gradient on the boundary such that the fracture propagates orthogonal to the outer boundary. By using dependencies,

$$m = m(\omega, u_{i,j}, u_{i,jk}) \,, \quad \mathfrak{f} = \mathfrak{f}(\omega, u_{i,j}, u_{i,jk}) \,, \quad (1.11)$$

after the variational formulation, $\delta\mathcal{A} = 0$, called the principle of least action, we obtain for variation of displacement and auxiliary variable two weak forms:

$$F_1 = \int_{\mathcal{B}} \left(\rho_0 g_i \delta u_i - \rho_0 \ddot{u}_i \delta u_i - \frac{\partial \mathfrak{f}}{\partial u_{i,j}} \delta u_{i,j} - \frac{\partial \mathfrak{f}}{\partial u_{i,jk}} \delta u_{i,jk} \right) \mathrm{d}V + \int_{\partial\mathcal{B}} \hat{t}_i \delta u_i \, \mathrm{d}A \,,$$

$$F_2 = \int_{\mathcal{B}} \left(-c_1 \dot\omega \delta\xi - m \delta\xi - c_2 \omega_{,i} \delta\xi_{,i} - c_3 \omega_{,ij} \delta\xi_{,ij} \right) \mathrm{d}V \,. \tag{1.12}$$

This formulation is general under the assumption that the LAGRANGEan density is correct. The first weak form, F_1, is the accustomed formulation for strain gradient theory in generalized mechanics, if we use an elastic or a stored energy density:

$${}^e\psi = \frac{1}{2} \varepsilon_{ij} C_{ijkl} \varepsilon_{kl} + \frac{1}{2} \varepsilon_{ij,k} D_{ijklmn} \varepsilon_{lm,n} + \varepsilon_{ij} G_{ijklm} \varepsilon_{kl,m} \,, \tag{1.13}$$

with the linearized strain, $\varepsilon_{ij} = (u_{i,j} + u_{j,i})/2$. Rank 4,5,6 tensors, $\boldsymbol{C}, \boldsymbol{G}, \boldsymbol{D}$, are material parameters for the bulk material modeled by the strain gradient (linear) material model. For the coupling to the damage variable, we utilize the degradation function, g, and the free energy density is analogous to the standard formulation given by $\mathfrak{f} = g\, {}^e\psi$. For seeing the relation to the standard formulation, consider $c_1 = c_3 = 0$ and insert

$$m = \frac{3G_c}{8\ell} + \frac{\partial g}{\partial \omega} {}^e\psi^* \,, \quad c_2 = \frac{3G_c \ell}{4} \,, \tag{1.14}$$

in the weak form, F_2. Now the generalization may be easily established by proposing

$$c_1 = \frac{1}{M} \,, \quad c_3 = \ell^2 c_2 \,, \tag{1.15}$$

in order to acquire adequate units. Indeed, M is called a mobility parameter used in the literature for controlling the crack propagation speed [56]. In this way, we have a possible generalized damage mechanics model by solving the following weak forms:

$$F_1 = \int_{\mathcal{B}} \Big(\rho_0 g_i \delta u_i - \rho_0 \ddot{u}_i \delta u_i - g\left(C_{ijkl} E_{kl} + G_{ijklm} E_{kl,m}\right) \delta u_{i,j}$$

$$- g\left(D_{ijklmn} E_{lm,n} + E_{lm} G_{lmijk}\right) \delta u_{i,jk} \Big) \mathrm{d}V + \int_{\partial\mathcal{B}} \hat{t}_i \delta u_i \, \mathrm{d}A \,,$$

$$F_2 = \int_{\mathcal{B}} \Big(-\frac{1}{M} \dot\omega \delta\xi - \Big(\frac{3G_c}{8\ell} + \frac{\partial g}{\partial \omega} {}^e\psi^* \Big) \delta\xi - \frac{3G_c \ell}{4} \omega_{,i} \delta\xi_{,i}$$

$$- \frac{3G_c \ell^3}{4} \omega_{,ij} \delta\xi_{,ij} \Big) \mathrm{d}V \,. \tag{1.16}$$

1.3 Strain Gradient Parameters

A microstructure delivers higher order parameters by means of a homogenization method [57]. Ample homogenization techniques are available in the literature, for the case of strain gradient elasticity, we refer to [58–63]. We follow the asymptotic analysis [64–67] that is applied in [68–71]. As a homogenization method, this approach is utilized in one-dimensional problems for composites [72, 73] and in two-dimensional continuum [74–78] by using numerical solutions. We employ a VOIGT-like notation by introducing A, B denoting $\{11, 22, 12\}$ and α, β indicating $\{111, 221, 121, 112, 222, 122\}$ for writing out all material parameters,

$$C_{AB} = \begin{pmatrix} C_{1111} & C_{1122} & C_{1112} \\ C_{2211} & C_{2222} & C_{2212} \\ C_{1211} & C_{1222} & C_{1212} \end{pmatrix},$$

$$G_{A\alpha} = \begin{pmatrix} G_{11111} & G_{11221} & G_{11121} & G_{11112} & G_{11222} & G_{11122} \\ G_{22111} & G_{22221} & G_{22121} & G_{22112} & G_{22222} & G_{22122} \\ G_{12111} & G_{12221} & G_{12121} & G_{12112} & G_{12222} & G_{12122} \end{pmatrix},$$

$$D_{\alpha\beta} = \begin{pmatrix} D_{111111} & D_{111221} & D_{111121} & D_{111112} & D_{111222} & D_{111122} \\ D_{221111} & D_{221221} & D_{221121} & D_{221112} & D_{221222} & D_{221122} \\ D_{121111} & D_{121221} & D_{121121} & D_{121112} & D_{121222} & D_{121122} \\ D_{112111} & D_{112221} & D_{112121} & D_{112112} & D_{112222} & D_{112122} \\ D_{222111} & D_{222221} & D_{222121} & D_{222112} & D_{222222} & D_{222122} \\ D_{122111} & D_{122221} & D_{122121} & D_{122112} & D_{122222} & D_{122122} \end{pmatrix}.$$

(1.17)

All parameters are calculated by using an equivalence of elastic energy at the microstructure length-scale (microscale) modeled with PLA and void compared with the elastic energy at the homogenized strain gradient continuum (macroscale). Since the microscale is a quadratic energy—only incorporating strains and positive material parameters—the energy is positive such that the combination of all parameters must be positive in determined parameters. The positive-definiteness in strain gradient parameters is of importance for a unique solution [79–81].

We strictly follow the method in [82–85] and use the parameters obtained in [86] by using a microstructure of 2.5 mm thick walls grid infill structure used in 3-D printing. As material we use PLA of YOUNG's modulus 3500 MPa and POISSON's ratio of 0.3 where an RVE is generated by 5.5×5.5 mm rectangle with 3×3 mm void rectangle generating 70% of infill ratio. Rank 4,5,6 tensors read

$$C_{AB} = \begin{pmatrix} 1987 & 475 & 0 \\ 475 & 1987 & 0 \\ 0 & 0 & 281 \end{pmatrix} \text{MPa},$$

$$G_{A\alpha} = \begin{pmatrix} 0 & 0 & 0 & 0 & 0 & 0 \\ 0 & 0 & 0 & 0 & 0 & 0 \\ 0 & 0 & 0 & 0 & 0 & 0 \end{pmatrix} \text{N/mm},$$

$$D_{\alpha\beta} = \begin{pmatrix} 469 & 634 & 1 & 3 & 1 & -75 \\ 634 & 4078 & 0 & 1 & -2 & 1473 \\ 1 & 0 & 1142 & 1473 & -75 & 0 \\ 3 & 1 & 1473 & 4079 & 636 & 1 \\ 1 & -2 & -75 & 636 & 469 & -1 \\ -75 & 1473 & 0 & 0 & 0 & 1143 \end{pmatrix} \text{N}.$$

(1.18)

For damage parameters, we use a realistic estimation of material parameters for the base PLA material as given in Table 1.1.

Table 1.1: Damage properties for the strain gradient model for a 2-D geometry under the assumption of a thin plate (plane stress).

Parameters	Notation	Value
Fracture toughness	K	$1\,\text{MPa}\sqrt{\text{m}}$
Modulus	E	$10\,\text{GPa}$
Critical energy release rate	G_c	$K^2/E \times 10^3\,\text{MPa}\,\text{mm}$
length-scale	ℓ	4 times the smallest discrete element length
Mobility parameter	M	$2000\,1/(\text{MPa}\,\text{s})$
Mass density	ρ	$5000 \times 10^{-12}\,\text{ton/mm}^3$

1.4 Numerical Implementation and Results

In space and time, we use discrete representations of the functions. For the time discretization, we employ the finite difference method also called EULER backward scheme for the damage variable as well as displacement, where the upper index $(\cdot)^0$ denotes the computed value one time step before and $(\cdot)^{00}$ the value two time steps before. Hence, we obtain

$$\dot{\omega} = \frac{\omega - \omega^0}{\Delta t}, \quad \ddot{u} = \frac{u - 2u^0 + u^{00}}{\Delta t^2}, \quad (1.19)$$

with a time step, Δt, which we choose constant throughout the simulation for the sake of a simpler algorithm. For the space discretization, we utilize the finite element method where the functions are represented as their nodal values and an interpolation between these nodal values. Dividing the computational domain into nodes is called triangulation and meshing that is done in Salome by using Netgen algorithm.

For discrete representation, we follow the HU–WASHIZU principle [87] and use a mixed space formulation such that the unknowns $\omega, \boldsymbol{u}$ are augmented by their derivatives, $\nabla\omega, \nabla\boldsymbol{u}$. In this way, necessary regularity is ensured since the gradient is modeled as an additional variable. Indeed, an additional constraint is added by a penalty method. For accuracy of this method in strain gradient problems, we refer to [88]. We use a discretization using LAGRANGE elements and generate piecewise continuous polynomials that are adequate for approximation in $\mathcal{H}^1$. This triangulation is denoted $\mathcal{T}$ and consists of non-overlapping triangles, τ. We use linear elements, $\mathcal{P}_1$, with a polynomial degree 1 in the case of the phase-field (damage variable) and quadratic elements, $\mathcal{P}_2$, with a polynomial degree of 2 for the displacement. For their gradients, we use one degree less. As is common in the GALERKIN approach, we use the same space for trial and test functions. Unknowns are displacement and damage variable, and their derivatives, constructing a mixed space in 2-D discrete representation of the continuum

$$\begin{aligned}\mathcal{V} = \Big\{ &\{u_i\} \in \left[\mathcal{H}^1(\Omega)\right]^2 : \{u_i\}\big|_\tau \in \mathcal{P}_2(\tau) \ \forall \tau \in \mathcal{T} \\
&\wedge \{\nabla u_{ij}\} \in \left[\mathcal{H}^1(\Omega)\right]^4 : \{u_i\}\big|_\tau \in \mathcal{P}_1(\tau) \ \forall \tau \in \mathcal{T} \\
&\wedge \{\omega\} \in \left[\mathcal{H}^1(\Omega)\right]^1 : \{\omega\}\big|_\tau \in \mathcal{P}_1(\tau) \ \forall \tau \in \mathcal{T} \\
&\wedge \{\nabla\omega_i\} \in \left[\mathcal{H}^1(\Omega)\right]^2 : \{\nabla\omega_i\}\big|_\tau \in \mathcal{P}_0(\tau) \ \forall \tau \in \mathcal{T} \Big\}. \end{aligned} \quad (1.20)$$

By adding constraints

$$\Lambda_1 = \int_\mathcal{B} \lambda^1_{ij} \big(\nabla u_{ij} - u_{i,j}\big) \, \mathrm{d}V \,, \quad \Lambda_2 = \int_\mathcal{B} \lambda^2_i \big(\nabla\omega_i - \omega_{,i}\big) \, \mathrm{d}V \,, \quad (1.21)$$

with constant multipliers, λ^1, λ^2, we acquire their variations

$$\delta\Lambda_1 = \int_\mathcal{B} \lambda^1_{ij} \big(\delta\nabla u_{ij} - \delta u_{i,j}\big) \, \mathrm{d}V \,, \quad \delta\Lambda_2 = \int_\mathcal{B} \lambda^2_i \big(\delta\nabla\omega_i - \delta\omega_{,i}\big) \, \mathrm{d}V \,. \quad (1.22)$$

By an analysis at their units, according to the HU–WASHIZU principle, we conclude to use

$$\lambda^1_{ij} = \frac{\partial f}{\partial u_{i,j}} \,, \quad \lambda^2_i = \frac{\partial {}^\mathrm{d}\psi}{\partial \omega_{,i}} \,. \quad (1.23)$$

For the numerical implementation of such a weak form, we use the open-source package collection SyFi developed under the FEniCS project [89, 90] by following the computational framework as in [91]. FEniCS offers assembly and solution by means of the finite element method with the chosen element type. It supports symbolic differentiation, which is exploited herein for "simply" implementing the weak form. All code is developed in Python.

As a material, we use the strain gradient material model with the elastic parameters as in Eq. (1.18) with the damage related parameters as in Table 1.1. Specifically the length-scale ℓ is chosen for 4 times the (minimum) element length in order to allow a smooth transition. Depending on the mesh the phase-field varies. However, the use of higher gradients ensures for a mesh independent displacement formulation. For showing this result, we have constructed an asymmetric double notch tensile test simulation as drawn in Fig. 1.1. The geometry is 20 mm × 5 mm with 2 notches of 0.1 mm thick and 1 mm deep.

We use a staggered scheme as similar to other works [92, 93]. The staggered method is based on solving two smaller problems, and hence, it is normally faster than a monolithic method based on solving a bigger problem. The reason is that the computation time is growing exponentially in greater problems [94, 95]. But the accuracy in a staggered scheme is less than a monolithic method, therefore, we apply many iterations in each time step until we obtain an error small enough. This approach is often used yet a convergence is not guaranteed. Especially in this formulation, there are numerical benefits because of the additional regularization owing to the higher gradients. Until the initiation of the fracture in Fig. 1.2, less than 2 iterations suffice but then we have limited to 10 iterations for the sake of computational time. A parallel computation (via MPI) has been performed by a computing node using Intel Xeon E7-4850, in total 64 cores each with the 40 MB cache, equipped with 256

Fig. 1.1: Double asymmetric notched tensile testing simulation, left boundary is clamped and right boundary is under uniaxial tensile loading by a given displacement, $\hat{u}$.

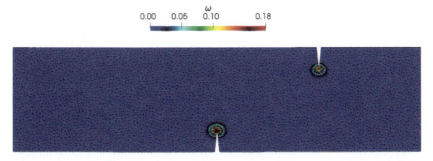

Fig. 1.2: Damage variable, ω, at $t = 0.64$ s, where the damage initializes around the notches.

GB memory in total, running Linux Kernel 5 Ubuntu 20.04. In this setting, the total computation takes around 3 hours for 50 time steps of 0.02 s.

For the implementation of the boundaries, for the displacement problem we use DIRICHLET boundary conditions for u and zero NEUMANN boundary condition for ∇u. For the damage problem, we use zero NEUMANN boundary condition for ω and $\nabla \omega$. In order to counterattack numerical problems for the purely NEUMANN boundaries, we use a line search algorithm based semi-smooth solver for variational inequalities based on NEWTON's method [96] from PETSc packages. Linearized problem is solved by GMRES iterative solver. For the displacement problem, we use a NEWTON–RAPHSON solver from PETSc packages by using a BICGSTAB iterative solver for the linearized problem. In both cases, we have utilized the same Incomplete Lower Upper (ILU) decomposition based preconditioner [97]. The problem solution method is highly scalable and we used all 64 cores during the simulation.

Additionally, we have implemented an updated LAGRANGEan technique with a remeshing strategy by using VEDO module [98]. An updated LAGRANGE is simply adding the displacement to the nodal positions in order to acquire the current configuration (one time step before) during the calculation. Especially in large deformations, this implementation allows to use the linearized strain measure with an adequate accuracy. Yet the element quality may decrease by displacing the nodes such that we use a remeshing if the aspect ratio in one of the elements is decreased less than 0.2. In this manner, the implementation may be used for any deformation without numerical problems and with high accuracy.

Displacement is controlled on both ends such that the crack propagation is stable. We demonstrate the final time step where the crack is developed on two notches and a displacement jump is visible in Fig. 1.3. We emphasize that the displacement field is a continuous function, yet the phase-field leads to a degradation resulting a sharp change in displacement. One remarkable result is that the phase-field is considered a diffusive approach such that the thickness of this distribution is far from a fracture interpretation. This issue is caused by the choice of ℓ being four times the element size. As a consequence, the mesh is chosen very fine to obtain a sharp change in displacement. Herein, we use a higher gradients enriched formulation in displacement and damage problem. Therefore, even if the mesh is coarse, the sought-after sharp displacement change is achieved.

1.5 Conclusion

We have developed a phase-field approach by using strain gradient elasticity theory in a consistent manner. The consistency is because of using the same order of the terms and therefore an adequate regularization in the numerical approach. Since the computational cost is high, we aim for a scalable approach that we have acquired owing to this regularization. In general, higher order models are numerically unstable and there are different special elements or alternative formulations. We have used standard elements in the FEM and performed a mixed mode fracture simulation

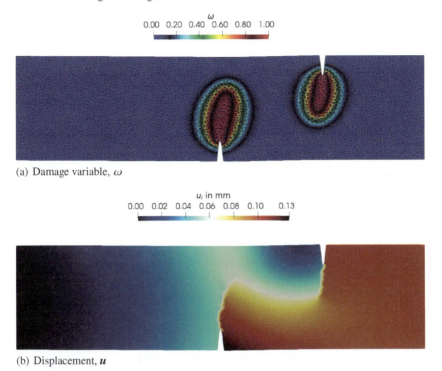

Fig. 1.3: Damage variable (a) and displacement (b) where the structural integrity is lost at $t = 1$ s, damage variable is distributed over 4 finite elements leading to a relatively thick phase-field band; however, displacement demonstrates the sharp contour as expected from a realistic model.

by using a material with a microstructure. The microstructure is chosen and all parameters are determined by an asymptotic homogenization approach. We interpret the methodology being successful since the determined parameters circumvents any numerical difficulty. For a transparent exchange, the numerical implementation is employed by open-source packages and publicly available. The Python code is using the FEniCS project available at http://www.fenicsproject.org/download. The computational implementation is available in [99] to be used under the GNU Public license [100].

References

[1] Murakami S (2012) Continuum Damage Mechanics: a Continuum Mechanics Approach to the Analysis of Damage and Fracture, vol 185. Springer Science & Business Media

2. Altenbach H, Sadowski T (eds) (2015) Failure and Damage Analysis of Advanced Materials, CISM International Centre for Mechanical Sciences, vol 560. Springer, Vienna
3. Altenbach H, Kolupaev VA (2015) Classical and non-classical failure criteria. In: Altenbach H, Sadowski T (eds) Failure and Damage Analysis of Advanced Materials, CISM International Centre for Mechanical Sciences, vol 560, Springer, pp 1–66
4. Öchsner A (2016) Continuum damage mechanics. In: Continuum Damage and Fracture Mechanics, Springer, pp 65–84
5. Kachanov L (1986) Introduction to Continuum Damage Mechanics, Mechanics of Elastic Stability, vol 10. Springer Science & Business Media
6. Lemaitre J, Desmorat R (2005) Engineering Damage Mechanics: Ductile, Creep, Fatigue and Brittle Failures. Springer Science & Business Media
7. Provatas N, Elder K (2011) Phase-Field Methods in Materials Science and Engineering. John Wiley & Sons
8. Miehe C, Welschinger F, Hofacker M (2010) Thermodynamically consistent phase-field models of fracture: Variational principles and multi-field FE implementations. International Journal for Numerical Methods in Engineering **83**(10):1273–1311
9. Kiendl J, Ambati M, De Lorenzis L, Gomez H, Reali A (2016) Phase-field description of brittle fracture in plates and shells. Computer Methods in Applied Mechanics and Engineering **312**:374–394
10. Kästner M, Hennig P, Linse T, Ulbricht V (2016) Phase-field modelling of damage and fracture—convergence and local mesh refinement. In: Naumenko K, Aßmus M (eds) Advanced Methods of Continuum Mechanics for Materials and Structures, Advanced Structured Materials, vol 60, Springer, pp 307–324
11. Teichtmeister S, Kienle D, Aldakheel F, Keip MA (2017) Phase field modeling of fracture in anisotropic brittle solids. International Journal of Non-Linear Mechanics **97**:1–21
12. Clayton J, Knap J (2016) Phase field modeling and simulation of coupled fracture and twinning in single crystals and polycrystals. Computer Methods in Applied Mechanics and Engineering **312**:447–467
13. Levitas VI (2018) Phase field approach for stress-and temperature-induced phase transformations that satisfies lattice instability conditions. Part I. General theory. International Journal of Plasticity **106**:164–185
14. Babaei H, Levitas VI (2018) Phase-field approach for stress-and temperature-induced phase transformations that satisfies lattice instability conditions. Part 2. Simulations of phase transformations Si I ↔ Si II. International Journal of Plasticity **107**:223–245
15. Amirian B, Jafarzadeh H, Abali BE, Reali A, Hogan JD (2022) Phase-field approach to evolution and interaction of twins in single crystal magnesium. Computational Mechanics **70**(4):803–818
16. Kuhn C, Müller R (2016) A discussion of fracture mechanisms in heterogeneous materials by means of configurational forces in a phase field fracture model. Computer Methods in Applied Mechanics and Engineering **312**:95–116

[17] Bilgen C, Weinberg K (2019) On the crack-driving force of phase-field models in linearized and finite elasticity. Computer Methods in Applied Mechanics and Engineering **353**:348–372

[18] Schreiber C, Kuhn C, Müller R, Zohdi T (2020) A phase field modeling approach of cyclic fatigue crack growth. International Journal of Fracture **225**(1):89–100

[19] Wolff M, Böhm M, Altenbach H (2018) Application of the Müller–Liu entropy principle to gradient-damage models in the thermo-elastic case. International Journal of Damage Mechanics **27**(3):387–408

[20] Amirian B, Jafarzadeh H, Abali BE, Reali A, Hogan JD (2022) Thermodynamically-consistent derivation and computation of twinning and fracture in brittle materials by means of phase-field approaches in the finite element method. International Journal of Solids and Structures **252**:111,789

[21] Ambati M, Gerasimov T, De Lorenzis L (2015) A review on phase-field models of brittle fracture and a new fast hybrid formulation. Computational Mechanics **55**(2):383–405

[22] Schröder J, Wick T, Reese S, Wriggers P, Müller R, Kollmannsberger S, Kästner M, Schwarz A, Igelbüscher M, Viebahn N, Bayat HR, Wulfinghoff S, Mang K, Rank E, Bog T, D'Angella D, Elhaddad M, Hennig P, Düster A, Garhuom W, Hubrich S, Walloth M, Wollner W, Kuhn C, Heister T (2021) A selection of benchmark problems in solid mechanics and applied mathematics. Archives of Computational Methods in Engineering **28**:713–751

[23] Yin B, Kaliske M (2020) Fracture simulation of viscoelastic polymers by the phase-field method. Computational Mechanics **65**:293–309

[24] Kamensky D, Moutsanidis G, Bazilevs Y (2018) Hyperbolic phase field modeling of brittle fracture: Part i—theory and simulations. Journal of the Mechanics and Physics of Solids **121**:81–98

[25] Forest S, Lorentz E (2004) Localization phenomena and regularization methods. In: Besson J (ed) Local Approach to Fracture, Presses de l'Ecole des Mines Paris, pp 311–371

[26] Carlsson J, Isaksson P (2019) Crack dynamics and crack tip shielding in a material containing pores analysed by a phase field method. Engineering Fracture Mechanics **206**:526–540

[27] Peerlings RH, de Borst R, Brekelmans WM, De Vree J (1996) Gradient enhanced damage for quasi-brittle materials. International Journal for numerical methods in engineering **39**(19):3391–3403

[28] Frémond M, Nedjar B (1996) Damage, gradient of damage and principle of virtual power. International Journal of Solids and Structures **33**(8):1083–1103

[29] Bažant ZP (2000) Size effect. International Journal of Solids and Structures **37**(1-2):69–80

[30] Zreid I, Kaliske M (2014) Regularization of microplane damage models using an implicit gradient enhancement. International Journal of Solids and Structures **51**(19-20):3480–3489

[31] Placidi L, Barchiesi E (2018) Energy approach to brittle fracture in strain-gradient modelling. Proceedings of the Royal Society A: Mathematical, Physical and Engineering Sciences **474**(2210):20170,878

[32] Placidi L, Barchiesi E, Misra A (2018) A strain gradient variational approach to damage: a comparison with damage gradient models and numerical results. Mathematics and Mechanics of Complex Systems **6**(2):77–100

[33] Mousavi S, Paavola J (2014) Analysis of plate in second strain gradient elasticity. Archive of Applied Mechanics **84**(8):1135–1143

[34] dell'Isola F, Seppecher P (1997) Edge contact forces and quasi-balanced power. Meccanica **32**:33–52

[35] Mumford DB, Shah J (1989) Optimal approximations by piecewise smooth functions and associated variational problems. Communications on pure and applied mathematics **42**(5):577–685

[36] Francfort GA, Marigo JJ (1998) Revisiting brittle fracture as an energy minimization problem. Journal of the Mechanics and Physics of Solids **46**(8):1319–1342

[37] Ambrosio L, Tortorelli VM (1990) Approximation of functional depending on jumps by elliptic functional via t-convergence. Communications on Pure and Applied Mathematics **43**(8):999–1036

[38] Wu JY (2017) A unified phase-field theory for the mechanics of damage and quasi-brittle failure. Journal of the Mechanics and Physics of Solids **103**:72–99

[39] Pham K, Amor H, Marigo JJ, Maurini C (2011) Gradient damage models and their use to approximate brittle fracture. International Journal of Damage Mechanics **20**(4):618–652

[40] Neff P, Ghiba ID, Madeo A, Placidi L, Rosi G (2014) A unifying perspective: the relaxed linear micromorphic continuum. Continuum Mechanics and Thermodynamics **26**(5):639–681

[41] Abali BE, Müller WH (2016) Numerical solution of generalized mechanics based on a variational formulation. Oberwolfach reports - Mechanics of Materials, European Mathematical Society Publishing House **17**(1):9–12

[42] Reiher JC, Bertram A (2020) Finite third-order gradient elastoplasticity and thermoplasticity. Journal of Elasticity **138**(2):169–193

[43] Naumenko K, Altenbach H, Kutschke A (2011) A combined model for hardening, softening, and damage processes in advanced heat resistant steels at elevated temperature. International Journal of Damage Mechanics **20**(4):578–597

[44] Placidi L, Misra A, Barchiesi E (2019) Simulation results for damage with evolving microstructure and growing strain gradient moduli. Continuum Mechanics and Thermodynamics **31**(4):1143–1163

[45] Natarajan S, Annabattula RK, Martínez-Pañeda E, et al (2019) Phase field modelling of crack propagation in functionally graded materials. Composites Part B: Engineering **169**:239–248

[46] Bilgen C, Kopaničáková A, Krause R, Weinberg K (2020) A detailed investigation of the model influencing parameters of the phase-field fracture approach. GAMM-Mitteilungen **43**(2):e202000,005

[47] Cuomo M, Contrafatto L, Greco L (2014) A variational model based on isogeometric interpolation for the analysis of cracked bodies. International Journal of Engineering Science **80**:173–188

[48] Aldakheel F, Wriggers P, Miehe C (2018) A modified Gurson-type plasticity model at finite strains: formulation, numerical analysis and phase-field coupling. Computational Mechanics **62**(4):815–833

[49] Bleyer J, Alessi R (2018) Phase-field modeling of anisotropic brittle fracture including several damage mechanisms. Computer Methods in Applied Mechanics and Engineering **336**:213–236

[50] Singh A, Das S, Altenbach H, Craciun EM (2020) Semi-infinite moving crack in an orthotropic strip sandwiched between two identical half planes. ZAMM-Journal of Applied Mathematics and Mechanics/Zeitschrift für Angewandte Mathematik und Mechanik **100**(2):e201900,202

[51] Welschinger FR (2011) A variational framework for gradient-extended dissipative continua: application to damage mechanics, fracture, and plasticity. PhD thesis, Universität Stuttgart

[52] Hansen-Dörr AC, Dammaß F, de Borst R, Kästner M (2020) Phase-field modeling of crack branching and deflection in heterogeneous media. Engineering Fracture Mechanics **232**:107,004

[53] Abali BE, Zohdi TI (2020) Multiphysics computation of thermal tissue damage as a consequence of electric power absorption. Computational Mechanics **65**:149–158

[54] Dittmann M, Aldakheel F, Schulte J, Schmidt F, Krüger M, Wriggers P, Hesch C (2020) Phase-field modeling of porous-ductile fracture in non-linear thermo-elasto-plastic solids. Computer Methods in Applied Mechanics and Engineering **361**:112,730

[55] Abali BE, Klunker A, Barchiesi E, Placidi L (2021) A novel phase-field approach to brittle damage mechanics of gradient metamaterials combining action formalism and history variable. ZAMM-Journal of Applied Mathematics and Mechanics/Zeitschrift für Angewandte Mathematik und Mechanik **101**(9):e202000,289

[56] Amirian B, Abali BE, Hogan JD (2023) The study of diffuse interface propagation of dynamic failure in advanced ceramics using the phase-field approach. Computer Methods in Applied Mechanics and Engineering **405**:115,862

[57] Mandadapu KK, Abali BE, Papadopoulos P (2021) On the polar nature and invariance properties of a thermomechanical theory for continuum-on-continuum homogenization. Mathematics and Mechanics of Solids **26**(11):1581–1598

[58] Abali BE, Yang H, Papadopoulos P (2019) A computational approach for determination of parameters in generalized mechanics. In: Altenbach H, Müller WH, Abali BE (eds) Higher Gradient Materials and Related Generalized Continua, Advanced Structured Materials, vol 120, Springer, Cham, chap 1, pp 1–18

[59] Yvonnet J, Auffray N, Monchiet V (2020) Computational second-order homogenization of materials with effective anisotropic strain-gradient behavior. International Journal of Solids and Structures **191**:434–448

[60] Altenbach H, Forest S (eds) (2016) Generalized Continua as Models for Classical and Advanced Materials, Advanced Structured Materials, vol 42. Springer, Cham

[61] Dos Reis F, Ganghoffer J (2012) Construction of micropolar continua from the asymptotic homogenization of beam lattices. Computers & Structures **112**:354–363

[62] Solyaev Y (2022) Self-consistent assessments for the effective properties of two-phase composites within strain gradient elasticity. Mechanics of Materials **169**:104,321

[63] Areias P, Melicio R, Carapau F, Carrilho Lopes J (2022) Finite gradient models with enriched RBF-based interpolation. Mathematics **10**(16):2876

[64] Bensoussan A, Lions JL, Papanicolaou G (1978) Asymptotic Analysis for Periodic Structures. North-Holland, Amsterdam

[65] Hollister SJ, Kikuchi N (1992) A comparison of homogenization and standard mechanics analyses for periodic porous composites. Computational Mechanics **10**(2):73–95

[66] Chung PW, Tamma KK, Namburu RR (2001) Asymptotic expansion homogenization for heterogeneous media: computational issues and applications. Composites Part A: Applied Science and Manufacturing **32**(9):1291–1301

[67] Temizer I (2012) On the asymptotic expansion treatment of two-scale finite thermoelasticity. International Journal of Engineering Science **53**:74–84

[68] Forest S, Pradel F, Sab K (2001) Asymptotic analysis of heterogeneous cosserat media. International Journal of Solids and Structures **38**(26-27):4585–4608

[69] Eremeyev VA (2016) On effective properties of materials at the nano-and microscales considering surface effects. Acta Mechanica **227**(1):29–42

[70] Ganghoffer JF, Goda I, Novotny AA, Rahouadj R, Sokolowski J (2018) Homogenized couple stress model of optimal auxetic microstructures computed by topology optimization. ZAMM-Journal of Applied Mathematics and Mechanics/Zeitschrift für Angewandte Mathematik und Mechanik **98**(5):696–717

[71] Turco E (2019) How the properties of pantographic elementary lattices determine the properties of pantographic metamaterials. In: Abali B, Altenbach H, dell'Isola F, Eremeyev V, Öchsner A (eds) New Achievements in Continuum Mechanics and Thermodynamics, Advanced Structured Materials, vol 108, Springer, pp 489–506

[72] Boutin C (1996) Microstructural effects in elastic composites. International Journal of Solids and Structures **33**(7):1023–105

[73] Barchiesi E, Dell'Isola F, Laudato M, Placidi L, Seppecher P (2018) A 1d continuum model for beams with pantographic microstructure: asymptotic micro-macro identification and numerical results. In: dell'Isola F, Eremeyev VA, Porubov A (eds) Advances in Mechanics of Microstructured Media and Structures, Advanced Structured Materials, vol 87, Springer, pp 43–74

[74] Bacigalupo A (2014) Second-order homogenization of periodic materials based on asymptotic approximation of the strain energy: formulation and validity limits. Meccanica **49**(6):1407–1425

[75] Boutin C, Giorgio I, Placidi L, et al (2017) Linear pantographic sheets: Asymptotic micro-macro models identification. Mathematics and Mechanics of Complex Systems **5**(2):127–162

[76] Placidi L, Andreaus U, Della Corte A, Lekszycki T (2015) Gedanken experiments for the determination of two-dimensional linear second gradient elasticity coefficients. Zeitschrift für angewandte Mathematik und Physik **66**(6):3699–3725

[77] Aydin G, Sarar BC, Yildizdag ME, Abali BE (2022) Investigating infill density and pattern effects in additive manufacturing by characterizing metamaterials along the strain-gradient theory. Mathematics and Mechanics of Solids p 10812865221100978

[78] Sarar BC, Yildizdag ME, Abali BE (2023) Comparison of homogenization techniques in strain gradient elasticity for determining material parameters. In: Altenbach H, Berezovski A, dell'Isola F, Porubov A (eds) Sixty Shades of Generalized Continua - Dedicated to the 60th Birthday of Prof. Victor A. Eremeyev, Advanced Structured Materials, vol 170, Springer International Publishing, Cham, pp 631–644

[79] Nazarenko L, Glüge R, Altenbach H (2021) Positive definiteness in coupled strain gradient elasticity. Continuum Mechanics and Thermodynamics **33**(3):713–725

[80] Nazarenko L, Glüge R, Altenbach H (2021) Uniqueness theorem in coupled strain gradient elasticity with mixed boundary conditions. Continuum Mechanics and Thermodynamics **34**(1):93–106

[81] Eremeyev VA (2021) Strong ellipticity conditions and infinitesimal stability within nonlinear strain gradient elasticity. Mechanics Research Communications **117**:103,782

[82] Abali BE, Barchiesi E (2021) Additive manufacturing introduced substructure and computational determination of metamaterials parameters by means of the asymptotic homogenization. Continuum Mechanics and Thermodynamics **33**:993–1009

[83] Vazic B, Abali BE, Yang H, Newell P (2022) Mechanical analysis of heterogeneous materials with higher-order parameters. Engineering with Computers **38**(6):5051–5067

[84] Yang H, Abali BE, Müller WH, Barboura S, Li J (2022) Verification of asymptotic homogenization method developed for periodic architected materials in strain gradient continuum. International Journal of Solids and Structures **238**:111,386

[85] Abali BE, Vazic B, Newell P (2022) Influence of microstructure on size effect for metamaterials applied in composite structures. Mechanics Research Communications **122**:103,877

[86] Aydin G, Yildizdag ME, Abali BE (2022) Strain-gradient modeling and computation of 3-D printed metamaterials for verifying constitutive parameters

determined by asymptotic homogenization. In: Giorgio I, Placidi L, Barchiesi E, Abali BE, Altenbach H (eds) Theoretical Analyses, Computations, and Experiments of Multiscale Materials, Advanced Structured Materials, vol 175, Springer, Cham, pp 343–357
[87] Washizu K (1982) Variational Methods in Elasticity and Plasticity, 3rd edn. Pergamon Press, New York
[88] Shekarchizadeh N, Abali BE, Bersani AM (2022) A benchmark strain gradient elasticity solution in two-dimensions for verifying computational approaches by means of the finite element method. Mathematics and Mechanics of Solids **27**(10):2218–2238
[89] Alnæs MS, Mardal KA (2010) On the efficiency of symbolic computations combined with code generation for finite element methods. ACM Transactions on Mathematical Software (TOMS) **37**(1):1–26
[90] Alnæs MS, Mardal KA (2012) Syfi and sfc: Symbolic finite elements and form compilation. In: Automated Solution of Differential Equations by the Finite Element Method, Springer, pp 273–282
[91] Abali BE (2017) Computational Reality, Solving Nonlinear and Coupled Problems in Continuum Mechanics. Advanced Structured Materials, Springer
[92] Barchiesi E, Yang H, Tran C, Placidi L, Müller WH (2021) Computation of brittle fracture propagation in strain gradient materials by the FEniCS library. Mathematics and Mechanics of Solids **26**(3):325–340
[93] Tangella RG, Kumbhar P, Annabattula RK (2022) Hybrid phase field modelling of dynamic brittle fracture and implementation in FEniCS. In: Krishnapillai S, Velmurugan R, Ha SK (eds) Composite Materials for Extreme Loading, Lecture Notes in Mechanical Engineering (LNME), Springer Nature, Singapore, pp 15–24
[94] Cheng P, Zhu H, Zhang Y, Jiao Y, Fish J (2022) Coupled thermo-hydro-mechanical-phase field modeling for fire-induced spalling in concrete. Computer Methods in Applied Mechanics and Engineering **389**:114,327
[95] Lu Y, Helfer T, Bary B, Fandeur O (2020) An efficient and robust staggered algorithm applied to the quasi-static description of brittle fracture by a phase-field approach. Computer Methods in Applied Mechanics and Engineering **370**:113,218
[96] Benson SJ, Munson TS (2006) Flexible complementarity solvers for large-scale applications. Optimization Methods and Software **21**(1):155–168
[97] Hysom D, Pothen A (2001) A scalable parallel algorithm for incomplete factor preconditioning. SIAM Journal on Scientific Computing **22**(6):2194–2215
[98] Musy M, Jacquenot G, Dalmasso G, de Bruin R, Pollack A, Claudi F, Badger C, Sullivan B, Hrisca D, Volpatto D, Schlömer N, Zhou Z (2021) vedo: A python module for scientific analysis and visualization of 3D objects and point clouds. Zenodo
[99] Abali BE (2020) Supply code for computations. http://bilenemek.abali.org/
[100] GNU Operating System (2007) GNU general public license. http://www.gnu.org/copyleft/gpl.html

Chapter 2
Creep-Damage Processes in Cyclic Loaded Double Walled Structures

Holm Altenbach, Dmytro Breslavsky, and Oksana Tatarinova

Abstract The paper presents an approach to determining the level of creep deformation and long-term strength of structural elements that operate under conditions of cyclic loading and heating. The method for solving the boundary - initial value problem is described. It is based on the combination of FEM and difference methods of integration for initial problems. The basis of the method is the developed and verified constitutive equations for modeling the cyclic creep-damage processes in the material. The main feature of the method is the transformation of the initial cyclic problem to a new at uniform loading and heating, but with constitutive equations of developed type. The case of the cycle stresses varying in a wide range, including in the conditions where they exceed the yield stress, as well as the case of creep when it is not exceeded by stresses, are considered. The numerical model of double-walled blade is considered and different cyclic creep modes of its operation were analyzed.

2.1 Introduction

Creep processes, which are accompanied by the accumulation of hidden damage, significantly limit the lifetime of various structural elements operating in high-temperature fields and high pressures. First of all, this applies to power machines, such as steam and gas turbines, gas turbine engines (GTE) etc. The modes of operation of their structural elements, leading to a complex form of boundary conditions, as

---

Holm Altenbach
Lehrstuhl für Technische Mechanik, Institut für Mechanik, Fakultät für Maschinenbau, Otto-von-Guericke-Universität Magdeburg, Universitätsplatz 2, 39106 Magdeburg, Germany,
e-mail: holm.altenbach@ovgu.de

Dmytro Breslavsky · Oksana Tatarinova
Department of Computer Modelling of Processes and Systems, National Technical University "Kharkiv Polytechnic Institute", UKR-61002, Kharkiv, Ukraine,
e-mail: dmytro.breslavsky@khpi.edu.ua, ok.tatarinova@gmail.com

well as the very complex geometric shape of these elements, led to the need to use numerical methods of calculation, primarily the Finite Element Method (FEM). By now, approaches and methods of FE analysis of creep under a complex stress state, implemented in modern versions of engineering calculation systems, can already be considered satisfactorily developed.

Methods for numerical FE analysis of creep-damage processes, as well as of the development of macroscopic defects resulting from the long-term action of these processes, are also being developed quite successfully [1–5], although they have not yet become standard for engineers. This is largely due to the complexity of constructing the defining constitutive and evolution equations, and, most importantly, obtaining the constants that are included in them, followed by verification of the resulting relationships. The choice of approach - the use of a scalar or tensor expression for the damage parameter also either limits the possibility of a more adequate description of the process of defect development in the material, or requires a very large amount of expensive and lengthy experimental investigations.

Recent years have been characterized by increased interest in the use of cooled blades operating at elevated temperatures in gas turbine engines (GTE). It is noted that double wall transpiration cooling (DWTC) systems allow to increase the operating temperature of gas turbines in comparison with a further increase in engine efficiency [6]. Creep calculations and analysis of the long-term strength of cooled and double-walled blades continue to be the focus of researchers [7–13].

Today thermomechanical stresses are one of the most serious problems in the implementation of these systems, and they must be taken into account, along with aerothermic characteristics, at the initial stages of design [8]. With the help of the proposed computational method, which combines both parts of the analysis, the modelling of the long-term behavior of double walled blades was performed. The calculated temperature distribution was used in thermomechanical FEA to determine the stresses in the double wall under thermal loading.

The fracture of aviation turbine blades at high temperatures was studied in [7]. Constitutive creep equations with temperature interpolation are constructed, and heat transfer is analyzed. The deformed state of the blade before failure is analyzed. The creep fracture time of the blades is determined to be 91 hours.

According to the data presented in [9], it can be concluded, that temperature differences have a greater impact on the service life of the blade than pressure differences. In [13], the effect of holes on the creep of samples with holes simulating a cooling blade was investigated. It was shown that the creep term was longer in thin-walled specimens with one central hole and shorter in specimens with multiple holes due to their interaction.

In [10–12], an analysis of the typical behavior of a cooled turbine blade was performed. The authors used an original approach in which the complex three-dimensional design of the turbomachine blade is represented by simplified two -and one dimensional models. The possibilities of analytical solution for the problem components, which represent such nonlinear deformation processes, as plasticity, ratchetting, creep, etc., were used.

GTE blades used on vehicles operate under conditions of complex temperature-force cyclic loading. For such conditions of their operation, it is known [14] that the supposition of varying components on constant load values or temperatures can significantly increase the creep rate and damage accumulation in the material. In this regard, the problem of an adequate description of the processes of cyclic deformation continues to be relevant [15–22].

Cyclic deformation processes are more complex than static ones. In this regard, experimental studies are carried out to construct constitutive equations for the description of cyclic deformation and to understand the processes taking place in the material. The processes of the interaction of creep and low-cycle fatigue are studied [15, 17] and the dependence of the main values on the strain, the strain rate, static recovery and the average stress ranges were experimentally verified [15]. The effects of previous cyclic loading on the creep of steel were studied [22]. For the case of the interaction of creep and cyclic plasticity, data on the change in the slope of the "creep strain rate - stress" curve when a certain stress value is reached, were determined [19]. For static load conditions with cyclic fatigue, the process of stress relaxation was studied [21]. Cyclic strengthening processes were studied, the influence of maximum plastic deformation due to preloading and ratcheting was analyzed [20].

The experimental results obtained in these and other studies are used to formulate and verify the constitutive equations to reflect all the main effects that occur during cyclic deformation - creep, plasticity, strengthening, damage, etc. The approaches of continuum mechanics and continuum damage mechanics, models of Hayhurst [22], Chaboche [14–16, 23], physical based and micromechanical models [17, 18, 21] are used now. The built constitutive equations are used in the simulation of a complex stress state in FEA.

The presented paper contains a description of the calculation method and constitutive equations for creep-damage processes under cyclic loads and heating, which is used for the analysis of a simplified DWTC system model. The large number of cooling channels and the complex geometric shape of the blades lead to the fact that direct numerical analysis often cannot be satisfactorily applied to elucidate the qualitative patterns of their deformation and damage accumulation leading to fracture. In this regard, a simplified two-dimensional model, includes a number of typical DWTC system operating modes are considered. FEM was used for numerical modeling, which allows transferring all the main approaches and algorithms to a general three-dimensional model.

2.2 Constitutive Equations

To carry out computational studies of creep using FE approaches, it is necessary to use creep-damage constitutive equations, as well as plasticity at the stages of forcing engines, which can be implemented to the general method and algorithms that support it. The incremental theories of creep and plasticity are used.

2.2.1 Static Loading

To determine plastic strains, we apply the flow rule with isotropic hardening [24] and use the Huber-von Mises plasticity condition:

$$f(\sigma_{ij}) = \frac{3}{2} S_{ij} S_{ij} - \left[\Phi\left(\int dp_i\right)\right]^2 \qquad (2.1)$$

where

$$S_{ij} = \sigma_{ij} - \delta_{ij}\sigma_{ii}$$

are the components of stress deviator,

$$\int dp_i$$

is the Odquist parameter, σ_v is von Mises equivalent stress. In this case, the components of plastic strain ε_{ij}^p increments are determined as follows:

$$d\varepsilon_{ij}^p = \frac{3}{2}\frac{dp_i}{\sigma_v} S_{ij}. \qquad (2.2)$$

Classical creep-damage laws (strain hardening or Norton creep, Kachanov-Rabotnov damage equation for scalar parameter, Arrhenius-type temperature function [3, 24]) are used

$$\dot{c}_{ij} = \frac{3}{2} B c_{vM}^{-\alpha} \frac{(\sigma_v)^{n-1}}{(1-\omega)^l} \exp\left(-\frac{Q_c}{T}\right) S_{ij}; \quad Q_c = \frac{U_c}{R}; \qquad (2.3)$$

$$\dot{\omega} = D \frac{(\sigma_{vd})^m}{(1-\omega)^l} \exp\left(-\frac{Q_d}{T}\right); \quad Q_d = \frac{U_d}{R}. \qquad (2.4)$$

Here c_{ij} are the components of creep strain tensor, c_{vM} is von Mises equivalent creep strain, U_c, U_d are the values of activation energies for creep and creep damage accumulation processes, R is universal constant. σ_{vd} is equivalent stress has to be estimated by use of strength criteria at the step of governing the conditions of hidden damage accumulation finishing. As is known [3], the values of material constants B, D, n, m, l, α included in (2.3) – (2.4), can be obtained by use of experimental data processing.

2.2.2 Cyclic Loading. Stresses Lower the Yield Limit

Let us consider the main equations for the case of cyclic varying of temperatures and stresses. In this case, by cyclicity we understand the alternation of long periods of gas turbine or turbine operation (hours in the first case and months in the other) with stop periods. We will demonstrate it by a simplified dependence of temperature and

2 Creep-Damage Processes in Cyclic Loaded Double Walled Structures

stress on time (Fig. 2.1), in which we neglect the regions of increase and decrease. Here t_1 is the operating time, t_2 is the dwell time. Further, in this approximation, we consider the periods T_c, defined as $T_c = t_1 + t_2$, to be the same for the entire time of further operation. In fact, the rectangular approximation is often quite sufficient, since the integration operation (see below) is used to obtain the equations, and the value of the corresponding area under the curve is decisive.

Let us use the constitutive equations (2.3) – (2.4), and for the sake of simplicity we use Norton's law and assume $\alpha = 0$. To perform an analysis of the behavior of structural elements operating under a complex stress state, it is first necessary to analyze the one-dimensional behavior of the materials from which they are made. Behavior patterns obtained at the same time can also be detected when solving two- and three-dimensional problems, which will facilitate their analysis.

First, we consider the one-dimensional case with the action of tension stress σ^u. For the case of stress cycling, it can be represented by the sum of constant σ and time-varying components σ^1: $\sigma^u = \sigma + \sigma^1$. Similarly, the temperature function has a constant part T and periodically varying T_1: $\tilde{T} = T + T_1$. The law of cyclic varying for above stress is represented by a polyharmonic law with a period T_p:

$$\sigma^u = \sigma + \sigma^1 = \sigma\left(1 + \sum_{k=1}^{\infty} M_k \sin\left(\frac{2\pi k}{T_p} t + \beta_k^\sigma\right)\right) \qquad (2.5)$$

where $M_k = \dfrac{\sigma_{ak}}{\sigma}$, σ_{ak} are the coefficients of stress function σ_1 expansions into Fourier series. First, we will assume that the temperature has a constant value T and the exponential factor in (2.3) – (2.4) will be equal to 1. We will expand the creep strain and damage parameter functions into an asymptotic series with a small parameter

$$\mu = \frac{T_p}{t_*},$$

where t_* is the time of finishing the hidden damage accumulation. We limit ourselves to two terms of the these series, which is the usual procedure of asymptotic methods [25, 26]:

$$c \cong c^{(0)}(t) + \mu c^{(1)}(\xi); \quad \omega \cong \omega^{(0)}(t) + \mu \omega^{(1)}(\xi) \qquad (2.6)$$

where $c^{(0)}(t), \omega^0(t), c^{(1)}(\xi), \omega^{(1)}(\xi)$ are the functions which reflect creep and damage processes in slow time (0) and fast time (1). Here we consider two time variables:

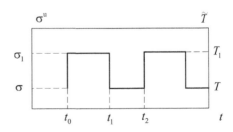

Fig. 2.1 Representation of simplified temperature and stress cycle.

a slow time t that varies from 0 to the time to the fracture time value t_* and fast time

$$\tau = \frac{t}{\mu} \quad \text{or} \quad \xi = \frac{\tau}{T_p}, 0 \leq \xi \geq 1.$$

Now let us substitute the asymptotic series (2.6) into equations (2.3) – (2.4) and average the obtained equations [25, 26] by the period of stress varying. After these transformations, we get the expressions of the creep strain and the damage parameter on the time interval:

$$\left\langle c^{(0)}(\xi) \right\rangle = \int_0^1 c^0(t) d\xi = c^0(t); \quad \left\langle c^{(1)}(\xi) \right\rangle = \int_0^1 c^1(\xi) d\xi \cong 0; \quad (2.7)$$

$$\left\langle \omega^{(0)}(\xi) \right\rangle = \int_0^1 \omega^0(t) d\xi = \omega^0(t); \quad \left\langle \omega^{(1)}(\xi) \right\rangle = \int_0^1 \omega^1(\xi) d\xi \cong 0. \quad (2.8)$$

The next step is to substitute (2.7) and (2.8) into the system of equations (2.3)-(2.4). The results change the basic system to the following

$$\dot{c} = Bg_n(M_k) \frac{\sigma^n}{(1-\omega)^l}; \quad \dot{\omega} = Dg_m(M_k) \frac{\sigma^m}{(1-\omega)^l}; \quad \omega(0) = \omega_0, \quad \omega(t_*) = 1; \quad (2.9)$$

$$g_n(M_k) = \int_0^1 \left(1 + \sum_{k=1}^{\infty} M_k \sin(2\pi k \xi)\right)^n d\xi;$$

$$g_m(M_k) = \int_0^1 \left(1 + \sum_{k=1}^{\infty} M_k \sin(2\pi k \xi)\right)^m d\xi.$$

Here, the functions $g_n(M_k)$ and $g_m(M_k)$ reflect the influence of the cyclicity of the processes of creep and damage accumulation.

After that, let us add to the consideration the cyclic temperature varying:

$$\tilde{T} = T + T^1 = T\left(1 + \sum_{i=1}^{\infty} M_i^T \sin\left(\frac{2\pi i}{T_T} t + \beta_i^T\right)\right); \quad M_i^T = \frac{T_i^a}{T} \quad (2.10)$$

where T_i^a are the coefficients of expansion the temperature function T_1 in Fourier series. We similarly use two time scales with a small parameter $\hat{\mu} = T_T/t_*$:

$$T \cong T^{(0)}(t) + \hat{\mu} T^{(1)}(\xi). \quad (2.11)$$

Using transformations similar to those described above for creep strain (see, for example, [27, 28]), it is possible to obtain expressions for the similar influence functions $g_T(T)$ for creep and $g_T^{\omega(T)}$ for damage equation:

$$g_T(T) = B \int_0^1 \exp\left(-\frac{Q_c}{T}\left(1+\sum_{i=1}^{\infty} M_i^T \sin(2\pi i\xi)\right)^{-1}\right) d\xi; \qquad (2.12)$$

$$g_T^{\omega}(T) = D \int_0^1 \exp\left(\frac{-Q_d}{T}\left(1+\sum_{i=1}^{\infty} M_i^T \sin(2\pi i\xi)\right)^{-1}\right) d\xi.$$

So, now the creep-damage laws, taking into account temperature and stress varying, can be written in the following form:

$$\dot{c} = g_n(M_k)g_T(T)\frac{\sigma^n}{(1-\omega)^l}, \quad c(0)=0; \qquad (2.13)$$

$$\dot{\omega} = g_m(M_k)g_T^{\omega}(T)\frac{\sigma^m}{(1-\omega)^l}; \quad \omega(0)=\omega_0, \; \omega(t_*)=1. \qquad (2.14)$$

System (2.13) – (2.14) can be considered as a new system of governing equations for the averaged cyclic creep-damage process. Its analysis shows that when using it, there is no need to integrate over the cycle.

Next, after passing to the general case of a complex stress state with the usual use of the corresponding invariants of the stress tensor or their combination, we obtain:

$$\dot{c}_{ij} = \frac{3}{2}g_T(T)g_n(M_k^{\sigma_v})\frac{\sigma_v^{n-1}}{(1-\omega)^l}S_{ij}; \quad \dot{\omega} = g_m(M_k^{\sigma_{vd}})g_T^{\omega}(T)\frac{\sigma_{vd}^m}{(1-\omega)^l}; \qquad (2.15)$$

$$\omega(0) = \omega_0, \omega(t_*) = 1;$$

$$g_T(T) = B \int_0^1 \exp\left(-\frac{Q_c}{T}\left(1+\sum_{i=1}^{\infty} M_i^T \sin(2\pi i\xi)\right)^{-1}\right) d\xi;$$

$$g_T^{\omega}(T) = D \int_0^1 \exp\left(\frac{-Q_d}{T}\left(1+\sum_{i=1}^{\infty} M_i^T \sin(2\pi i\xi)\right)^{-1}\right) d\xi; \quad M_i^T = \frac{T_i^a}{T};$$

$$g_n(M_k^{\sigma_v}) = \int_0^1 \left(1+\sum_{k=1}^{\infty} M_k^{\sigma_v} \sin(2\pi k\xi)\right)^n d\xi; \quad M_k^{\sigma_v} = \frac{\sigma_v^{ak}}{\sigma_v}:$$

$$g_m(M_k^{\sigma_{vd}}) = \int_0^1 \left(1+\sum_{k=1}^{\infty} M_k^{\sigma_{vd}} \sin(2\pi k\xi)\right)^m d\xi; \quad M_k^{\sigma_{vd}} = \frac{\sigma_{vd}^{ak}}{\sigma_v}.$$

2.2.3 Cyclic Load. Overloading with Transition to Plastic Deformation

As was noted above, under forced operating modes of a gas turbine engine, in addition to the already developing creep strains, plastic strains can occur in the material of its structural elements. First, we also consider uniaxial deformation.

Let us formulate the problem. The uniaxial sample is instantly loaded in the elastic area, then deformed by creep with stress σ during time t_1. Then, for time t_2, a load is added incrementally, which realizes the stress σ_1, and its value exceeds the value of the yield limit σ_y for the given temperature. After that, the sample is also gradually unloaded to the value σ. Next, the process of loading and unloading is repeated (Fig. 2.1).

It is known [29] that an adequate description of the step load during creep can be implemented using the strain hardening theory. Let us consider it. With a stepped load from σ to σ_1, the strain rate is determined by the angle of inclination of the tangent to the strain curve with stress σ_1 at a point that can be found by parallel transfer to it along the time axis of the point from the curve constructed at stress σ. After the onset of stress σ_1, the creep strain increases according to the law corresponding to the law of its varying at the mentioned point. In the case when the stress σ_1 exceeds the yield point, we assume that the total strain also increases by stepped law by addition of plastic part. Its value can be determined by the deformation curve $(\sigma - \varepsilon)$.

Let us consider creep deformation of the rod made from high-chromium corrosion-resistant foundry heat-resistant nickel based alloy (Ni 57%, Cr 16%, Co 11%, W 5%) and heated evenly to temperature 950 C [30]. The creep curves of this material for 4 stress values for a deformation time of 1h are presented in Fig. 2.2. Note that the first two curves correspond to the deformation in which the stress values exceed the yield strength of this alloy at the given temperature $\sigma_y = 390$ MPa, the other two are obtained during the initial elastic deformation. As can be seen, the curves do not

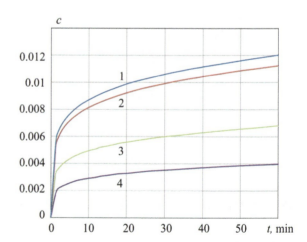

Fig. 2.2 Creep curves of heat-resistant nickel based alloy. Static loading. Stress values: 1 - 450 MPa, 2 - 420 MPa, 3 - 250 MPa, 4 - 145 MPa.

differ qualitatively. In this regard, they were processed using one dependency (2.3). It was done by use of strain hardening law. For the case of creep without damage, the following values of constants were obtained after calculations:

$$B = 5.26 \cdot 10^{-27} \text{MPa}^{-n}/\text{h}, \quad n = 5.508, \quad \alpha = 4.678.$$

Let us apply the obtained constants to Eq. (2.3) for the analysis of deformation processes with a step varying of stresses and a uniaxial stress state. To do this, we will conduct a numerical simulation of the creep process with cyclic loadings, using calculations based on strain hardening theory. A number of calculations were carried out with different input data, and below we present typical results. First, consider the loading due to Program 1 corresponding to stress varying that do not exceed the yield limit at this temperature.

2.2.3.1 Program 1

Initial value of stress $\sigma = 250$ MPa, the greater value $\sigma_1 = 350$ MPa, $t_0 = 0.166$ h (10 min), $t_1 = 0.083$ h (5 min), $t_2 = 0.25$ h. Loading time is equal to 1 h. The results are presented in Fig. 2.3, where the dependence of the total strain ε ($\varepsilon = \varepsilon^{elast} + c$) on time is given. Here, curve 1 and 3 correspond to the static load at $\sigma = 350$ MPa, curve 3 – at $\sigma = 250$ MPa. Curve 2 is built for data from cyclic loading according to program 1. As can be seen from the figure, the curve for cyclic loading is similar to

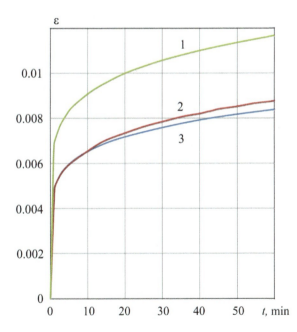

Fig. 2.3 Creep curves of heat-resistant nickel based alloy. Static (curves 1 - 350 MPa and 3 - 250 MPa) and cyclic (curve 2) loading.

the curves for static loading, therefore, it is possible to use the ratio of type (2.13) to describe the averaged process.

2.2.3.2 Program 2

Now consider the load according to program 2, in which the load in the plastic zone is cyclically added to the creep caused by the stress, which exceeds the yield strength. We analyse the case of deformation with hardening of the material in each cycle of additional loading, when the yield stress changes due to hardening process, as shown in Fig. 2.4.

Initial value of stress $\sigma = 370$ MPa, the greater value $\sigma_1 = 420$ MPa, $t_0 = 0.25$ h, $t_1 = 0.083$ h, $t_2 = 0.083$ h. Loading time is equal to 1 h. The results are shown in Fig. 2.5. As can be seen from the comparison of the location of curves 2 in Figs. 2.3 and 2.5, they are qualitatively different, which is due to the instantaneous plastic additional loading in the cycle of program 2. With a rather small difference between the stress values σ and σ_1 in 50 MPa, we observe jumps in deformation during additional loading. During unloading, there is an elastic reduction, but the plastic strain accumulated during the cycle remains and is added to the full value. It is also possible to see that with each cycle as the yield strength increases, the amount

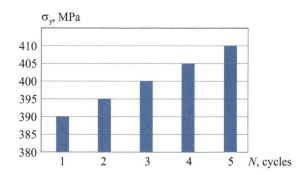

Fig. 2.4 Dependence of the yield limit on the number of the loading cycle.

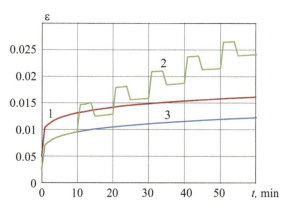

Fig. 2.5 Creep curves of heat-resistant nickel based alloy. Static (curves 1 - 420 MPa, and 3 - 370 MPa) and cyclic cyclic step loading according to program 2 (curve 2).

of reduction in total strain also decreases. This means that when loading has a lot of cycles, the shape of the strain curve will approach a smoother one. A similar conclusion is confirmed by the shape of the strain curve with 10 similar cycles of program 2, which is presented in Fig. 2.6. A loading time of 2 h was set.

From the analysis of the curve in Fig. 2.6, built according to program 2 of the step loading, it can be seen that when the current value reaches the yield strength value of the acted stress $\sigma_1 = 420$ MPa, the deformation begins to proceed similarly to the process with initially elastic stresses (as according to program 1). Such loading processes are similar to ratcheting processes [31], but with continued growth of creep strains.

The method of obtaining the averaged equations discussed in the previous section cannot be used directly to obtain the averaged equation in the case when there is an overloading in the cycle, which leads to the occurrence of plastic strains with material hardening. This is due to the different nature of curves for the cyclic creep with plastic strains creep under static loading. In this regard, an approach is proposed that allows obtaining an approximate form of such an averaged equation, for its use in numerical modeling of structural elements of turbomachines.

Let us assume that due to experiments or numerical modeling using strain hardening law (it is this that makes it possible to calculate the case of additional loading [29]) a set of uniaxial creep curves under purely static loading and a corresponding set of curves for the case of cyclic overloads is obtained (Fig. 2.7). The set should consist of as many calculated curves as possible (three are shown in the figure to better understand the appropriate arrangement). As can be seen after analyzing the curves of the Fig. 2.7, the cyclic overload curves in the case under consideration with stress values from 390 to 420 MPa are of the same type. They are characterized by higher values of strains in the cycle at higher values of overload stresses. After the overload is completed, the strains follow the same segments of the curves corresponding to the main static load. This means that the function of irreversible cyclic deformation does not depend on the value of the overload stresses, but only on the stress of basic loading.

Next, for each curve of cyclic creep with overloads, using approximation procedures, we obtain averaged curves, i.e., curves that correspond exactly to the irreversible

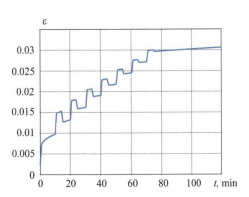

Fig. 2.6 Cyclic creep curve of heat-resistant nickel based alloy (program 2).

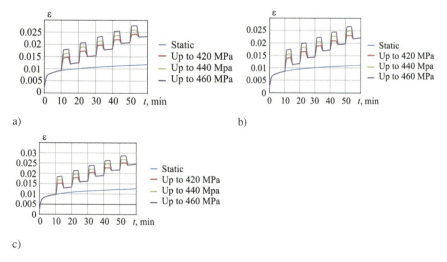

Fig. 2.7: Creep curves of heat-resistant nickel based alloy for static stresses (a) 330 MPa, b) 350 MPa, c) 370 MPa) and cyclic overloading up to 420, 440 and 460 MPa.

strain accumulated in the sample. This is shown in Fig. 2.8, where the cyclic creep curve with the overload amplitude σ_1 (curve 1), the averaged curve 2 and the curve under static loading (3) are presented for one set of applied stress values. Points used for approximation are marked with circles.

Using classical methods of processing the static creep curves, we find the values of the constants B, n, α, included in the equation for the static creep strain rate function:

$$\dot{c} = B c^{-\alpha} \sigma^n \qquad (2.16)$$

Further, by integrating Eq. (2.16), we obtain an expression for the dependence of creep strains on time:

$$c_j = b \sigma_j^r t^a, \quad j = 1 \ldots N, \qquad (2.17)$$

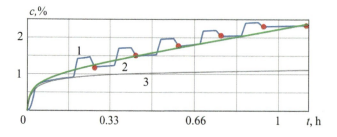

Fig. 2.8: Schematic presentation of the cyclic creep curve (1), averaged curve (2) and static creep curve (3).

where B, r, a are the constants, N is the number of curves obtained. According to the procedure described above for obtaining averaged curves of irreversible strain (curve like curve 2 in Fig. 2.8), we obtain approximation dependencies for these functions $f_j(t;\sigma_j), j = 1\ldots N$. Than it is possible to determine the influence function in the form of the additional coefficient $k_j(t;\sigma_j)$, multiplied by which the value of the function of static creep strains (curve 3 in Fig. 2.8) the demanded value on the averaged curve 2 will be obtained:

$$k_j(t;\sigma_j) = \frac{f_j(t;\sigma_j)}{b\sigma_j^r t^a}, \quad j = 1\ldots N, \tag{2.18}$$

Next, using the values of the obtained functions $k_j(t;\sigma_j)$ at N points on the plane (t,σ) for each of the set of points (t_i,σ_j) $(i = 1\ldots M)$ we obtain the values of the function $K(t_i,\sigma_j)$, which reflects the effect of cyclic loading on creep. With the help of approximation procedures in the two-dimensional domain, we obtain the expression of the function $K(t,\sigma)$ for the all possible values of times and stresses. It is already possible to include it in calculations for cyclic loading. For the function of cyclic creep strains, we obtain

$$c = K(\sigma,t)b\sigma^r t^a, \tag{2.19}$$

or for the cyclic creep strain rate function

$$\dot{c} = \dot{K}(\sigma,t)b\sigma^r t^a + aK(\sigma,t)b\sigma^r t^{a-1} \tag{2.20}$$

For the case of a complex stress state, we get:

$$\dot{c}_{ij} = \frac{3}{2}b\sigma_v^{r-1}\left(\dot{K}(\sigma_v,t)t^a + aK(\sigma_v,t)t^{a-1}\right)S_{ij} \tag{2.21}$$

The analysis of expression (2.21) shows that it provides the possibility of simulating creep processes with overload stresses in the cycle exceeding the yield limit, using only the formulation of the creep problem under static loading, but with an governing equation of the type (2.21).

To illustrate the method of obtaining an equation of type (2.21), which allows calculations for a wide range of stresses of the main loading process σ, let us continue the analysis of the deformation of considered alloy at a temperature of 950°C. As an example, consider curves 2 and 3 of Fig. 2.5. The value of the static stress $\sigma = 370$ MPa. We will use curve 2 to obtain a new curve that will correspond to curve 2 in Fig. 2.8, which is an averaged curve that collectively describes the development of irreversible strains during creep with overloads. Such a curve is constructed - it is curve 1 in Fig. 2.9. The points show the values taken from curve 2 of Fig. 2.5 at moments of partial unloading.

Next, it was necessary to construct a function that would approximate above mentioned curve 1 from Fig. 2.9. After a number of mathematical experiments, it was found that the hyperbolic function best satisfies the conditions of approaching the experiment in the first cycles and reaching the asymptote at larger time values. It

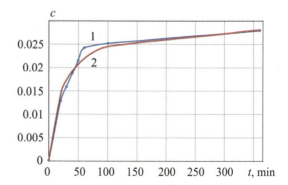

Fig. 2.9 The averaged deformation curve during creep with overloads (1) and the approximated curve (2).

was accepted in the following form:

$$c = \frac{t}{b_0 + b_1 t} \qquad (2.22)$$

The values of constants are included in (2.22): $b_0 = 742.61$ min, $b_1 = 33.82$. The graph of this function is represented by the curve 2 in Fig. 2.9. As can be seen from the comparison of the values presented on curves 1 and 2 of this figure, the worst difference that occurs in the area after the fourth unloading does not exceed 10%. On other sections, and most importantly, on larger time values, the differences are 2-3%. This is satisfactory for calculations.

Next, let us consider the deriving of function $K(t,\sigma)$. For this, it was necessary to carry out similar actions with the curves obtained for other stress values like presented above for the case of $\sigma = 370$ MPa. The stress values were taken in the range of 295-370 MPa with a step of 15 MPa. For each of these creep curves with overloads, an approximation algorithm was performed and constants to Eq. (2.22) were found. They were close enough to presented in (2.22). After that, it was possible to determine the values of the coefficients $k_j(t;\sigma_j)$ for expression (2.18). The calculations were carried out for the above stress values and time of 100 min with a step of 20 min. Thus, for each of the 36 points, the coefficients $k_j(t;\sigma_j)$ were found. They are represented by the surface $K(t,\sigma)$ in Fig. 2.10.

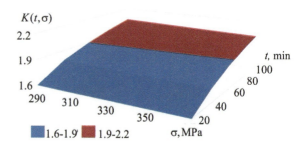

Fig. 2.10 Dependence of the function $K(t,\sigma)$, reflects the influence of the cyclic overloads on creep rates, from the stress values of the main process and time.

After that, the obtained digital values function $K(t,\sigma)$ of two coordinates in the plane, time and stresses, were used for the numerical determination using the approximation algorithm of the function of two variables. The analytical expression of the function $K(t,\sigma)$ was obtained as follows:

$$K(t,\sigma) = V_0 \sigma^{v_1} t^{v_2} \qquad (2.23)$$

where numerical constants included have the following values:

$$V_0 = 0.785 \text{MPa}^{-v_1} \text{h}^{-v_2}, \quad v_1 = -0.32, \quad v_2 = 0.15.$$

To use this function when calculating a complex stress state, it is necessary to make a transition to equivalent stresses and strains due to (2.21) according to relations (2.23) as well as to obtain the expression of creep strain rate. We obtain:

$$\dot{c}_{ij} = \frac{3}{2} b V_0 (a+v_2) \sigma_v^{r-1+v_1} t^{v_2+a-1} S_{ij} \qquad (2.24)$$

2.3 Problem Statement

Let us consider the general mathematical formulation of the boundary initial value problem of the creep of deformed solids for the volume V with isotropic properties in the Cartesian coordinate system x_i ($i = 1, 2, 3$). It us supposed, that non-varied in time displacement values are known in the part of solid's surface S_1 $u_i|_{S_1} = \bar{u}_i$. Another surface part S_2 is loaded by traction p with constant in time $p_i^0(x)$ and cyclically varied in time $\Phi_i(x,t)$ components:

$$p_i(x,t) = p_i^0(x) + \Phi_i(x,t), \quad x \in S_2 \qquad (2.25)$$

where

$$\Phi_i(x,t) = p_i^a(x)\Phi(t) = p_i^a(x) \sum_{k=1}^{\infty} A_k \sin\left(\Omega_k t + \beta_k^p\right) \qquad (2.26)$$

are the periodical expansions with period T_p;

$$p_i^a(x), \quad A_k = \sqrt{a_k^2 + b_k^2}, \quad \Omega_k = 2\pi k/T_p, \quad \beta_k^p = \arctan(a_k/b_k)$$

are known values.

The solid V is an inhomogeneous temperature field, which is set on the surface S by the sum of the constant T and periodically varying T_1 components:

$$\tilde{T}(x,t) = T(x) + T^1(x,t), \quad x \in S, \qquad (2.27)$$

where

$$T^1(x,t) = T^a(x)\theta(t) = T^a(x)\sum_{k=1}^{\infty} A_k^T \sin\left(\Omega_k^T t + \beta_k^T\right), \qquad (2.28)$$

are the periodical expansions with period T_T;

$$T^a(x), \quad A_k^T = \sqrt{\left(a_k^T\right)^2 + \left(b_k^T\right)^2}, \quad \Omega_k^T = 2\pi k/T_T, \quad \beta_k^T = \arctan\left(a_k^T/b_k^T\right)$$

are known values.

Due to Lagrange approach we consider the small strains and displacements that usually occur in the structural elements of power engineering. The following notations are used: **u** for displacement vector with components $u_i(x,t)$; $\boldsymbol{\sigma}$, $\boldsymbol{\varepsilon}$ are the stress and strain tensors with components $\sigma_{ij} = \sigma_{ji}(x,t)$ and $\varepsilon_{ij} = \varepsilon_{ji}(x,t)$. Both of them are functions of co-ordinates x_i ($i = 1, 2, 3$) and time t. Let us assume that at any time the strain tensor is the sum of elastic and temperature strain tensors, tensors of irreversible plasticity and creep strains:

$$\varepsilon_{ij} = \varepsilon_{ij}^e + \varepsilon_{ij}^T + \varepsilon_{ij}^p + c_{ij}, \qquad (2.29)$$

where $\boldsymbol{\varepsilon}^e, \boldsymbol{\varepsilon}^T$ are the elastic and thermal strain tensors with components $\varepsilon_{ij}^e(x)$, $\varepsilon_{ij}^T(x)$; $\boldsymbol{\varepsilon}^p$ is plastic strain tensor with components $\varepsilon_{ij}^p = \varepsilon_{ij}^p(x)$; **c** is creep strain tensor with components $c_{ij} = c_{ji}(x,t)$, $c_{ij}(x,0) = 0$, $(i,j = 1,2,3)$.

For thermal strains, let's limit ourselves to the generalized law of thermoelasticity of homogeneous isotropic solids [24], known as the Duhamel-Neumann law, so that at any time the relationship between stresses, strains and temperature is written as follows:

$$\sigma_{ij} = \lambda\varepsilon_0\delta_{ij} + 2G(\varepsilon_{ij} - \varepsilon_{ij}^p - c_{ij}) - (3\lambda + 2G)\varepsilon_{ij}^T; \qquad (2.30)$$

$$\varepsilon_0 = \varepsilon_{km}\delta_{km}; \quad \lambda = \frac{\nu E}{(1+\nu)(1-2\nu)}; \quad G = \frac{E}{2(1+\nu)}; \quad \varepsilon_{ij}^T = \alpha_{Te}\tilde{T}\delta_{ij}$$

where λ, G are Lamé parameters; α_{Te}, δ_{ij} are coefficient of thermal expansion and Kronecker delta.

As in [25–27], we present the basic system of equations for determining the stress-strain state of the solid during creep under the conditions of a known temperature field $\tilde{T}(x,t)$

$$\sigma_{ij,j} = \rho\ddot{u}_i; \quad x_i \in V; \quad \sigma_{ij}n_j = p_i^0(x) + \Phi_i(x,t), \quad x_i \in S_2; \qquad (2.31)$$

$$\varepsilon_{ij} = \frac{1}{2}\left(u_{i,j} + u_{j,i}\right), x_i \in V; \quad u_i|_{S_1} = \bar{u}_i, \quad x_i \in S_1;$$

$$\sigma_{ij} = \lambda\varepsilon_0\delta_{ij} + 2G(\varepsilon_{ij} - \varepsilon_{ij}^p - c_{ij}) - (3\lambda + 2G)\alpha_{Te}\tilde{T}\delta_{ij}$$

where, in addition to the previously defined notations, **n** is a unit vector with components $n_i, i = 1,2,3$ of the external normal to the solid's surface.

The system of differential equations (2.31), which should be specified by adding to it the constitutive equations of the material (2.15), will describe the general mathematical formulation of boundary- initial value creep-damage problem at periodically varying temperatures and stresses. To apply the constitutive equations, let us transform the system of differential equations (2.30) using the method of two time scales and averaging over the period of the cyclic varying the components of temperature and stress.

Let us assume that the time of the creep process until the completion of the hidden damage is much longer than the periods of the cyclic components of stress and temperature $t_* \gg \max(T_p, T_T)$, and limit ourselves to the first approximation of the asymptotic expansions for the components $\mathbf{u}$, $\boldsymbol{\varepsilon}$ and $\boldsymbol{\sigma}$ with a small parameter $\mu = \min[(t_*/T_p)^{-1}, (t_*/T_T)^{-1}]$, $\mu \ll 1$:

$$u_i \cong u_i^{(0)}(x,t) + \mu u_i^{(1)}(x,\xi), \quad \varepsilon \cong \varepsilon^{(0)}(x,t) + \mu \varepsilon^{(1)}(x,\xi);$$

$$\varepsilon_{ij}^T \cong \varepsilon_{ij}^{T(0)}(x,t) + \mu \varepsilon_{ij}^{T(1)}(x,\xi), \quad \varepsilon_{ij}^P \cong \varepsilon_{ij}^{P(0)}(x,t) + \mu \varepsilon_{ij}^{P(1)}(x,\xi); \quad (2.32)$$

$$\sigma_{ij} \cong \sigma_{ij}^{(0)}(x,t) + \mu \sigma_{ij}^{(1)}(x,\xi), \quad c_{ij} \cong c_{ij}^{(0)}(x,t) + \mu c_{ij}^{(1)}(x,\xi)$$

where x, t and ξ are formally independent variables. Then, after performing the transformations that can be found in [25, 26], we obtain two systems of equations - the main (2.33) and auxiliary (2.34).

System (2.33) describes the motion of a system of material points during irreversible deformation on a slow time scale:

$$\sigma_{ij,j} = 0, \quad x_i \in V; \quad \sigma_{ij}n_j = p_i^0, \quad x_i \in S_2; \quad (2.33)$$

$$\varepsilon_{ij} = (u_{i,j} + u_{j,i})/2, \quad x_i \in V; \quad u_i|_{S_1} = \bar{u}_i, \quad x_i \in S_1;$$

$$\sigma_{ij} = \lambda \varepsilon_0 \delta_{ij} + 2G(\varepsilon_{ij} - \varepsilon_{ij}^P - c_{ij}) - (3\lambda + 2G)\alpha_T T \delta_{ij}.$$

Here, and in following text, the superscript "0", that describe the creep processes of heterogeneously heated solids that occur on a slow time scale, is omitted in the functions.

The system of equations (2.33) should be supplemented with the creep-damage constitutive equations (2.15). This system describes the general mathematical formulation of the boundary initial value creep-damage problem at periodically varying temperatures and stresses.

To specify the system of equations (2.33), which must be applied with use of constitutive equations (2.15), it is necessary to define the stress fields σ_{ij}^{ak}, $k = 1, 2, \ldots$ as well as temperature fields T_i^a, $i = 1, 2, \ldots$, which describe the periodically varying processes of the stress-strain state and temperature over time ξ, $0 \leq \xi \leq 1$. Auxiliary systems of equations are intended for this purpose. A system of equations is obtained for time scale $(0 \leq \xi \leq 1)$ [26, 27]:

$$\sigma_{ij,j}^{(1)} = \rho \mu^{-3} u_{i,\xi\xi}^{(1)}; \quad \sigma_{ij}^{(1)} n_j = \mu^{-1} \Phi_i, \quad x_i \in S_2; \quad (2.34)$$

$$\varepsilon_{ij}^{(1)} = (u_{i,j}^{(1)} + u_{j,i}^{(1)})/2, \quad x_i \in V; \quad u_i^{(1)}\big|_{S_1} = 0, \quad x_i \in S_1;$$

$$\sigma_{ij}^{(1)} = \lambda\varepsilon_0^{(1)}\delta_{ij} + 2G\varepsilon_{ij}^{(1)} - (3\lambda + 2G)\alpha_{Te}T^1 - \sigma_{ij}^{(1)p}.$$

The system of equations (2.34) corresponds to the equations of thermoelasticity (thermoplasticity in the case of the presence of plastic overloads in the cycle) of the solid at given periodically varying loading with frequencies which is significantly lower from lower solid's natural frequency Ω_0: ($\Omega^p = 2\pi/T_p$, $\Omega^T = 2\pi/T_T$) $\ll \Omega_0$. The traction and temperature functions are specified on the corresponding parts of the solid's surface:

$$\Phi_i(x,\xi) = p_i^a(x)\Phi(\xi) = p_i^a(x)\sum_{k=1}^{\infty}\Phi_k(\xi), \quad \Phi_k(\xi) = A_k\sin\left(\mu\Omega_k\xi + \beta_k^p\right), \quad (2.35)$$

$$T^1(x,\xi) = T^a(x)\theta(\xi) = T^a(x)\sum_{k=1}^{\infty}\theta_k(\xi), \quad \theta_k(\xi) = A_k^T\sin\left(\mu\Omega_k^T\xi + \beta_k^T\right).$$

This allows us to consider system (2.34) as corresponding to the non-stationary deformation of the body under the action of harmonically varying pressure on the surface S_2 $p_{ik}^{(1)} = p_i^a(x)\varphi_k(\xi)$, where $\varphi_k(\xi) = \Phi_k(\xi)/\mu$, $\varphi_{k,\xi\xi} = -\mu^2\Omega_k^2\varphi_k(\xi)$ (as a harmonic function). If the periods of traction and temperature are assumed to coincide, then, as is known, in (2.34) it is possible to separate the variables by coordinates and time

$$\sigma_{ij}^{(1)} = \sum_k \sigma_{ij}^{ak}(x)\varphi_k(\xi), \quad (2.36)$$

$$u_i^{(1)} = \sum_{k=1}^{\infty} u_i^{ak}(x)\varphi_k(\xi),$$

$$T^1 = \sum_{k=1}^{\infty} T^{ak}(x)\varphi_k(\xi).$$

Boundary value problems for amplitude values of unknowns periodically varying on a fast time scale will have the following form ($k = 1, 2, \ldots$):

$$\sigma_{ij,j}^{ak} = -\Omega_k^2\mu^{-1}\rho u_i^{ak}, \quad \sigma_{ij}^{ak}n_j = p_i^a(x)A_k, \quad x \in S_2; \quad (2.37)$$

$$\varepsilon_{ij}^{ak} = 1/2\left(u_{i,j}^{ak} + u_{j,i}^{ak}\right); \quad x \in V; \quad u_i^{ak} = 0, \quad x \in S_1;$$

$$\sigma_{ij}^{ak} = \lambda\varepsilon_0^{ak}\delta_{ij} + 2G\varepsilon_{ij}^{ak} - (3\lambda + 2G)\alpha_{Te}T^a(x)A_k^T - \sigma_{ij}^{pak}.$$

Inertial components in the first equations of system (2.37) are formally preserved when deriving the equations, meanwhile, they can be neglected in calculations, under the conditions of considering the processes with far from resonant frequencies $\Omega^p = \Omega^T \ll \Omega_0$. Under this assumption, systems (2.37) can be considered as a static problem of thermo-elasticity (thermo-plasticity).

Amplitude values of unknowns periodically varying on a fast time scale are calculated after solving the problem of non-stationary thermal conductivity for the heating-cooling cycle and determining the value of the function $T^{(1)}$. This makes it possible to solve systems of Eqs. (2.37).

According to the algorithm proposed above, after solving the initial-boundary value problem (2.33), the stress-strain state of the solid's creep-damage process with cyclic varying of temperature and external force fields is determined. Its integration is carried out until the finish of hidden damage accumulation or until the time for the analysis of the deformation process, which is specified by the problem conditions.

2.4 Comparison Between Data of Direct Approach and Use of Averaged Function K

Next, let us proceed to the analysis of the results of solving the problem of creep under intermittent overloads, which lead to the occurrence of plastic deformation. In this case, the traction p varyings according to the time law, which is similar to the presented in Fig. 2.1.

Let us consider the creep of a long plate (beam) subjected to bending in its plane by a traction part p_b=1 MPa and tensed by an another traction part p_e, which has a constant value of 25 MPa up to the middle of the plate, and then increases linearly to 75 MPa. Such a load can be considered as simulating the behavior of a turbomachine blade, which is loaded by surface pressure from the working body (bending) and tensed by centrifugal forces. The left edge of the plate is rigidly fixed.

Plate dimensions: length 100 mm, width 12 mm, thickness 1 mm. The plate is periodically loaded with both types of load in such a way that plastic strains occur in a certain part of it. At the same time, p_b reaches the value p_{bo}=1.45 and p_{eo} = $1.45 p_e$. It has a constant value of 36.25 MPa up to the middle of the plate, and then increases linearly to 109 MPa. The material of the plate is the above considered nickel based alloy at a temperature of 950°C. The considered conditions correspond to a two-dimensional plane stress state.

For the direct numerical modeling of considered cyclic creep process, it was necessary to solve a number of boundary and initial-boundary problems, namely:

1. creep under tensile load p_b and p_e during t_0 =0.25 h;
2. instantaneous elastic-plastic loading with load $p_{bo} + p_{eo}$;
3. creep under load $p_{bo} + p_{eo}$ during t_1 =0.25 h to 0.5 h;
4. unloading to $p_b + p_e$;
5. creep under load $p_b + p_e$ during t_2 =0.25 h to 0.75 h;
6. instantaneous elastic-plastic loading to load $p_{bo} + p_{eo}$;
7. creep under load $p_{bo} + p_{eo}$ during t_1 =0.25 h to 1 h.

After studies of the convergence of the solutions, a FE mesh with 600 elements and 357 nodes was involved in the calculations. Computer simulation was carried out and the results of sequential data calculation of these 7 problems were obtained. The results in the form of the final distribution of von Mises strains in the beam are shown in (Fig. 2.11a). The areas of the beam where irreversible plastic strains take place are marked with ovals. As you can see, they occupy a very limited area near the fixed side.

Fig. 2.11: Distributions of von Mises equivalent strain in a beam area. a) - direct solution of 7 creep and plastic problems; b) – calculations with use of Eq. (2.24).

Next, the same elastic-plastic-creep behavior of the beam under consideration was modeled using the obtained Eq. (2.24), which uses the influence function $K(\sigma_{vM}, t)$ and reflects the effect of cyclic overloads. The result in the form of a similar distribution of von Mises strains in the beam is shown in (Fig. 2.11b). Comparing the distributions obtained by direct integration (Fig. 2.11a) and by using the influence function (Fig. 2.11b), we conclude that calculations based on the equivalent rate of irreversible strains qualitatively and quantitatively correctly determine the location and level of maximum strains. In the rest beam area, the deviations between the distributions are also minimal. Somewhat larger strains occur around the zones near the fixed side, which are about 0.5%. Such a deviation can be considered as satisfactory, taking into account the fact that when determining long-term strength based on calculations of accumulated damage, the area of maximum stresses and strains is decisive. The obtained conclusion regarding the satisfactory degree of accuracy when solving problems with overloading in the cycle, which leads to the occurrence of plastic strains, allows us to use the proposed approach for the modeling the deformation of more complex structural elements.

2.5 Numerical Simulation of the Cyclic Creep-damage in DWTC System Model

As noted in the literature review, cooled systems and blades (common name Double-wall transpiration cooling system - DWTC system) are now widely used in practice -

both in gas turbines and in turbines [6, 8]. Real blades are three-dimensional objects with a rather complex geometric shape, with a large number of different cooling channels. Correct numerical simulation of such objects by FEM using requires the meshes with a very large number of finite elements (of the order of 10^6) as well as large amounts of computer resources. In the presence of such data volumes, their processing in order to identify qualitative patterns of deformation and damage is not a simple task, and errors in solving physically nonlinear problems that accumulate when using models with a large number of elements can lead to incorrect conclusions.

In this regard, the papers have been published, in which the behavior of DWTC systems is simulated using simpler models (for example, rod models in [12]). Using analytical and approximate methods for such a model, it was possible to obtain a number of qualitative regularities of the behavior of the systems under consideration. But the important factors of non-homogeneous temperature and stress distributions, stress redistribution during physically nonlinear deformation remained unconsidered.

However, it is proposed to develop this approach in the direction of, on the one hand, the complication of the model, so that it reflects the main features inherent in a complex stress state, and on the other hand, so that the correct numerical simulation can take place in an acceptable time frame (the calculation time of one variant should not exceed 20-30 min). Also, this model should be built based on those approaches that are used in the engineering analysis of structural elements, namely the FEM approaches. Therefore, in this paper, the construction of a simplified two-dimensional model of a cooled turbine blade is proposed. At the same time, the main components of the thermal load inherent in the deformation of such blades will be taken into account: non-uniform temperature distribution due to the effect of the cooling, non-uniform pressure on the blade and the contribution of centrifugal forces. There remains the possibility of expanding the field of research, for example, due to the addition of new cooling channels to the model, the possibility of making a blade from different materials, etc.

2.5.1 Description of the Calculation Model

A model of a structural element with a plane cross-section, which has two walls and a bridge between them, simulating the effect of a blade's shelf (Fig. 2.12), is involved in the simulation. The following dimensions were set: total height $h = 0.3$ m, shelf length $L = 0.09$ m, shelf height $h_1 = 0.06$ m, blade height $h_2 = 0.235$ m, blade wall thickness $h_w = 0.024$ m, cooling channel width $a = 0.018$ m. The lower surface is rigidly fixed. The load consist of pressure from the gas flow, action of centrifugal loads.

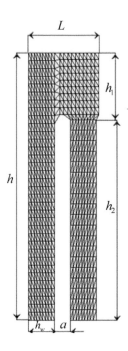

Fig. 2.12 FE scheme and main dimensions of the blade model.

2.5.2 Determination of the Temperature and Stress Field in the Blade

In this modelling cycle, we will consider the steady-state temperature field that is created in the blade during its steady-state cooling. For calculations, we use the developed research program *FEM Temperature 2d*. Let us use the boundary conditions of the 1st kind and set the temperature distribution along the model's boundaries.

Let us consider several boundary conditions of the same type, which differ by the level of blade heating. On the left edge (see Fig. 2.12) the temperature T_l is set, inside the model, on the edges of the cooling channel T_{in}, on the right side is the temperature that varies from T_l on the shelf to T_{rmin} on the blade.

A special preprocessor program for generating FE meshes has been developed for numerical modelling. An example of its operation for a model with 18 FE in wall thickness, a total of 1170 elements, is shown in Fig. 2.12. The results of solving the stationary thermal conductivity problem in the form of temperature distribution in finite elements are shown in the figures in the corresponding subsections.

Let us discuss the problem of the refinement of stress state. In the case of solving the problem of thermo-elasticity at the initial stage of integration, the considered calculation scheme has a defect associated with the presence of large unphysical values of stresses in the boundary. In the real design, the blade is further continued by the root, which is in contact with the locking joint. To clarify the stress-strain state in the proposed model, a real complex analysis of the stress-strain state of such a blade was performed. The used calculation scheme is presented in Fig. 2.13.

The results of the thermoelastic analysis of the stress-strain state for the three-dimensional model of the blade are presented in Fig. 2.14 a) (temperature distribution) and 2.14 b) (distribution of von Mises equivalent stress). From the analysis of the stress state calculation, it can be seen that there are increased stress values in the area of the transition from the blade to the root.

In this regard, for the creep-damage simulation in the blade model using the calculation scheme presented in Fig. 2.12, stress values in a von Mises stress range from 360 MPa to 260 MPa are considered in the area of the model's fixation.

2.5.3 Creep Calculations for a Two-dimensional Model of a Blade Made of Nickel Based Alloy

Let us consider the results of the cycle of calculations of the stress-strain state accumulation in the blade model given in Fig. 2.12. Blade material is high-chromium corrosion-resistant foundry heat-resistant nickel based alloy (Ni 57%, Cr 16%, Co 11%, W 5%) [30]. Temperature range: 950-850°C. $T_1 = 950°C$, $T_{in} = 850°C$, $T_{rmin} = 900°C$. The distribution of the temperature field along the cross section of the model, which is consistent with that obtained for the 3D model (2.14) a), is given in Fig. 2.15.

Let us consider the case of loading the lateral face of the blade with pressure from the gas flow, which varies according to the functional dependence on the height of the blade as a set of linear sections. The graph of this dependence is shown in Fig. 2.16 a).

The action of centrifugal loads with an intensity of 8 MPa is given. Temperature stresses are taken into account. The results of the calculations in the form of distributions of the von Mises stress fields along the cross section of the model are given in Fig. 2.16. Figure 2.16 b) contains data obtained for the case of pressure acting

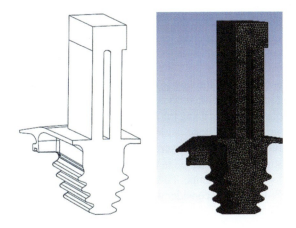

Fig. 2.13 A sketch of a simplified blade model and its FE mesh.

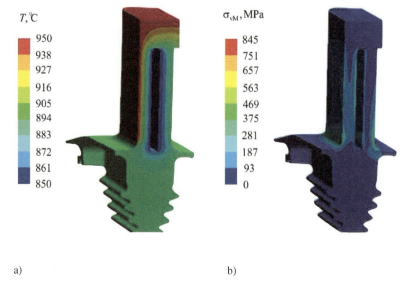

Fig. 2.14: Calculation results in a three-dimensional statement: a) – temperature distribution; b) – distribution of von Mises equivalent stresses. $t = 0$ h.

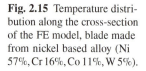

Fig. 2.15 Temperature distribution along the cross-section of the FE model, blade made from nickel based alloy (Ni 57%, Cr 16%, Co 11%, W 5%).

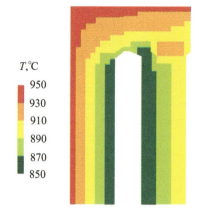

according to the law given in Fig. 2.16 a), as well as Fig. 2.16 b) for the case of constant pressure with a value of 13 MPa, which corresponds to the average value of pressure when using the law presented in Fig. 2.16 a). Comparing these distributions, we come to a conclusion about the practical closeness of the results. In this regard, further calculations were performed for the case of constant pressure. Note that the stress values obtained in the thermoelastic calculation do not exceed the yield strength for the given temperature of 950°C, $\sigma_y = 390$ MPa.

First, let us consider the results of numerical modeling of the blade behavior under the action of only static loads. The constitutive equations (2.3), (2.4) are applied. Let

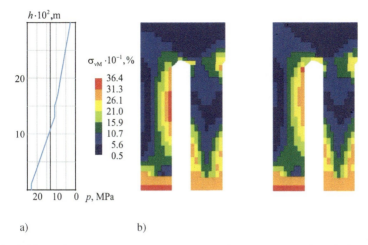

Fig. 2.16: Distributions of pressure (a) and of von Mises stress (b). Blade made from nickel based alloy (Ni 57%, Cr 16%, Co 11%, W 5%). Static loading. $t = 0$ h..

us present in addition to described creep constants of considered alloy the values of constants are included in damage evolution equation (2.4): $D = 1.18 \cdot 10^{-17}$ MPa^{-m}/h, $m = l = 5.69$. The result of calculations show that the time of hidden damage accumulation is equal to 24.16 h. The obtained distributions of von Mises strains (a) and damage parameter (b) for this time value are presented in Fig. 2.17.

It can be seen from the obtained data that the largest strains are in the left side of a blade not so far to the transition zone to the blade root. The maximum value reaches 1.7%. Also, the similar strain values occur in the area of stress concentration near

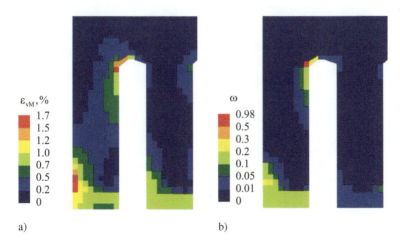

Fig. 2.17: Distribution of the von Mises strain (a) and damage parameter (b), blade made from nickel based alloy (Ni 57%, Cr 16%, Co 11%, W 5%). Model 1, static loading. $t = 24.16$ h.

the blade shelf. The fracture occurs in this place, but traditional type of the fracture in the blade fixed area is possible due to fairly high values of damage parameter (0.2-0.3) here. This area is characterized by joint action of temperature stresses and gas pressure.

Further let us present the data of calculation with considering cyclic overloading of blade with plastic deformation through a cycle. The influence function $K(\sigma_v, t)$ (2.24) was used in simulation. Similar distributions of strains and damage parameter are presented in Fig. 2.18.

An analysis of the resulting strain distribution shows that, due to an increase in the rate of accumulation of irreversible strain in the case of cyclic overloads, the level of strain accumulated by the time of failure in the latter case is much higher, by about 20–30%. The maximum strain increases from 1.7 to 2.2%. From the analysis of the distributions, it can be seen that the zones with maximum deformation remain practically the same as in the case of static loading.

The fracture time was changed insignificantly, this was only due to processes of stress varying due to more intensified creep. However, it is possible to stress the expansion of the zones of possible fracture: now it is practically equally likely in both places with maximum damage - both near the fixed side of the blade and in the area of the shelf.

The obtained results cannot be considered completely satisfactory from the point of view of design demands for blades with the necessary long-term strength. The lifetime of 22-24 hours corresponds to approximately the same number of GTE work cycles, which is insufficient. In this regard, we will demonstrate the possibility of making corrections to the model by increasing the thickness of the blade walls by 25%, and the height of the shelf by 15%. The developed FE preprocessor allows you to quickly switch to a new model. Further in the text, we will refer to the model

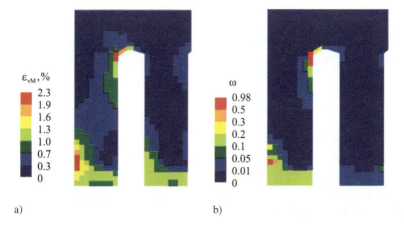

Fig. 2.18: 17 Distribution of the von Mises strain (a) and damage parameter (b), blade made from nickel based alloy (Ni 57%, Cr 16%, Co 11%, W 5%). Model 1, cyclic loading. $t = 22.47$ h.

considered above as model number 1, and the new one with increased dimensions as model number 2.

For model 2, a similar cycle of calculations was carried out, as presented above for model 1. Additionally, the influence of the pressure level from the gas flow on the deformation and damage in the blade was studied. The value of constant pressure in one of the options was reduced by 4 times, to 3.3 MPa. This version of the calculation model will be given the number 2.2, and the model in which the pressure value of 13 MPa is used will have the number 2.1. Note that due to the increase in the thickness of the walls and the size of the shelf, the overall level of stresses in the problems presented by models 2.1 and 2.2 is 10-15% lower.

The results of the calculations are given in Figs. 2.19-2.22, to model 2.1 a) and 2.2 b), respectively. Figures 2.19 and 2.20 show the distributions of von Mises strains, as well as Figs. 2.21 and 2.22 built for damage parameter distributions. All results are given for time t = 57.1 h, which precedes the moment of completion of hidden fracture.

Based on the analysis of the obtained results, it is possible to draw the following conclusions. The general level of strains remains approximately the same for all three analyzed variants. The maximum values for static load reach 1.6-1.7%, for cyclic loading, when rates are higher, 2.2-2.3%. The same strain level in this particular case is due, most likely, to the compensation of the higher strain rate in model 1 (higher stress level) and twice the time of deformation until the completion of hidden fracture for models 2.1 and 2.2. A significant strain level in all cases occurs in the area of the fixed side. In model 2.1, in which a higher level of pressure is set, during cyclic deformation, a significant strain level, up to 1%, also occurs on the inner side of the first wall. When the lateral pressure decreases (model 2.2), this distribution practically disappears. In general, it is possible to note that the impact of load cyclicity is reflected only on the general increased strain level while preserving the main areas of more intensive deformation.

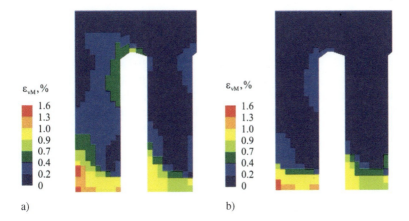

Fig. 2.19: Distribution of the von Mises strain, blade made from nickel based alloy (Ni 57%, Cr 16%, Co 11%, W 5%). Model 2.1 (a) and Model 2.2 (b), static loading. $t = 57.1$ h.

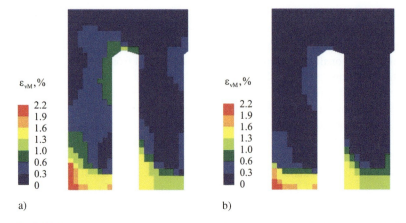

Fig. 2.20: Distribution of the von Mises strain, blade made from nickel based alloy (Ni 57%, Cr 16%, Co 11%, W 5%). Model 2.1 (a) and Model 2.2 (b), cyclic loading. $t = 57.1$ h.

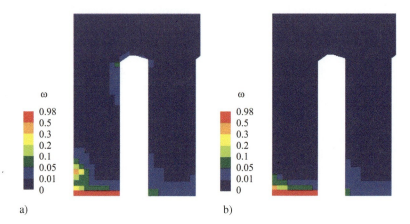

Fig. 2.21: Distribution of the damage parameter, blade made from nickel based alloy (Ni 57%, Cr 16%, Co 11%, W 5%). Model 2.1 (a) and Model 2.2 (b), static loading. $t = 57.1$ h.

When moving from model 1 to model 2, there is a qualitative change in the nature of the place of completion of hidden fracture. In model 2, the failure occurs in the area near fixed side, where the maximum stresses occur under elastic loading. Such a change is due to a decrease in the load on the first wall of the blade, which leads to a lower rate of accumulation of damage. Reducing the pressure on the blade (model 2.2) leads to the fact that practically all the damage is localized in the area of the fixed side.

Due to the non-linearity of the processes of stress redistribution during creep, it is not possible to draw a conclusion about the place of failure in advance, this is determined by the composition of the stress level. With the help of numerical simulation of creep and damage, as shown by the given results, it is possible to make

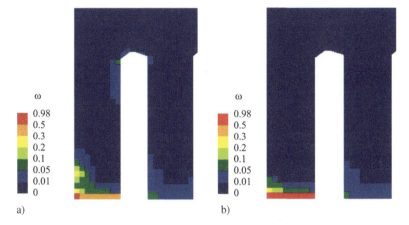

Fig. 2.22: Distribution of the damage parameter, blade made from nickel based alloy (Ni 57%, Cr 16%, Co 11%, W 5%). Model 2.1 (a) and Model 2.2 (b), cyclic loading. $t = 57.1$ h.

such a conclusion. That is, in the case under consideration, even a slight change in the thickness of the blade walls can lead to qualitative changes in the nature of the fracture. Note that the nature of the fracture in model 2, i.e. in the places of transition to the blade root, is more acceptable from the point of view that the real stress level in the three-dimensional model of the blade, which takes into account the contact interaction between the blade and the rotor, will be lower.

2.5.4 Creep Calculations for a Two-dimensional Model of a Blade Made of an Inconel X Alloy

As a second example of the behaviour the considered double-walled blade is creep-damage analysis in more suitable from the point of view of possibility of deformation and fracture occurrence range of temperature-loading conditions. The case of stresses which do not exceed the yield limit is considered. Dimensions are equal to presented for model 2 in previous section.

Blade material is Inconel X alloy, temperature range - 730-830°C. $T_l = 830°C$, $T_{in} = 730°C$, $T_{rmin} = 780°C$. The load is pressure of 0.66 MPa from the gas flow, assumed to be constant over the height of the blade, the action of centrifugal forces with an equivalent intensity that varies linearly from 12 MPa on the outer wall to 8 MPa on the inner wall. The same operations for the stress determining near blade root in 3D statement were done.

For the Inconel X alloy, the creep curves given in [32] were considered at a constant temperature of 1003K and stresses of 168.8 MPa and 211 MPa to determine the Norton creep law material parameters B, n and the long-term strength curve at stresses of 168.8 MPa, 211 MPa and 235 MPa for finding the material damage parameters D, m, l.

For a temperature of 1088 K, curves at the same stress values were used to find the material parameters. After their determining at two different temperatures, them were found for the total temperature range T = 1003 K-1088 K: $B = 1.07 \cdot 10^3 \text{MPa}^{-n}/\text{h}$, $n = 6.33, D = 1.1 \cdot 10^8 \text{MPa}^{-m}/\text{h}, m = 4.86, l = 1.054, Q_c = Q_d = 5.083 \cdot 10^4$ K.

First, we consider the results obtained for the case of a purely static load and temperatures that do not varying over time. The distribution of temperatures in the cross-section of the FE model is presented in Fig. 2.23; von Mises stresses under thermoelastic initial loading: Fig. 2.24; the values of the von Mises strains in this time (a) and damage parameter before the end of the process of hidden damage accumulation (b), $t_* = 61.1$ h: Fig. 2.25.

Next, consider the case of cyclic loading of the blade under consideration, in which the stresses do not exceed the yield limit. In this case, it is possible to apply the constitutive equations obtained using the methods of two time scales and averaging over the period (2.15) when solving the boundary – initial value creep problem (2.33). Let us take for modeling the case with five overloads in a cycle within one hour (Fig. 2.25): $t_1 = 7$ min, $t_2 = t_3 = 5$ min, $T_c = 1$ h. This cycle simulates the operation of a gas turbine for one hour.

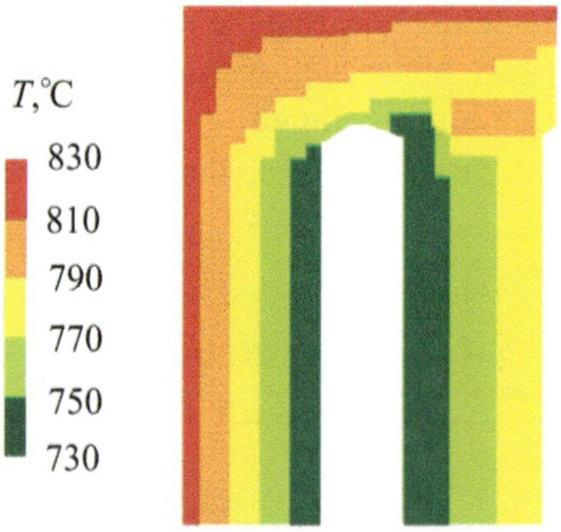

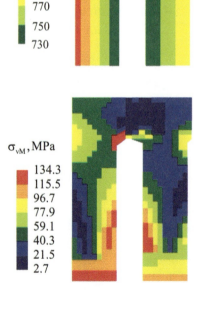

Fig. 2.23 Distribution of temperatures through the cross-section of the FE model, blade made of alloy Inconel X.

Fig. 2.24 Distribution of von Mises stresses through the cross-section of the FE model, blade made of alloy Inconel X, $t = 0$.

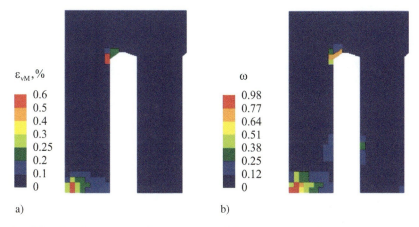

Fig. 2.25: Distribution of von Mises strains a) and damage parameter b) through the cross-section of the FE model, blade made of alloy Inconel X, static loading $t = 61.1$ h.

After performing the transformations and expanding the stress and temperature functions into Fourier series, the expressions for the influence functions of the type included in Eq. (2.15) were obtained:

$$g_n\left(M_k^{\sigma_i}\right) = \int_0^1 \left(a_0\left(M_k^{\sigma_i}\right) + \sum_{k=1}^{\infty} M_k^{\sigma_i} \cos(2\pi k\xi)\right)^n d\xi, \qquad (2.38)$$

$$g_m\left(M_k^{\sigma_e}\right) = \int_0^1 \left(a_0\left(M_k^{\sigma_e}\right) + \sum_{k=1}^{\infty} M_k^{\sigma_e} \cos(2\pi k\xi)\right)^m d\xi,$$

$$g_T\left(M_k^T\right) = \int_0^1 \exp\left(-\frac{Q_c}{T_m}\left(a_0\left(M_k^T\right) + \sum_{k=1}^{\infty} M_k^T \cos(2\pi k\xi)\right)^{-1}\right) d\xi,$$

$$M_k^{\sigma_i} = \frac{(\sigma_a)_i}{(\sigma_m)_i}, M_k^{\sigma_e} = \frac{(\sigma_a)_e}{(\sigma_m)_e}, M_k^T = \frac{T_a}{T_m},$$

where

$$a_0(M_k) = \frac{1}{T_p}\left(0.5T_c + (5M_k - 2.5)t_2 + 2t_3\right),$$

$$a_k(M_k) = \frac{1}{\pi k}\left[2M_k \sin\left(\frac{\pi k}{T_c}t_2\right) \cdot \left(\cos\left(\frac{\pi k}{T_c}(2t_1 + t_2)\right) + \right.\right.$$

$$+ \cos\left(\frac{\pi k}{T_c}(2t_1 + 3t_2 + 2t_3)\right) + \cos\left(\frac{\pi k}{T_c}(2t_1 + 5t_2 + 4t_3)\right) +$$

$$
\begin{aligned}
&+\cos\left(\frac{\pi k}{T_c}(2t_1+7t_2+6t_3)\right)+\cos\left(\frac{\pi k}{T_c}(2t_1+9t_2+8t_3)\right)+\\
&+2\sin\left(\frac{\pi k}{T_c}t_3\right)\cdot\left(\cos\left(\frac{\pi k}{T_c}(2t_1+2t_2+t_3)\right)+\cos\left(\frac{\pi k}{T_c}(2t_1+4t_2+3t_3)\right)+\right.\\
&+\cos\left(\frac{\pi k}{T_c}(2t_1+6t_2+5t_3)\right)+\cos\left(\frac{\pi k}{T_c}(2t_1+8t_2+7t_3)\right)\Big)+\\
&+\frac{1}{t_1}\left(\frac{T_c}{2\pi k}\left(\cos\left(\frac{2\pi k}{T_c}t_1\right)-1\right)+t_1\sin\left(\frac{2\pi k}{T_c}t_1\right)\right)-\\
&-\frac{1}{T_c-(t_1+5t_2+4t_3)}\left(\frac{T_c}{2\pi k}\left(1-\cos\left(\frac{2\pi k}{T_c}(t_1+5t_2+4t_3)\right)\right)+\right.\\
&+\left.(T_c-(t_1+5t_2+4t_3))\sin\left(\frac{2\pi k}{T_c}(t_1+5t_2+4t_3)\right)\right)\Big].
\end{aligned}
$$

The representation of the coefficients a_k does not look very simply, but from a practical point of view, the implementation of constitutive equations with influence functions of the type (2.38) is not a problem: this expression is added to just one small function of the software tool.

Let us consider the results of the calculation analysis of cyclic creep processes, which is accompanied by damage. We consider the case when the pressure on the blade and its heating-cooling occur according to the law presented in Fig. 2.25. The relationship between the components of the stress tensor during additional loading and temperatures during heating is considered to be as follows

$$\sigma_{ij}^a = (1+L)\sigma_{ij}^m, \quad T^a = (1+H)T^m$$

where $L < 1, H < 1$ are coefficients of the overloading and heating.

First, we present the numerical simulation data for different values of L and $H = 0$ m. They are presented in Fig. 2.26 and in Table 2.1. Analyzing the table, we come to the conclusion that the increase in cycle stress values leads to a reduction in the lifetime values and an increase in strains. Similar results were obtained earlier with other cycle parameters [25–28] and simple geometry.

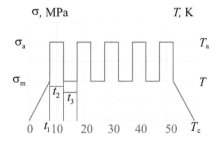

Fig. 2.26 Dependence of stress and temperature from time through the loading cycle.

2 Creep-Damage Processes in Cyclic Loaded Double Walled Structures

Table 2.1: Dependence of the time to fracture and maximum von Mises strains on the coefficient of additional loading.

L	0	0.16	0.25	0.33	0.5
Time to fracture, h	0.6	54.95	53.37	51.65	47.88
Maximum von Mises strain,%	1.6	1.2	1.3	1.35	1.4

As an example, Fig. 2.27 shows the distribution of the damage parameter and the von Mises strains at $L = 0.5$ along the cross-section of the blade. For other values of L, the distributions are qualitatively similar.

Finally, we present the results of the numerical simulation taking into account the cyclical varying of both loads and temperatures. As an example, consider the case $L = H = 0.25$. The results are shown in Fig. 2.28, where the distributions of the von Mises strains (a) and the damage parameter (b) along the cross-section of the blade are presented. The time to complete the hidden fracture was 44.45 hours.

Comparing the results of calculations taking into account the cyclical effect of temperatures (Fig. 2.28) and without it (Fig. 2.27) for the same value of load increase in the cycle $L = 0.25$, it is possible to conclude that, despite the fact that the value of the time to completion hidden fracture did not change significantly, only for three hours, in the area of the transition to the shelf, a higher value of the damage parameter was obtained, which may indicate the occurrence of an additional fracture place. The level of damage in the fixed side area changes slightly.

Analyzing the change in the strain level, we see that the additional cyclical varying in temperature increases it almost twice, the maximum values from 1.4% increase to 2.7%. Such a difference between varying in strains and the level of damage can be

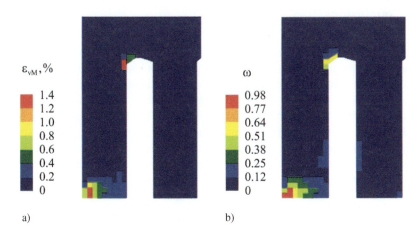

Fig. 2.27: Distribution of von Mises strains a) and damage parameter b) through the cross-section of the FE model, blade made of alloy Inconel X, cyclic loading $t = 47.88$ h.

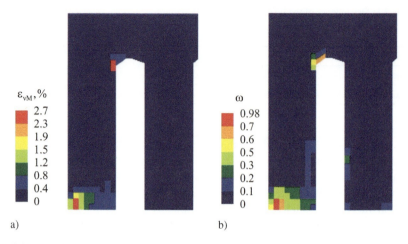

Fig. 2.28: Distribution of von Mises strains a) and damage parameter b) through the cross-section of the FE model, blade made of alloy Inconel X, cyclic loading and heating $t = 44.45$ h.

explained by a greater dependence of the creep rate on stresses for a given material and temperature range than the dependence of the damage parameter on them. This is reflected in the difference of approximately 1.5 times between the values of exponents n and m in the corresponding constitutive equations.

Calculations at other values of L and H provide the corresponding results of the intensification of the level of strains and the reduction of the lifetimes when the values of these coefficients increase.

2.6 Conclusions

An approach to determining the deformation level and long-term strength of structural elements that are under conditions of cyclic loading and heating and in the material of which creep develops is presented. The method for solving the boundary - initial value problems is described. It is based on the traditional combination of FEM and difference methods of integration for initial problems. The basis of the method is the developed and verified constitutive equations for describing the creep and damage of the material. The cases of the cycle stresses varying in a wide range, including the conditions where they exceed the yield stress, as well as the creep when it is not exceeded by stresses, are considered.

The basis of the calculation method is the formulation and description of equivalent creep processes, which allows you to significantly reduce the calculation time due to the absence of the need for direct integration by cycle, which, if there is a sufficiently large number of them, makes it impossible to effectively analyze options during design. Also, cycle integration is a rather complex procedure that cannot ensure in all cases the convergence of results with average computing resources. Finally, the transition to the modeling of averaged processes provides an opportunity to use modern engineering software complexes, which effectively implement the methods of creep analysis under static loading.

In the case when the cycle stress values do not exceed the yield stress, due to the similarity of the strain and damage accumulation curves under static and cyclic loading, it is possible to apply asymptotic methods together with period averaging and to formulate the form of a new boundary- initial value problem with new averaged constitutive equations. However, in the case of cyclic overloads with exceeding the yield stress, the creep curves under static and cyclic deformation are different. For the latter case, the paper proposes an approach that allows, based on the approximation of cyclic creep curves in a wide range of stresses, to obtain an expression for a new function that reflects the effect of creep acceleration due to load cyclicity.

The proposed approaches and solution methods were used to analyze the creep-damage processes in a model of a GTE blade with double walls. DWTC systems are currently being intensively developed, but due to the complexity of the geometry and significant three-dimensionality of the problem, direct computational analysis of the regularities of long-term high-temperature processes developing in their material is difficult and requires a large amount of resources and time. Due to this, the use of such system's models is one of the effective ways to better understand the processes that take place in them.

This paper proposes an approach to construct a simplified model of a blade with double walls, which, on the one hand, takes into account all types of temperature-force influences and stress levels, and on the other hand, thanks to the transition to a two-dimensional scheme, provides the possibility of both rapid modeling and visual displaying the results in one plane. It is clear that in the future it is necessary to move the main conclusions to three-dimensional modeling case.

The main conclusions obtained during numerical modeling can be summarized as follows. As with the analysis of cyclic creep and damage in simpler implementations of the geometry of structural elements obtained earlier [25–28], it was demonstrated that the addition of cyclic loads and temperatures in comparison with the corresponding static process leads to an increase in the rate of creep and damage accumulation, which is reflected by a greater level of accumulated strains and a shorter time to fracture.

The paper demonstrates the possibilities of the proposed approach to the analysis of the stress-strain state and long-term strength, taking into account the impact of load cyclicity by reducing of the problem dimension. It is clear that for a comprehensive analysis of the behavior of such complex systems as the considered double-wall blade, similar studies should be continued to take into consideration other important influencing factors, such as corrosion, multi-cycle fatigue, thermal shock, and others.

Acknowledgements Dmytro Breslavsky acknowledges the support by the Volkswagen Foundation within the programme "Visiting research program for refugee Ukrainian scientists" (Az. 9C184).

References

[1] Bathe KJ (2014) Finite Element Procedures, 2nd edn. Prentice Hall
[2] Murakami S (2012) Continuum Damage Mechanics - A Continuum Mechanics Approach to the Analysis of Damage and Fracture, Solid Mechanics and Its Applications, vol 28. Springer
[3] Naumenko K, Altenbach H (2016) Modeling High Temperature Materials Behavior for Structural Analysis - Part I: Continuum Mechanics Foundations and Constitutive Models, Advanced Structured Materials, vol 28. Springer
[4] Simo JC, Hughes TJR (1998) Computational Inelasticity. Springer, New York
[5] Zienkiewicz OC, Taylor RL, Fox DD (2013) The Finite Element Method for Solid and Structural Mechanics, 7th edn. Butterworth-Heinemann
[6] Murray AV (2019) Advanced gas turbine cooling: double-wall turbine cooling technologies in turbine ngv/blade applications. Phd thesis, University of Oxford, Oxford
[7] Li Z, Wen Z, Pei H, Yue X, Wang P, Ai C, Yue Z (2022) Creep life prediction for a nickel-based single crystal turbine blade. Mechanics of Advanced Materials and Structures **29**(27):6039–6052
[8] Murray AV, Ireland PT, Rawlinson AJ (2017) An integrated conjugate computational approach for evaluating the aerothermal and thermomechanical performance of double-wall effusion cooled systems. In: Turbo Expo: Power for Land, Sea, and Air, American Society of Mechanical Engineers, vol 5B: Heat Transfer, p V05BT22A015
[9] Rezazadeh Reyhani M, Alizadeh M, Fathi A, Khaledi H (2013) Turbine blade temperature calculation and life estimation - a sensitivity analysis. Propulsion and Power Research **2**(2):148–161
[10] Skamniotis C, Cocks AC (2021) 2d and 3d thermoelastic phenomena in double wall transpiration cooling systems for gas turbine blades and hypersonic flight. Aerospace Science and Technology **113**:106,610
[11] Skamniotis C, Cocks AC (2021) Creep-plasticity-fatigue calculations in the design of porous double layers for new transpiration cooling systems. International Journal of Fatigue **151**:106,304
[12] Skamniotis C, Cocks AC (2022) Analytical shakedown, ratchetting and creep solutions for idealized twin-wall blade components subjected to cyclic thermal and centrifugal loading. European Journal of Mechanics - A/Solids **95**:104,652
[13] Wen Z, Liang J, Liu C, Pei H, Wen S, Yue Z (2018) Prediction method for creep life of thin-wall specimen with film cooling holes in ni-based single-crystal superalloy. International Journal of Mechanical Sciences **141**:276–289
[14] Altenbach H, Brünig M (eds) (2015) Inelastic Behavior of Materials and Structures Under Monotonic and Cyclic Loading, Advanced Structured Materials, vol 57. Springer, Cham
[15] Barrett PR, Hassan T (2020) A unified constitutive model in simulating creep strains in addition to fatigue responses of haynes 230. International Journal of Solids and Structures **185-186**:394–409

[16] Chaboche JL, Kanouté P, Azzouz F (2012) Cyclic inelastic constitutive equations and their impact on the fatigue life predictions. International Journal of Plasticity **35**:44–66
[17] Ding B, Ren W, Zhong Y, Yuan X, Zheng T, Shen Z, Guo Y, Li Q, Liu C, Peng J, Brnic J, Gao Y, Liaw PK (2022) Comparison of the creep-fatigue cyclic life saturation effect for three different superalloys. Materials Science and Engineering: A **842**:143,086
[18] Dong Y, Zhu Y, Wu F, Yu C (2022) A dual-scale elasto-viscoplastic constitutive model of metallic materials to describe thermo-mechanically coupled monotonic and cyclic deformations. International Journal of Mechanical Sciences **224**:107,332
[19] Meng L, Chen W (2022) A new thermodynamically based model for creep and cyclic plasticity. International Journal of Mechanical Sciences **214**:106,923
[20] Ohno N, Nakamoto H, Morimatsu Y, Okumura D (2021) Modeling of cyclic hardening and evaluation of plastic strain range in the presence of pre-loading and ratcheting. International Journal of Plasticity **145**:103,074
[21] Wang C, Xuan FZ, Zhao P, Guo SJ (2021) Effect of cyclic loadings on stress relaxation behaviors of 9–12%cr steel at high temperature. Mechanics of Materials **156**:103,787
[22] Zhang W, Wang X, Gong J, Jiang Y, Huang X (2017) Experimental and simulated characterization of creep behavior of p92 steel with prior cyclic loading damage. Journal of Materials Science & Technology **33**(12):1540–1548
[23] Chaboche JL (2002) Damage mechanics. In: Comprehensive Structural Integrity, Pergamon Press, vol 2, pp 213–282
[24] Lemaitre J, Chaboche JL (1994) Mechanics of Solid Materials. Cambridge University Press, Cambridge
[25] Breslavskii DV, Morachkovskii OK (1998) Nonlinear creep and the collapse of flat bodies subjected to high-frequency cyclic loads. International Applied Mechanics **34**(3):287–292
[26] Breslavsky DV, Morachkovsky OK, Tatarinova OA (2008) High-temperature creep and long-term strength of structural elements under cyclic loading. Strength of Materials **40**(5):531–537
[27] Breslavsky D, Morachkovsky O, Tatarinova O (2014) Creep and damage in shells of revolution under cyclic loading and heating. International Journal of Non-Linear Mechanics **66**:87–95
[28] Breslavs'kyi DV, Korytko YM, Morachkovs'kyi OK (2011) Cyclic thermal creep model for the bodies of revolution. Strength of Materials **43**(2):134–143
[29] N RY (1969) Creep Problems in Structural Members. North-Holland Series in Applied Mathematics and Mechanics, North-Holland Publishing Company, Amsterdam
[30] Maksyuta II, Klyass OV, Kvasnitskaya YG, Myalnitsa GF, Mikhnyan EV (2014) Technological features of high chrome nickel allay, complex-alloyed rhenium and tantalum (in Russ.). Electrometallurgy Today (1):41–48

[31] Maugin GA (1999) The Thermomechanics of Nonlinear Irreversible Behaviors - An Introduction, World Scientific Series on Nonlinear Science Series A, vol 27. World Scientific
[32] Guarnieri G (1954) The creep-rupture properties of aircraft sheet alloys subjected to intermittent load and temperature. In: Frey D (ed) Symposium on Effect of Cyclic Heating and Stressing on Metals at Elevated Temperatures ASTM STP No. 165-EB, ASMT, pp 105–147

Chapter 3
Creep Mechanics – Some Historical Remarks and New Trends

Holm Altenbach, Johanna Eisenträger, Katharina Knape, and Konstantin Naumenko

Abstract Creep mechanics is a branch of continuum mechanics that began to develop in the late 19th century. In the 1930s the first theories were developed that allowed the analysis of structures and the description of material behavior. The theory of viscoelasticity introduced approaches that could be linked to rheological models. Therefore, rheological models and their perspectives will be discussed in the concluding part.

3.1 Starting Point - the Early Period of Creep Mechanics

The starting point for the development of creep mechanics were a number of accidents in the 19th century, some of which had a tragic outcome. The analysis of the accidents showed that the machines and systems were under relatively low mechanical loads, i.e. failure occured at stresses that were significantly below the yield stress of the materials used. At the same time, all accidents and the damage observed as a result had in common that the temperature level, i.e., the operating temperatures, was elevated.

The main items of the history of creep mechanics are given, for example, in [1–4]. The first publications on creep problems were published in the second half of the 19th century. However, the main results were summarized for the first time in [5, 6]. The first constitutive law for creep problems was the Norton-Bailey law [7, 8]. This one-dimensional power law (in the sense of mathematics) contains two material parameters: the magnitude and the creep exponent. Consequently, the material description in the creep range requires more experimental effort in comparison to elastic behavior. At the beginning, energy machine design was the main

---

Holm Altenbach · Johanna Eisenträger · Katharina Knape · Konstantin Naumenko
Lehrstuhl für Technische Mechanik, Institut für Mechanik, Fakultät für Maschinenbau, Otto-von-Guericke-Universität Magdeburg, Universitätsplatz 2, 39106 Magdeburg, Germany,
e-mail: holm.altenbach@ovgu.de, johanna.eisentraeger@ovgu.de, katharina.knape@ovgu.de, konstantin.naumenko@ovgu.de

application field of creep mechanics. For example in 1933, Stodola reported about gas turbine applications [9]. However, mechanical loads are multiaxial such that the stress and the strain in the one-dimensional case should be substituted by their tensorial counterparts. Odqvist [10, 11] and Bailey [8] introduced a corresponding theory for isotropic material behavior using invariants of the stress and the strain tensors. A consistent tensorial description was privided by Prager [12] and Reiner [13], which includes in additional anisotropy. Mismatches with experimental results resulted in further modifications of the creep equations, e.g., the strain hardening theory presented by Nadai [14] and Soderberg [15] and discussed, for example, by Rimrott [16]. Stability problems considering creep were presented by Hoff [17, 18], and elements of the geometrically nonlinear theory had to be developed. The increasing use of polymer materials resulted in the development of new theories, partly based on analogies between viscoelasticity and creep. However, viscoelastic behavior is often mathematically described with the help of integral equations, see [19] among others. The simplest models were suggested using rheological models [20]. Finally, another application field should be mentioned - the creep in concrete which was studied, for example, by N.K. Arutunyan [21]. The last one was the starting point for the development of a new branch of mechanics of deformable bodies - mechanics of growing solids [22]. It must be pointed out that from the end of 1950ies special attention was paid to tertiary creep as final stage of creep in materials and structures. This branch of creep mechanics is also called creep-damage mechanics or only damage mechanics. The development of this research field started with pioneering works of Kachanov [23] and Rabotnov [24]. This pure phenomenological approach was later more and more founded upon knowledge of material scientists and we have from the beginning of 1980ies, mechanism-based approaches started to develop [25–28].

Nowadays, an extensive of textbooks and monographs on creep mechanics reporting on established research results is available. Most authors prefer the inductive approach. In many cases, creep equations in the one-dimensional form are introduced based on experimental observations. The generalization is given step by step applying mathematical approaches of tensor calculus. It seems that there is no book that represents creep mechanics as strong as books on continuum mechanics. Such books are only for the elasticity or plasticity, see, for example, [29, 30]. For creep mechanics, the following books can be recommended: [31–44]. Whereas theories for static (better quasi-static) applications in the case of monotonous loads and under isothermal conditions are well established, cyclic loads and their consequences for the creep behavior are under discussion and require further research.

3.2 IUTAM Symposia and Other Events Devoted to Problems in Creep Mechanics

Creep mechanics is the object of scrutiny of several scientific organizations worldwide. Various conferences including presentations about the topics and activities devoted to creep problems were organized. The first conference series with attention on modelling and simulation of creep in materials and applications to structures was established by the International Union for Theoretical and Applied Mechanics (IUTAM). The idea to provide IUTAM symposia *Creep in Structures* was born in the late fifties of the last century. Since 1960, every 10 years such conferences were organized:

- 1960 - Stanford (U.S.A.) [45],
- 1970 - Göteborg (Sweden) [46],
- 1980 - Leicester (U.K.) [47],
- 1990 - Kraków (Poland) [48] and
- 2000 - Nagoya (Japan) [49].

The rather long time interval between two symposia can be explained as the main focus was on creep in metals and structures made of metals and alloys. The duration of time-dependent processes resulting in significant data and their verification needs long-term tests in the time range of years.

Due to the arising interest of creep in other materials like polymers and composites, shorter time ranges for experiments are realistic (this was the result of discussions during the Nagoya meeting in 2000). However, the next symposium after the meeting in Nagoya was only held in Paris/France in 2012 [50]. At the same time, the title of the symposium was changed to *Advanced Materials Modelling for Structures*. This year's symposium comes 11 years after the Paris Symposium. The traditional name was returned, so that in 2023 *Creep in Structures VI* is organized.

There are a lot of other conferences devoted to selected creep and creep-damage problems, for example [51, 52]. The International Association for Applied Mathematics and Mechanics (GAMM) offered plenary lectures during their annual conferences and in 1998 and 2001 the topic of plenary lectures were related to creep mechanics [53, 54]. In 1999, the International Centre for Mechanical Sciences had organized a special course on creep problems [55].

3.3 Research Directions and Magdeburg's Contributions

3.3.1 Kachanov-Rabotnov Approach and Mechanism-Based Models

Since 1993, the creep and creep-damage behavior of materials and thin-walled structures has bin investigated [56–61]. The first papers were based on classical creep mechanics and the damage approach of Kachanov and Rabotnov. In [62], the damage activation and deactivation was considered, and in [63] the shear correction factors in creep-damage analysis of beams, plates, and shells were presented. Edge effects were investigated in [64] and an interesting application - creep in a multipass weld - was discussed in [65]. Reference [66] included non-classical effects and in [67], some mathematical aspects were discussed. Additionally, considering engineering applications, problems of thin-walled structures were analyzed using both geometrical linear and non-linear formulations. Since 1997, instead of a purely phenomenological approach, mechanism-based models were applied [68]. The classical theory was extended to creep behavior with respect to a wide ranges of stress levels [69, 70].

3.3.2 Non-Classical Creep

The creep theory for materials exhibiting non-classical creep behavior (for example, different tension and compression behavior [32, 71], the equivalent behavior under tension, compression and torsion is different [32, 71], Poynting-Swift effect [72–74] or second order effects [75]) was in the spotlight of research activities from 1989 until 1996. This non-classical behavior is presented, e.g., in [76, 77]. The development of this specialization of creep mechanics was connected with the use of some light alloys, gray cast irons, polymers, ceramics, composites, and other materials whose creep behavior depends on the type of loading. In addition, the damage behavior of some materials features non-classical effects [78, 79]. Most of the approaches in this field are related to the generalization of the Rabotnov concept [32] about creep and a material damage parameter. In [80], isotropic and anisotropic constitutive equations reflecting creep behavior depending on the loading type are presented and applied to shells of revolution. Furthermore, a numerical example for the aluminium alloy AK4-1T is discussed. In [81], an isotropic creep law for non-classical creep behavior was suggested and the basic test for the estimation of the constitutive parameters was described. The model was applied to pure copper M1E (Cu 99%) at a temperature of 573 K. In [82–84], the energetic variant of creep constitutive equations based on the dissipation energy was discussed and applied to the titanium alloy OT-4 at 748 K. Finally, in [85], the behavior of polymers with loading type depending behavior was analyzed. The main results concerning the non-classical models in elasticity, plasticity, creep, strength criteria and fatigue were summarized in the monograph [86].

3.3.3 Benchmark Tests for Creep Problems

In [87], the accuracy of creep-damage predictions was investigated. It was shown that the accuracy of the creep-damage finite element predictions in beams and plates (bending problems) can be compared with solutions based on the Ritz method. The mesh sensitivity of the long-term predictions for beams and plates was investigated while accounting for different element types. The main conclusions were that the approximations or meshes justified for the elastic solutions using displacement based variational methods cannot be used for the creep damage analysis and the discretization established based on convergent steady state creep solutions can be used for the continuum damage mechanics analysis of thin-walled structures in bending. In

[88], the use of solid and shell elements for creep-damage problems in thin-walled structures was discussed.

3.3.4 Rheological Models

Rheological modelling is a widely used tool in material modelling and simulation. The history of rheological modelling is presented in [89, 90]. The "official beginning" was in the twenties of the last century, but the first rheological models were suggested by Robert Hooke and other scientists at the same time. With the increasing use of plastics in the fifties of the last century, the method became very popular. The focus was on the time-dependent behavior of plastics and their phenomenological description. However, in the last years, rheological models were used also in other cases with applications to metals, concrete, geomaterials, etc. Starting with [91] the creep behavior of a material with a hard and soft phase was investigated using a rheological model. Further developments of this approach are presented in [92–95] for alloys and in [96] for a polymer.

3.3.5 Thesis

The following theses were finished in the Magdeburg's team until summer 2023:

- PhD theses

    1. Jewgenij Kostenko: Zur Modellierung und Berechnung mehrschichtiger Flächentragwerke unter Einbeziehung des Werkstoffkriechens (December 1992),
    2. Konstantin Naumenko: Modellierung und Berechnung der Langzeitfestigkeit dünnwandiger Flächentragwerke unter Einbeziehung von Werkstoffkriechen und Schädigung (February 1996),
    3. Yevgen Gorash: Development of a creep-damage model for non-isothermal long-term strength analysis of high-temperature components operating in a wide stress range (July 2008),
    4. Sergii Kozhar: Festigkeitsverhalten der Al-Si-Gusslegierung AlSi12CuNiMg bei erhöhten Temperaturen (March 2011),
    5. Oksana Ozhoga-Maslovskaja: Microscale modeling grain boundary damage under creep conditions (January 2014)
    6. Mykola Ievdokymov: Identification technique of mechanism-based constitutive model for cast iron under thermo-mechanical loads (May 2015),
    7. Thomas Hanke: Viskoelastische Beschreibung des Langzeit-Kriechverhaltens von Ethylen-Tetraflurethylen (ETFE) Folien für Membrankissen-Konstruktionen (March 2016, in cooperation with Fraunhofer-Institut für Mikrostruktur von Werkstoffen und Systemen Halle),

8. Johanna Eisenträger: A Framework for Modeling the Mechanical Behavior of Tempered Martensitic Steels at High Temperatures (May 2018),
9. Vansssa Hammerschmidt: Entwicklung eines Prüfkonzeptes für thermomechanisch hoch beanspruchte Bereiche von Zylinderköpfen und numerische Abbildung der thermischen Belastungszyklen (June 2018, in cooperation with Volkswagen AG),
10. Steffen Mittag: Mechanismenbasierte probabilistische Bewertung der Ermüdungslebensdauer von Metallen unter Berücksichtigung der Streuung der temperaturabhängigen Eigenschaften (June 2019, in cooperation with Hochschule Offenburg),
11. Andreas Jilg Development and Implementation of a Cyclic Plasticity Model with Thermal Softening for Hot Work Tool Steel (June 2019, in cooperation with Hochschule Offenburg),
12. Carl Fischer: Schädigungsentwicklung und mechanismenbasierte Lebensdauermodellierung von Aluminiumgusslegierungen unter thermomechanischen Ermüdungsbelastungen (October 2021, in cooperation with Hochschule Offenburg)

- DSc theses

  1. Konstantin Naumenko: Modeling of High-Temperature Creep for Structural Analysis Applications (April 2006, published as [97])

3.4 Outlook

Further impulses for creep research are mainly expected from the following areas: power plant construction, aircraft construction and microsystem technology. The first two items are related to traditional application fields. The performance and efficiency improvement leads partly to an increase of exploitation temperature, creep and creep-damage became more important. As the component size decreases, the temperature increases and failure can be avoided only after intensive investigation of this problem.

References

[1] Odqvist FKG (1981) Historical survey of the development of creep mechanics from its beginnings in the last century to 1970. In: Ponter ARS, Hayhurst DR (eds) Creep in Structures, Springer Berlin Heidelberg, Berlin, Heidelberg, pp 1–12
[2] Naumenko K, Altenbach H (2016) Modeling High Temperature Materials Behavior for Structural Analysis. Part I: Continuum Mechanics Foundations and Constitutive Models, Advanced Structured Materials, vol 28. Springer

[3] Altenbach H, Eisenträger J (2020) Introduction to creep mechanics. In: Altenbach H, Öchsner A (eds) Encyclopedia of Continuum Mechanics, Springer Berlin Heidelberg, Berlin, Heidelberg, pp 1337–1344
[4] Altenbach H, Knape K (2020) On the main directions in creep mechanics of metallic materials. Proceedings of the National Academy of Sciences of Armenia: Mechanics **73**(3):24–43
[5] da Costa Andrade EN (1910) On the viscous flow in metals, and allied phenomena. Proceedings of the Royal Society of London Series A, Containing Papers of a Mathematical and Physical Character **84**(567):1–12
[6] da Costa Andrade EN (1914) The flow in metals under large constant stresses. Proceedings of the Royal Society of London Series A, Containing Papers of a Mathematical and Physical Character **90**(619):329–342
[7] Norton FH (1929) Creep of Steel at high Temperatures. McGraw-Hill, New York
[8] Bailey RW (1935) The utilization of creep test data in engineering design. Proceedings of the Institution of Mechanical Engineers **131**(1):131–349
[9] Stodola A (1933) Die kriecherscheinungen, ein neuer technisch wichtiger aufgabenkreis der elastizitätstheorie. ZAMM - Journal of Applied Mathematics and Mechanics / Zeitschrift für Angewandte Mathematik und Mechanik **13**(2):143–146
[10] Odqvist FKG, Hult J (1962) Kriechfestigkeit metallischer Werkstoffe. Springer, Berlin u.a.
[11] Odqvist FKG (1974) Mathematical Theory of Creep and Creep Rupture. Oxford University Press, Oxford
[12] Prager W (2004) Strain Hardening Under Combined Stresses. Journal of Applied Physics **16**(12):837–840
[13] Reiner M (1945) A mathematical theory of dilatancy. American Journal of Mathematics **67**(3):350–362
[14] Nadai A (1938) The influence of time upon creep. The hyperbolic sine creep law. In: Contributions to the Mechanics of Solids dedicated to Stephen Timoshenko by his friends on the occasion of his sixties birthday anniversary, Macmillian, pp 155–170
[15] Soderberg CR (1938) Plasticity and creep in machine design. In: Contributions to the Mechanics of Solids dedicated to Stephen Timoshenko by his friends on the occasion of his sixties birthday anniversary, Macmillian, pp 97–210
[16] Rimrott FPJ (????) The hyperbolic sine creep law in engineering practice. In: Rimrott FJP, Schwaighofer J (eds) Mechanics of the Solid State, University of Toronto Press, Toronto, pp 168–180
[17] Hoff NJ (2021) The Necking and the Rupture of Rods Subjected to Constant Tensile Loads. Journal of Applied Mechanics **20**(1):105–108
[18] Hoff NJ (1957) Buckling at high temperature. The Aeronautical Journal **61**(563):756–774
[19] Rabotnov YN (1980) Elements of Hereditary Solid Mechanics. Mir, Moscow
[20] Reiner M (1969) Deformation and Flow. An Elementary Introduction to Rheology, 3rd edn. H.K. Lewis & Co., London

[21] Arutyunyan NK (1966) Some Problems in the Theory of Creep. Pergamon Press, Oxford
[22] Arutiunian NK (1977) Boundary value problem of the theory of creep for a body with accretion. Journal of Applied Mathematics and Mechanics **41**(5):804–810
[23] Kachanov LM (1958) O vremeni razrusheniya v usloviyakh polzuchesti (on the time to rupture under creep conditions, in russ.). Izv AN SSSR Otd Tekh Nauk (8):26 – 31
[24] Rabotnov YN (1959) O mechanizme dlitel'nogo razrusheniya (A mechanism of the long term fracture, in Russ.). Voprosy prochnosti materialov i konstruktsii, AN SSSR pp 5 – 7
[25] Trampczynski WA, Hayhurst DR, Leckie FA (1981) Creep rupture of copper and aluminium under non-proportional loading. J Mech Phys Solids **29**:353 – 374
[26] Kowalewski ZL, Hayhurst DR, Dyson BF (1994) Mechanisms-based creep constitutive equations for an aluminium alloy. The Journal of Strain Analysis for Engineering Design **29**(4):309 – 316
[27] Othman AM, Dyson BF, Hayhurst DR, Lin J (1994) Continuum damage mechanics modelling of circumferentially notched tension bars undergoing tertiary creep with physically-based constitutive equations. Acta Metallurgica et Materialia **42**(3):597 – 611
[28] Dyson BF, McLean M (1998) Microstructural evolution and its effects on the creep performance of high temperature alloys. In: Strang A, Cawley J, Greenwood GW (eds) Microstructural Stability of Creep Resistant Alloys for High Temperature Plant Applications, Cambridge University Press, Cambridge, pp 371 – 393
[29] Bertram A (2012) Elasticity and Plasticity of Large Deformations, 3rd edn. Springer, Berlin
[30] Lurie A (2010) Theory of Elasticity. Foundations of Engineering Mechanics, Springer
[31] Hult JA (1966) Creep in Engineering Structures. Blaisdell Publishing Company, Waltham
[32] Rabotnov YN (1969) Creep Problems in Structural Members. North-Holland, Amsterdam
[33] Kraus H (1980) Creep Analysis. Wiley & Sons, New York
[34] Malinin NN (1981) Raschet na polzuchest' konstrukcionnykh elementov (in Russ., Creep Calculations of Structural Elements). Mashinostroenie, Moscow
[35] Boyle JT, Spence J (1983) Stress Analysis for Creep. Butterworth, London
[36] Lemaitre J, Chaboche J-L (1990) Mechanics of Solid Materials. University Press, Cambridge
[37] Skrzypek JJ (1993) Plasticity and Creep. CRC Press, Boca Raton
[38] Penny RK, Mariott DL (1995) Design for Creep. Chapman & Hall, London
[39] Skrzypek J, Ganczarski A (1998) Modelling of Material Damage and Failure of Structures. Foundation of Engineering Mechanics, Springer, Berlin
[40] Betten J (2008) Creep Mechanics, 3rd edn. Springer, Berlin

[41] Hyde T, Sun W, Hyde C (2013) Applied Creep Mechanics. McGraw-Hill Education
[42] Lokoshchenko AM (2016) Polzuchest' i dlitel'naya prochnost' metallov (in Russ., Creep and Long-term Strength of Metals). FIZMATLIT, Moscow
[43] Naumenko K, Altenbach H (2016) Modeling High Temperature Materials Behavior for Structural Analysis. Part I: Continuum Mechanics Foundations and Constitutive Models, Advanced Structured Materials, vol 28. Springer
[44] Naumenko K, Altenbach H (2020) Modeling High Temperature Materials Behavior for Structural Analysis. Part II: Solution Procedures and Structural Analysis Examples, Advanced Structured Materials, vol 112. Springer
[45] Hoff NJ (ed) (1962) Creep in Structures. Springer, Berlin
[46] Hult J (ed) (1972) Creep in Structures. Springer, Berlin
[47] Ponter ARS, Hayhurst DR (eds) (1981) Creep in Structures. Springer, Berlin
[48] Zyczkowski M (ed) (1991) Creep in Structures. Springer, Berlin
[49] Murakami S, Ohno N (eds) (1991) Creep in Structures. Kluwer, Dordrecht
[50] Altenbach H, Kruch S (eds) (2013) Advanced Materials Modelling for Structures, Advanced Structured Materials, vol 19. Springer, Berlin · Heidelberg
[51] Skrzypek JJ, Ganczarski A (eds) (2003) Anisotropic Behaviour of Damaged Materials, Lecture Notes in Applied and Computational Mechanics, vol 9. Springer, Berlin Heidelberg
[52] N O, T U (eds) (2007) Engineering Plasticity and Its Applications from Nanoscale to Macroscale, Lecture Notes in Applied and Computational Mechanics, vol 9. Trans. Tech. Publications Ltd., Stafa-Zürich
[53] Betten J (1998) Anwendungen von Tensorfunktionen in der Kontinuumsmechanik anisotroper Materialien. ZAMM-Journal of Applied Mathematics and Mechanics/Zeitschrift für Angewandte Mathematik und Mechanik: Applied Mathematics and Mechanics **78**(8):507–521
[54] Altenbach H (2002) Creep analysis of thin-walled structures. ZAMM - Journal of Applied Mathematics and Mechanics / Zeitschrift für Angewandte Mathematik und Mechanik **82**(8):507–533
[55] Altenbach H, Skrzypek JJ (eds) (1999) Creep and Damage in Materials and Structures, CISM International Centre for Mechanical Sciences, vol 399. Springer
[56] Altenbach H, Naumenko K (1995) Geometrically nonlinear small strain creep theory for thin plates including isotropic damage. Mathematical Modelling and Scientific Computing **5**(2):89–99
[57] Altenbach H, Morachkovsky O, Naumenko K, Sichov A (1996) Zum Kriechen dünner Rotationsschalen unter Einbeziehung geometrischer Nichtlinearität sowie der Asymmetrie der Werkstoffeigenschaften. Forschung im Ingenieurwesen **62**(3):47–57
[58] Altenbach J, Altenbach H, Naumenko K (1997) Lebensdauerabschätzung dünnwandiger Flächentragwerke auf der Grundlage phänomenologischer Materialmodelle für Kriechen und Schädigung. Technische Mechanik **17**(4):353–364
[59] Altenbach H, Morachkovsky O, Naumenko K, Sychov A (1997) Geometrically nonlinear bending of thin-walled shells and plates under creep-damage conditions. Archive of Applied Mechanics **67**(5):339–352

[60] Altenbach H, Naumenko K (1997) Creep bending of thin-walled shells and plates by consideration of finite deflections. Computational Mechanics 19(6):490–495
[61] Altenbach H (1999) Creep-damage behaviour of plates and shells. Mechanics of Time-Dependent Materials 3(2):103–123
[62] Altenbach H, Huang C, Naumenko K (2001) Modelling of the creep-damage under reversed stress states by considering damage activation and deactivation. Technische Mechanik 21(4):273–282
[63] Altenbach H, Naumenko K (2002) Shear correction factors in creep-damage analysis of beams, plates and shells. JSME International Journal Series A Solid Mechanics and Material Engineering 45(1):77–83
[64] Altenbach J, Altenbach H, Naumenko K (2004) Edge effects in moderately thick plates under creep-damage conditions. Technische Mechanik 24(3):254–263
[65] Naumenko K, Altenbach H (2005) A phenomenological model for anisotropic creep in a multipass weld metal. Archive of Applied Mechanics 74(11):808–819
[66] Altenbach J, Altenbach H, Naumenko K (1999) A phenomenological creep theory including nonclassical effects and creep-damage coupling. Appl Mech and Engng 4(-):11–16
[67] Altenbach H, Deuring P, Naumenko K (1999) A system of ordinary and partial differential equations describing creep behaviour of thin-walled shells. Zeitschrift für Analysis und ihre Anwendungen 18(4):1003–1030
[68] Altenbach H, Altenbach J, Naumenko K (1997) On the prediction of creep damage by bending of thin-walled structures. Mechanics of Time-Dependent Materials 1(2):181–193
[69] Altenbach H, Gorash Y, Naumenko K (2008) Steady-state creep of a pressurized thick cylinder in both the linear and the power law ranges. Acta Mechanica 195(1-4):263–274
[70] Naumenko K, Altenbach H, Gorash Y (2009) Creep analysis with a stress range dependent constitutive model. Archive of Applied Mechanics 79(6-7):619–630
[71] Lucas GE, Pelloux RMN (1981) Texture and stress state dependent creep in Zircaloy-2. Metallurgical Transactions A 12(7):1321–1331
[72] Swift W (1947) Length changes in metals under torsional overstrain. Engineering 163:253–257
[73] Billington EW (1977) Non-linear mechanical response of various metals. III. Swift effect considered in relation to the stress-strain behaviour in simple compression, tension and torsion. Journal of Physics D: Applied Physics 10(4):553–569
[74] Billington E (1985) The Poynting-Swift effect in relation to initial and post-yield deformation. International Journal of Solids and Structures 21(4):355–371
[75] Truesdell C (1964) Second-order effects in the mechanics of materials. In: Reiner M, Abir D (eds) Second Order Effects in Elasticity, Plasticity and Fluid Dynamics, Pergamon Press, Oxford, pp 1–50
[76] Murakami S, Yamada Y (1974) Effects of hydrostatic pressure and material anisotropy on the transient creep of thick-walled tubes. International Journal of Mechanical Sciences 16(3):145–160

[77] Betten J (1981) Creep Theory of Anisotropic Solids. Journal of Rheology **25**(6):565–581
[78] Krajcinovic D (1984) Continuum damage mechanics. Appl Mech Review **37**(1):1–6
[79] Murakami S (1988) Mechanical Modeling of Material Damage. Journal of Applied Mechanics **55**(2):280–286
[80] Altenbach H, Zolochevsky A (1991) Kriechen dünner schalen aus anisotropen werkstoffen mit unterschiedlichem zug-druckverhalten. Forschung im Ingenieurwesen **57**(6):172–179
[81] Altenbach H, Schieße P, Zolochevsky AA (1991) Zum Kriechen isotroper Werkstoffe mit komplizierten Eigenschaften. Rheologica Acta **30**(4):388–399
[82] Altenbach H, Zoločevskij AA (1992) Zur Anwendung gemischter Invarianten bei der Formulierung konstitutiver Beziehungen für geschädigte anisotrope Kontinua. ZAMM - Zeitschrift für angewandte Mathematik und Mechanik **72**(8):375–377
[83] Altenbach H, Zolochevsky A (1992) Energy version of creep and stress-rupture strength theory for anisotropic and isotropic materials which differ in resistance to tension and compression. Journal of Applied Mechanics and Technical Physics **33**(1):101–106
[84] Altenbach H, Zolochevsky AA (1994) Eine energetische Variante der Theorie des Kriechens und der Langzeitfestigkeit für isotrope Werkstoffe mit komplizierten Eigenschaften. ZAMM - Zeitschrift für angewandte Mathematik und Mechanik **74**(3):189–199
[85] Altenbach H, Altenbach J, Zolochevsky A (1995) A generalized constitutive equation for creep of polymers at multiaxial loading. Mechanics of Composite Materials **31**(6):511–518
[86] Altenbach H, Altenbach J, Zolochevsky A (1995) Erweiterte Deformationsmodelle und Versagenskriterien der Werkstoffmechanik. Deutscher Verlag für Grundstoffindustrie
[87] Altenbach H, Kolarov G, Morachkovsky OK, Naumenko K (2000) On the accuracy of creep-damage predictions in thinwalled structures using the finite element method. Computational Mechanics **25**(1):87–98
[88] Altenbach H, Kushnevsky V, Naumenko K (2001) On the use of solid- and shell-type finite elements in creep-damage predictions of thinwalled structures. Archive of Applied Mechanics **71**(2):164–181
[89] Altenbach H, Eisenträger J (2021) Rheological modeling in solid mechanics from the beginning up to now. Lecture Notes of TICMI **22**:13–29
[90] Altenbach H (2023) Rheological modeling- historical remarks and actual trends in solid mechanics. In: Altenbach H, Kaplunov J, Lu H, Nakada M (eds) Advances in Mechanics of Time-Dependent Materials, Advanced Structured Materials, vol 188, Springer International Publishing, Cham, pp 1–16
[91] Naumenko K, Altenbach H, Kutschke A (2011) A combined model for hardening, softening, and damage processes in advanced heat resistant steels at elevated temperature. International Journal of Damage Mechanics **20**(4):578–597

[92] Eisenträger J, Naumenko K, Altenbach H, Gariboldi E (2017) Analysis of temperature and strain rate dependencies of softening regime for tempered martensitic steel. The Journal of Strain Analysis for Engineering Design **52**(4):226–238
[93] Eisenträger J, Naumenko K, Altenbach H (2018) Numerical implementation of a phase mixture model for rate-dependent inelasticity of tempered martensitic steels. Acta Mechanica **229**(7):3051–3068
[94] Eisenträger J, Naumenko K, Altenbach H (2018) Calibration of a phase mixture model for hardening and softening regimes in tempered martensitic steel over wide stress and temperature ranges. The Journal of Strain Analysis for Engineering Design **53**(3):156–177
[95] Eisenträger J, Naumenko K, Altenbach H (2019) Numerical analysis of a steam turbine rotor subjected to thermo-mechanical cyclic loads. Technische Mechnik **19**(3):261–281
[96] Altenbach H, Girchenko A, Kutschke A, Naumenko K (2015) Creep behavior modeling of polyoxymethylene (POM) applying rheological models. In: Altenbach H, Brünig M (eds) Inelastic Behavior of Materials and Structures Under Monotonic and Cyclic Loading, Advanced Structured Materials, vol 57, Springer, pp 1–15
[97] Naumenko K, Altenbach H (2007) Modeling of Creep for Structural Analysis. Foundations of Engineering Mechanics, Springer

Chapter 4
Various State-of-the-Art Methods for Creep Evaluation of Power Plant Components in a Wide Load and Temperature Range

Eike Blum, Yevgen Kostenko, and Konstantin Naumenko

Abstract Many power plant components are exposed to high temperature environments and complex loading conditions over long period of operation. An important part in the life-time assessment is the reliable prediction of strain/stress state using robust creep modeling to avoid possible integrity or functionality issues and failures in such components. The goal of this work is to apply different state-of-art creep models including the empirical Norton-Bailey, modified Garofalo equations and the advanced constitutive visco-plastic model KORA to the analysis of typical high-temperature power plant components in a wide range of loads and temperatures. Among other things, an advantage of each model and its robustness is discussed, which should reflect both inelastic deformation and stress relaxation. The material parameters were identified from experimental data for 10%CrMoV heat resistant steels in the creep range. The results of non-linear Finite Element Analysis (FEA) were used for the subsequent integrity assessment of benchmark examples as well of the practical example of the steam turbine component. The material laws were implemented in the commercial software NX CAE. The results for long-term assessment of real steam turbine component are presented and discussed. In addition, an outlook on further developments of the material modeling and assessment procedure is also provided.

Eike Blum · Yevgen Kostenko
Siemens Energy Global GmbH & Co. KG, Mülheim/Ruhr, Germany,
e-mail: eike-marcel.blum.ext@siemens-energy.com,
yevgen.kostenko@siemens-energy.com

Konstantin Naumenko
Lehrstuhl für Technische Mechanik, Institut für Mechanik, Fakultät für Maschinenbau, Otto-von-Guericke-Universität Magdeburg, Universitätsplatz 2, 39106 Magdeburg, Germany,
e-mail: konstantin.naumenko@ovgu.de

4.1 Introduction

Steam turbine components as auxiliary pipes, rotors, turbine and valve casings etc. are operating under complex multi-axial loading conditions at high temperature for a long period of time. Structural analysis of such components requires the consideration of inelastic deformation processes at high temperature. Inelastic structural analysis can be performed with FEA for the given component geometry, boundary/contact conditions and kinematic constraints.

The main part of such analysis is a definition of material properties. There are different ways to describe the material behavior using empirical or constitutive material models. The parameter calibration for both methods is a very important part of modeling, and its determining effort depends on the complexity of the model.

A key step is to develop and to calibrate a material model which reflects basic features of inelastic material response for the specified loading and temperature profiles. In the last decades many examples of structural analysis of creep are performed and presented in the open literature, e.g. [1–3]. Two approaches to the modeling of creep behavior are usually applied.

The first is based on pure empirical equations (e.g., Norton-Bailey, Garofalo etc.) to describe creep curves under constant stress and temperature levels. One example is the power law creep equipped with an Arrhenius type function of temperature and a strain hardening function. This approach is widely used in the practice since the material parameters in the creep equation can be identified directly from experimental data and identification procedures are well established. Furthermore, creep equations are available in commercial finite element codes and can be applied for the structural analysis.

The second approach introduces internal state variables and corresponding evolution equations (e.g., Chaboche type) to capture changes in the material microstructure. Examples include backstress tensors, dislocation densities as well as softening and damage variables [2, 4]. This approach can be applied to capture the material behavior under complex varying thermo-mechanical loading paths, e.g. [5]. Furthermore, critical damage zones can be predicted, as shown in [6]. However, general procedures to identify material parameters are not well-established. Furthermore, a user material subroutine should be developed and implemented inside the commercial finite element code. The subroutine must include not only a set of constitutive and evolution equations but also time step and iteration procedures. As a rule, the developed subroutines are tested and applied to analyze specific components and loading conditions. In a general-purpose case including contact conditions, singular points, like edges, etc. the developer does not guarantee the convergence of the solution. Therefore, this approach has found only restricted application in the industry.

In this paper we address the creep analysis of heat resistant steels for a wide stress range including low and moderate stress values. Based on the available experimental data, different creep equations are discussed. The continuous response functions are introduced to fit creep behavior to a wide range of temperatures as well. All discussed models are incorporated into the NX CAE finite element code by use of user defined creep models. To verify the model assumptions a benchmark model of cube-one-element, pipe with analytical solution as well as an example for a real power plant component steam turbine valve casing are presented. The results of creep analysis under isothermal conditions illustrate the quality and accuracy of different approaches to reflect basic features of stress redistribution in the structural component.

The main target of this work is to carry out an inventory of available models (state-of-the-art) with a focus on assessment regarding robustness and computing performance required by industrial application. Such assessment shows the needs of modeling adjustment for commercial use and provides an overview of further necessary steps in material model adaptations in academic research groups.

4.2 Applied Creep Models

4.2.1 Norton-Bailey Equation

The empirical Norton-Bailey creep equation, e.g. [1] can be used to represent primary and secondary creep

$$\dot{\varepsilon}_{cr} = A\sigma^n t^m \quad (4.1)$$

In Eq. (4.1) $\dot{\varepsilon}_{cr}$ is the creep strain rate, σ the effective stress and t the time. A, n, m are material parameters. The Norton exponent n is stress dependent and can only be fitted for a limited stress range at one time. For example, diffusion creep occurs at low stresses ($n = 1$) and dislocation creep ($n \approx 5$) at higher stresses as shown in Fig. 4.1.

4.2.2 Modified Garofalo Equation

The empirical modified Garofalo equation, see e.g. [7] has separate terms for primary, secondary and tertiary creep

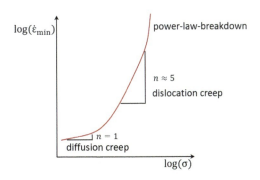

Fig. 4.1 Stress dependence of Norton exponent

$$\varepsilon_{cr} = \varepsilon_{cr,\max} H(t) + \dot{\varepsilon}_{cr,\min} t + K_3(T) \left(\frac{t}{t_{23}}\right)^f \qquad (4.2)$$

In Eq. (4.2) $\varepsilon_{cr,\max}$ is the maximum of the primary creep strain. $H(t)$ is a time-dependent hardening function increasing from 0 to 1. Secondary creep is represented by the minimum creep strain rate $\dot{\varepsilon}_{cr,\min}$ multiplied by time t. The tertiary creep component is determined by the f power of time t referred to transition time t_{23} and the temperature-dependent factor $K_3(T)$. A description of Eq. (4.2) and the associated parameters can be found in [8].

Equation (4.2) contains a non-linear dependence between time and creep strain. In the user creep subroutine, for strain hardening the time is substituted by Newton iterations.

4.2.3 Constitutive Model

Constitutive models, suitable to calculate stresses, deformations and material damage in tensor notation are widely discussed in literature, e.g. [2]. The viscoplastic KORA-model, which was originally proposed by [9], has been continuously improved and extended mainly to focus on assessing the lifetime of components which undergo a generalized creep-fatigue loading. The in this work discussed approach, also outlined e. g. in [10, 11], mainly represents a Chaboche-type model, cp. [4] including some adaptions to describe a time dependent visco-plastic behavior.

The following paragraph summaries briefly the basics of the approach discussed in [12], starting with the strain tensor

$$E = E_{el} + E_{pl} \qquad (4.3)$$

as sum of elastic and plastic parts. The stress tensor

$$T = (1-D) C \cdot\cdot E_{el} \qquad (4.4)$$

results from the Hooke's law in tensor notation, where C is the fourth-rank elasticity tensor. The scalar D represents a calculative material damage, which can be understood as a measure of material degradation.

The connection between backstress tensor $\boldsymbol{\xi}$ and backstrain tensor Y is described as follows

$$\boldsymbol{\xi} = (1-D) c Y, \qquad (4.5)$$

where c is a constant. The von Mises equivalent stress is

$$f = \sqrt{\frac{3}{2} \frac{(T-\boldsymbol{\xi})^{\text{dev}}}{\sqrt{1-D}} \cdot\cdot \frac{(T-\boldsymbol{\xi})^{\text{dev}}}{\sqrt{1-D}}}, \qquad (4.6)$$

The variable

$$F = \frac{f}{\sqrt{1-D}} - k_0 \qquad (4.7)$$

indicates, whether elastic or plastic deformation becomes active. The latter is the case when the von Mises stress f, augmented by the damage D, is larger than the flow limit k_0, a material constant. The accumulation of plastic equivalent strain

$$\dot{s} = \frac{\langle F \rangle^m}{\eta} \exp\left(a \langle F \rangle^d\right) \qquad (4.8)$$

is modeled by a Norton-like power law approach with an additional exponential function, making it possible to reproduce the stress dependence of the creep rate in a wide stress range (Fig. 4.1). The brackets $\langle \ \rangle$ indicate the use of Föppl's convention: plastic strain is accumulated for positive stresses but never pushed back for negative stresses. The evolution of the plastic strain tensor

$$\dot{E}_p = \frac{3}{2} \frac{(T-\xi)^{\text{dev}}}{\sqrt{1-D}f} \dot{s} \qquad (4.9)$$

is linked to the flow rule, using the deviatoric stress tensor.

Following [13], the nonlinear dependence of the backstrain from the accumulated plastic is modelled by

$$\dot{Y} = \dot{E}_{pl} - B(s) b \sqrt{1-D} \dot{s} Y - p \left\| c \sqrt{1-D} Y \right\|^{w-1} Y \qquad (4.10)$$

and

$$B(s) = B_1 + (1-B_1) \exp(-B_2 s), \qquad (4.11)$$

where B_1 and B_2 are material parameters. The evolution of damage assumed as the sum which sums up to fatigue $\dot{D}_A$ and creep damage $\dot{D}_t$ terms

$$\dot{D} = \dot{D}_A + \dot{D}_t \qquad (4.12)$$

The kinetic law for $\dot{D}_A$ is associated with the accumulated plastic strain $\dot{s}$

$$\dot{D}_A = \frac{\dot{s}}{A_A}, \qquad (4.13)$$

where A_A is the material parameter. The kinetic law for the creep part is assumed as follows

$$\dot{D}_t = \left(\frac{\sigma^*}{A_t}\right)^{k_t} (1-D)^{-r_t}, \qquad (4.14)$$

where A_t, k_t and r_t are material parameters and σ^* is the damage equivalent stress. Further details regarding this model and nomenclature are detailed described in [12].

4.3 Structural Analysis

4.3.1 Verification of the Creep Models Based on Creep Tests

Figure 4.2 illustrates the calculated curves of the different creep models in comparison to three measurement curves at 550 °C for 10%Cr-Steel. The Norton-Bailey and the modified Garofalo curves are calculated analytically. The constitutive model curves are computed numerically as a one-cube-element model (3D hexahedron element). The applied Norton-Bailey model provides an acceptable quality of the measured data at low stresses, up to the beginning of the tertiary creep. The tertiary creep cannot be described at all. At high stresses, the Norton-Bailey curve is far below the measured data. This is due to the stress dependence of the power law exponent n. At higher stresses, dislocation creep starts, and the stress dependence of the creep strain becomes larger. Therefore, it is hardly possible to map a large stress range with one parameter set. At low stresses with a low $n \approx 1$ diffusion creep is dominant.

The modified Garofalo creep law is able to present the primary, secondary and tertiary creep with more accuracy. The material parameters used, represent the measurement curves at 120, 240 and 400 MPa. At the stress of 400MPa, the curve is also underestimated. However, the high stress range is matched better with the

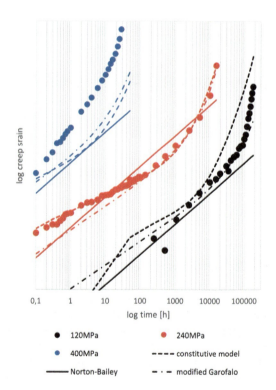

Fig. 4.2 Assessment of different creep models based on experimental data for 10%Cr-Steel at 550 °C.

4 Various State-of-the-Art Methods for Creep Evaluation

modified Garofalo than with the Norton-Bailey model. The tertiary creep prediction at the stress of 400MPa is underestimated as well.

In comparison to previous models the constitutive model reproduces well the creep curve at 240 MPa. However, at the stress of 400 MPa, the curve is suited also too low. At the low stress of 120 MPa, there is a slight drop to observe in the curve at about 50 hts clearly too early at 120 MPa.

4.3.2 Relaxation Test with Cube-one-Element Model

The relaxation test shown in Fig. 4.3 is a uniaxial tensile test with a total constant strain of 0.2% at 550°C. The experimental measured stress varies in the range of 150 to 280MPa, where all creep models reproduce good results. The empirical creep models are calculated in the strain hardening variant. Generally, the results of all creep models fit the measured data. The best stress relaxations result is obtained with the modified Garofalo and with the constitutive model.

4.3.3 Pipe Benchmark FE Model

The pipe model, often considered as benchmark, will be used to assess the results and computation time of the different creep routines. The pipe model is a 90° section of a thick wall hollow cylinder under internal pressure, see Fig. 4.4. The dimensions were selected with inner radius of 200 mm, outer radius of 500 mm and length of 3000 mm. The creep simulations are performed at a temperature of 550°C for a duration

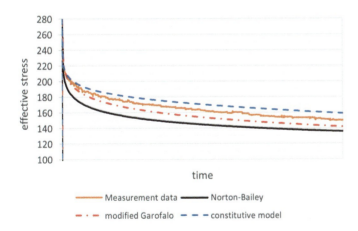

Fig. 4.3: Relaxation test at 550°C and constant total strain of 0.2%.

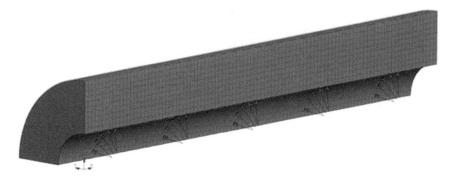

Fig. 4.4: 3D meshed view of the pipe FE model.

of to 200,000 h. For the pipe under internal pressure, an analytical solution exists for the steady-state case according to [14]. The steady-state circumferential stress is given by

$$\sigma_\theta = p \left(\frac{-(1-2/n)r^{-2/n} + b^{-2/n}}{a^{-2/n} - b^{-2/n}} \right) \quad (4.15)$$

Here p is the internal pressure, r the radial position, a the inner radius, b the outer radius und n is the stress exponent of the Norton-Bailey creep model.

In the analytical solution, a plane stress state is assumed. Therefore, in the FE model, all nodes are fixed in axial direction. The thermal expansion of the material is not considered. The two cut surfaces are fixed in the normal direction at symmetry plane.

The following results are generated with 30mm meshed hexahedron elements with quadratic shape function. The inner pressure is chosen to 100 MPa and the calculation time is 200,000 h.

The circumferential stress according to equation 15 is highest at the inner radius before creep. Due to creep, a stress redistribution takes place. The stress maximum shifts into the direction of the outer radius. Figure 4.5 shows the circumferential stress as a function of radius after 200000 h. The analytical circumferential stress will be calculated using the power law exponent n of the Norton-Bailey model. Therefore, the analytical solution and the Norton-Bailey FEA solution fit well together. According to the modified Garofalo model, the circumferential stress relaxes much more pronouncedly at the inner radius than in the Norton-Bailey model. This can be explained by additional modeling of tertiary creep effects, which occur mainly at the inner radius. For the modified Garofalo solution, further stress redistribution is expected due to the modeled tertiary creep.

Less stress relaxation is observed with the constitutive model than with the empirical models. This agrees with the results of the stress relaxation test given in Fig. 4.3, where the stresses are slightly above the measured curve. This is also due to the special parameter identification for the constitutive equation, as can be seen in Fig. 4.2.

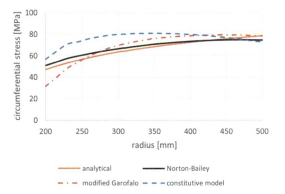

Fig. 4.5 Circumferential stress after 200000 h.

4.3.4 Performance Evaluation of User-Creep Routines

The computational complexity of the modified Garofalo creep routine is higher than that of the Norton-Bailey routine, because the strain hardening is solved numerically with Newton iterations. The computation times on the benchmark models of Norton-Bailey and modified Garofalo creep routine are compared in Fig. 4.6. For this purpose, the standard non-linear control parameters of the NX Nastran are used. The *automatic time stepping* option is therefore activated. The modified Garofalo simulations takes about two to three times longer. On the one hand, the time difference can be related to the higher numerical effort for strain hardening with Newton iterations. On the other hand, the difference in computing times is greater with increasing internal pressure. As the load increases, the creep strain increases in the modified Garofalo equation due to tertiary creep. The larger creep strain increments can cause smaller time steps. For an inner pressure of 100MPa, the solver uses 399 iterations for Norton Bailey and 472 iterations for modified Garofalo. This means one iteration takes about two seconds with the Norton Bailey solution. The modified Garofalo solution takes about four seconds per iteration, which is twice as long.

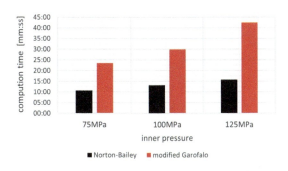

Fig. 4.6 Comparison of computation times of Norton-Bailey and modified Garofalo.

In the FE solution with the constitutive model, convergence problems exist in some cases. Convergence can be improved with the solver variable *Acceptable Error for Creep Integration Accuracy (CRLIMR)*. This leads to significant longer run times. Particularly with complex 3D models, convergence and computation performance differences are noticeable in relation to the empirical models. Further investigation of this area is necessary.

4.3.5 Temperature Interpolation for Norton-Bailey Creep Equation

It is important for computation accuracy to use the correct temperature interpolation. If the temperature of a creep simulation is located between temperature interpolation points of the specified material parameters, errors can occur during temperature interpolation. Using the Norton-Bailey creep routine as an example, different interpolation variants are tested on a hexahedron element. The creep strain is evaluated for a constant stress of 140 MPa and temperature range from 475 to 600°C in 25 °C steps. The interpolation points are at 450, 500, 550 and 600°C.

In NX CAE, material parameters are linear interpolated by default. This is fine for the Norton-Bailey parameters n and m in Eq. (4.1). However, the parameter A has a logarithmic temperature correlation. If the logarithmic relationship is taken in consideration during the interpolations, realistic results between the temperature support points in Fig. 4.7 are obtained. With only linear parameter interpolation, there are large deviations from the expected creep strain between the interpolation points. Another interpolation option is creep strain increment interpolation. Here, the creep strain increment is calculated for the temperature point below and above and interpolated to the current temperature. With the linear creep strain increment interpolation there are also deviation leaps in the creep curve. If the creep strain increment is interpolated double logarithmical

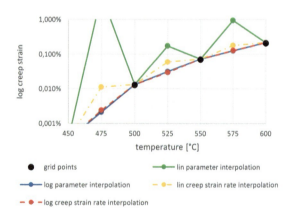

Fig. 4.7 Different temperature interpolation variants for Norton-Bailey creep law.

$$\Delta\varepsilon_{cr}(T) = \exp\left(\ln\left(\Delta\varepsilon(T_0)\right) + \frac{\ln\left(\Delta\varepsilon(T_1)\right) - \ln\left(\Delta\varepsilon(T_0)\right)}{\ln(T_1) - \ln(T_0)}\left(\ln(T) - \ln(T_0)\right)\right)$$
(4.16)

the results are also accurate.

In the case of the Norton-Bailey creep equation, partial logarithmic parameter interpolation or logarithmic creep strain increment interpolation can give good results. However, the material parameters must be appropriate for parameter interpolation. A disadvantage of the creep strain increment interpolation is that the derivatives of the creep strain increment to creep strain and the equivalent stress for the user creep routine cannot be calculated analytically directly.

4.3.6 Isothermal Steam Turbine Valve FE Model with a Constant Loading

The following paragraph deals with the analysis of an idealized thick-walled valve casing of a high-pressure turbine made from 10%CrMoV heat resistant steel. Such a component has to stop and/or to control the steam flow into a turbine. Due to the confidentiality of geometry and of material parameters, the data for the dimensions and the relevant loading as well as the results for the absolute values of stresses and the time-range were artificially normalized.

Therefore, the results should be considered as just qualitative. It is assumed that the valve casing is uniformly heated and loaded by internal pressure. The temperature and the pressure are assumed constant over time. Transient effects during start-ups and shut-downs are not considered. The main part of the analyzed casing is simplified as symmetrical.

Figures 4.8 and 4.9 show distribution of the von Mises stress as a result of loading and heating. We observe that the maximum stress levels appear on the inner side of the casing. As a result of creep, an essential stress redistribution takes place, leading to an increase in the stress value on the outer side of the casing. Figures 4.10 and 4.11 show corresponding strain values, which are lower than the allowable characteristic strain. Nevertheless, damage processes associated with voids and micro-cracks are expected on both the inner and the outer surfaces of the casing. It is to note that the Garofalo law results show by factor two higher strains than the Norton-Bailey law. Field experiences reveal that the reality is to be expected in between both solutions.

4.4 Conclusions

In this paper, the empirical Norton-Bailey and the modified Garofalo creep models as well as the constitutive KORA model are discussed. All models are implemented in the FE software NX CAE using user-defined creep subroutines. The results are compared using different computational models including a real steam turbine

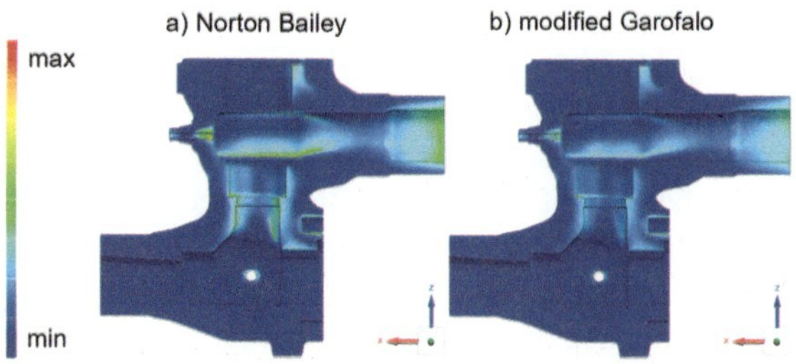

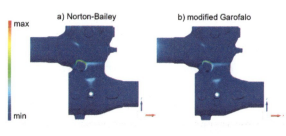

Fig. 4.8 Stress distribution normalized by creep rupture limit (inside). a) Norton-Bailey law, b) modified Garofalo law.

Fig. 4.9 Stress distribution normalized by creep rupture limit (outside). a) Norton-Bailey law, b) modified Garofalo law.

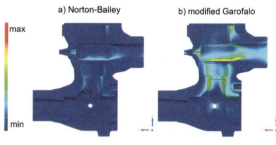

Fig. 4.10 Equivalent creep strain inside the valve. a) Norton-Bailey law, b) modified Garofalo law.

Fig. 4.11 Equivalent creep strain outside the valve. a) Norton-Bailey law, b) modified Garofalo law.

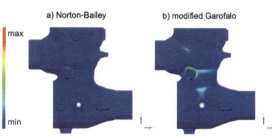

component. Using the discussed models, a detailed FEA is performed to predict creep in structural components. The example of an idealized high-pressure stop and control valve casing demonstrates that these methods are capable of reproducing basic features of creep in engineered structures, including time-dependent deformation, stress redistributions, and the formation of critical zones of creep damage.

The Norton-Bailey model is already incorporated into most commercial FE programs. Therefore, the use of user-defined creep subroutine is usually not necessary. The determination of the material parameters can usually be done by the customer.

However, the creep strain and stress relaxation can be predicted closer to reality using the modified Garofalo model, due to the consideration of tertiary creep. The associated user-defined creep subroutines for empirical models are robust, and the computational performance is acceptable.

Nevertheless, the empirical models are generally unable to reproduce the complex non-monotonic loading profiles, LCF influence, and macroscopic material response. This could be achieved by combining continuum mechanics approaches with a qualitative analysis of microstructural behavior using constitutive models.

The use of the constitutive model in the listed examples does not show significant advantages in the benchmark. Its perspective use can be advantageous when non-monotonic loading or superimposed LCF analysis is required. Computational time increases enormously with complex 3D models of real components, and issues with convergence also increase. For example, within the framework, it was not possible to compute a real component with a constitutive model without simplifying the material behavior. In addition, the use of user-defined creep subroutines of such a model does not currently meet industry requirements for robustness. It is the responsibility of academic research groups to adapt such models for use by customers in future projects. It is expected that the use of such models will allow the consideration and combination of different damage mechanisms such as plasticity, creep and LCF. At present, this only works in isolated cases. In addition, the determination of the material constants of such models is very time-consuming.

Summarized the main outputs of the work are as following:

1. Empirical models are simple, robust and sufficient for practical use.
2. Constitutive models are not robust and not adopted for needs of industry at present.
3. There is a need to continue and promote the development of constitutive models, focusing on robustness and usability.

References

[1] Betten J (2008) Creep Mechanics. Springer Science & Business Media
[2] Naumenko K, Altenbach H (2016) Modeling High Temperature Materials Behavior for Structural Analysis. Part I: Continuum Mechanics Foundations and Constitutive Models, Advanced Structured Materials, vol 28. Springer
[3] Eisenträger J, Naumenko K, Kostenko Y, Altenbach H (2019) Analysis of a power plant rotor made of tempered martensitic steel based on a composite model of inelastic deformation. In: Advances in Mechanics of High-Temperature Materials, Springer, pp 1–34
[4] Chaboche JL (1989) Constitutive equations for cyclic plasticity and cyclic viscoplasticity. International journal of plasticity **5**(3):247–302
[5] Naumenko K, Kutschke A, Kostenko Y, Rudolf T (2011) Multi-axial thermo-mechanical analysis of power plant components from 9–12% Cr steels at high temperature. Engineering Fracture Mechanics **78**(8):1657–1668

[6] Naumenko K, Kostenko Y (2009) Structural analysis of a power plant component using a stress-range-dependent creep-damage constitutive model. Materials Science and Engineering: A **510**:169–174
[7] Scholz A, Wang Y, Linn S, Berger C, Znajda R (2009) Modeling of mechanical properties of alloy CMSX-4. Materials Science and Engineering: A **510**:278–283
[8] Kloos K, Granacher J, Oehl M (1993) Beschreibung des Zeitdehnverhaltens warmfester Stähle. Teil 1: Kriechgleichungen für Einzelwerkstoffe. Materialwissenschaft und Werkstofftechnik **24**(8):287–295
[9] Reckwerth D, Tsakmakis C (2003) The principle of generalized energy equivalence in continuum damage mechanics. Deformation and failure in metallic materials pp 381–406
[10] Wang P, Cui L, Lyschik M, Scholz A, Berger C, Oechsner M (2012) A local extrapolation based calculation reduction method for the application of constitutive material models for creep fatigue assessment. International journal of fatigue **44**:253–259
[11] Kontermann C, Scholz A, Oechsner M (2014) A method to reduce calculation time for fe simulations using constitutive material models. Materials at High Temperatures **31**(4):334–342
[12] Kontermann C, Linn S, Oechsner M (2019) Application concepts and experimental validation of constitutive material models for creep-fatigue assessment of components. In: Turbo Expo: Power for Land, Sea, and Air, American Society of Mechanical Engineers, vol 58714, p V008T29A026
[13] Armstrong PJ, Frederick C, et al (1966) A Mathematical Representation of the Multiaxial Bauschinger Effect, vol 731. Berkeley Nuclear Laboratories Berkeley, CA
[14] Odqvist FKG, Hult J (1962) Kriechfestigkeit metallischer Werkstoffe. Springer, Berlin u.a.

Chapter 5
Creep and Irradiation Effects in Reactor Vessel Internals

Dmytro Breslavsky and Oksana Tatarinova

Abstract The paper is devoted to the presentation of the calculation method for determining the stress-strain state and long-term strength of reactor vessel internals (RVI) and the description of the results obtained with its help. The method is based on a complete mathematical formulation of the boundary- initial value problems of creep accompanied by irradiation effects. Deformation and damage accumulation caused by irradiation effects in the material when interacting with the effects caused by thermal creep, can significantly limit the safe operation of RVI. Elastic, thermoelastic, thermal and irradiation creep, irradiation swelling strains, creep damage and fracture are considered. The numerical solution of the boundary value problems is performed by the FEM, and the initial value problems are solved by time integration. To estimate cyclic deformation and fracture, the procedures of asymptotic methods and averaging over cycle periods are used. As an examples of the use of this calculation method, the results of creep modelling of fuel element, T-joint of tubes and notched plates are given. The issues of interaction of stresses, strains and damages of different nature under complex stress state are discussed.

5.1 Introduction

Creep accompanied by the accumulation of hidden damage is a complex phenomenon. In nuclear reactor vessel internals (RVI) it is also accompanied by effects associated with the action of irradiation. Deformation and damage accumulation caused by irradiation effects in the material, such as irradiation creep, irradiation swelling, when interacting with the effects caused by thermal creep, can significantly limit the safe operation of RVI [1, 2].

Dmytro Breslavsky · Oksana Tatarinova
Department of Computer Modelling of Processes and Systems, National Technical University "Kharkiv Polytechnic Institute", UKR-61002, Kharkiv, Ukraine,
e-mail: dmytro.breslavsky@khpi.edu.ua, ok.tatarinova@gmail.com

Creep, which occurs as a result of radiation and is accompanied by high plasticity of fuel and structural materials, is called irradiation creep. It is known that the rate of irradiation creep can be several times higher than the rate of creep of the same material at the same stress level [3–5]. In order to distinguish the mechanisms of creep that occurs during radiation exposure and under the action of only temperature-force fields, the term 'thermal' creep will be used for the latter.

Radiation exposure under conditions of long-term operation leads to another effect - irradiation swelling, which is volumetric instability caused by the formation of pores and bubbles in the material and the accumulation of inert gas. At the same time, the increase in volume is accompanied by a decrease in the density of the material. Irradiation swelling limits the operational properties of fuel rods and other elements of the reactor during operation [4].

Like as in traditional creep-damage studies [6], the role of numerical simulation of the RVI is no less important. Experimental studies are not only very expensive due to their duration, but also require, firstly, unique laboratory equipment, and, secondly, are associated with the need to observe strict safety measures due to possible irradiation damage. If it is still possible to conduct experiments with a uniaxial stress state, which are absolutely necessary to determine the properties of the material under radiation exposure, then there are very few data on deformation measurements under a complex stress state.

As with the numerical simulation of creep and fracture of structural elements, FEM approaches are used to analyze the stress-strain state of RVI using both standard engineering software such as ANSYS, ABAQUS [7–9], and specially developed ones [10–12]. The basis for numerical modeling of the deformation processes in structural elements subjected to the joint action of thermal-force and radiation fields is the use of the hypothesis of strains additivity (in most cases, the problems are solved using the Lagrange approach). Constitutive equations are formulated for each physically nonlinear component of the total strain tensor (plasticity, thermal and irradiation creep, irradiation swelling, etc.). It is clear that the adequacy and accuracy of numerical solutions depends on their adequacy. If classical, proven constitutive equations are used for plasticity and creep strains description, then the predicting the irradiation swelling strains, which are often quite significant and dangerous [4, 5], is still far from classical completion. Irradiation swelling strains have a volumetric nature, and their description in Solid Mechanics is carried out using the equations corresponding, for example, to the one used in the calculation of thermal strains. When formulating the functional dependence used to describe the process of strain accumulation during irradiation swelling, information is used regarding its dependence on the integral neutron flux, fluency or accumulated radiation dose, temperature, and time. However, further such dependencies are built directly using experimental distributions, which are processed by the method of least squares [13]. Unfortunately, in most publications, dependencies are formulated precisely for the expressions of swelling strains, while for the correct formulation of the boundary- initial value problems, a formulation in rates is required. Reformulation of these state equations obtained for the strain values to an incremental form can lead to significant inaccuracies in calculations [14].

The irradiation creep of metals and alloys is often found to be described by linear dependencies on the applied stresses [4]. Sometimes, when processing the results of experimental studies, it is possible to formulate the dependence of the irradiation creep function on swelling [5], but such equations still require further verification.

As shown by experimental and numerical studies [1–5], the addition of radiation exposure significantly, quantitatively and qualitatively, changes the characteristics of creep processes in structural elements. As for hidden damage accumulation processes, experimental data in most cases also indicate their intensification when radiation exposure is added. Damage calculations of RVI elements using Continuum Damage Mechanics (CDM) approaches are extremely few. This is primarily due to the difficulty of obtaining experimental data on long-term behavior during radiation exposure, and possibly to the lack of appropriate units in many FE software complexes.

Due to the fact that the main structural materials of the active zone of nuclear reactors were steels and alloys, which in most cases did not show significant anisotropy of physical and mechanical properties, the calculations were carried out assuming isotropy of deformation properties and accumulation of damage. By the way, a number of light alloys are used for the manufacture of RVI elements, which are characterized by significant anisotropy of both classical physical and mechanical properties [6] and deformation during radiation exposure [15–17]. Their numerical modeling requires the verification and formulation of more complex constitutive equations, and the use of tensor models to predict damage accumulation.

The operation of nuclear reactors is not constant over time, they are periodically shut down, and many RVI elements are subjected to cyclical varying in temperature and pressure. This mode of their work is associated with the need to reformulate the boundary initial value problems with the involvement of new constitutive equations, which reflect the influence of cyclic loading, heating-cooling, and irradiation on creep and damage accumulation. Investigations in this direction are just beginning to be carried out, modeling of non-uniform loading is the subject of few publications [18, 19]. Yang et al. [19] demonstrate the results of calculations of irradiation swelling were carried out using ANSYS procedures, in which physical and mechanical properties were determined as functions of time. Dubyk et al. [18] also used the developed additional software for the ANSYS to model the reactor baffle taking into account shutdowns.

This paper is devoted to the presentation of the main approaches, methods and constitutive equations used for numerical modeling of creep-damage processes in RVI elements made of materials that exhibit isotropic or anisotropic properties and can be cyclically loaded. The presented numerical examples demonstrate the capabilities of the calculation method for the purpose of determining the laws of deformation and fracture.

5.2 Problem Statement and Description of Solution Approaches.

Let us consider a solid with a volume V fixed on a part of the surface S_1 and loaded with volume forces $\boldsymbol{f}$ and traction $\boldsymbol{P}$ and on a part of the surface S_2 (Fig. 5.1). In the coordinate system $\boldsymbol{x} = (x_1, x_2, x_3)$ the motion of the continuum of material points will be described by the displacement vector $\boldsymbol{u}$, tensors of stresses $\boldsymbol{\sigma} = \sigma(x_i, t)$, $\sigma_{ij} = \sigma_{ji}$ and strains $\boldsymbol{\varepsilon} = \boldsymbol{\varepsilon}(x_i, t)$, $\varepsilon_{ij} = \varepsilon_{ji}$, $(i, j = 1, 2, 3)$, which are functions of coordinates and time t. Irreversible creep strains are represented by a tensor $\boldsymbol{c} = \boldsymbol{c}(x_i, t)$, $c_{ij} = c_{ji}$. The tensor connection for them with the stress tensor components and time is determined by the accepted constitutive equations.

An inhomogeneous temperature field $T(x_i, t)$ acts on the solid, the effect of which causes thermal strains $\boldsymbol{\varepsilon}^T = \boldsymbol{\varepsilon}^T(x_i, t)$, with components $\varepsilon_{ij}^T = \varepsilon_{ji}^T$. The components of the total initial strain e_{ij} consist of elastic e_{ij}^e and thermoelastic strains ε_{ij}^T. The presence of radiation exposure leads to the development of irradiation creep and irradiation swelling strains with components c_{ij}^r and ε_{ij}^{sw}, respectively.

Let us specify the nature of the external load field. Acting external forces can be divided into two components - main and oscillating action. To the first we include volume forces $\boldsymbol{f}(\boldsymbol{x}, t)$, $\boldsymbol{x} \in V$, and the part of traction $\boldsymbol{P}^0(\boldsymbol{x}, t)$, $\boldsymbol{x} \in S_2$, which slowly varies over time or remain unchanged. The second part include tractions and forces that change over time according to the harmonic law with amplitude P_i^a and period T_c

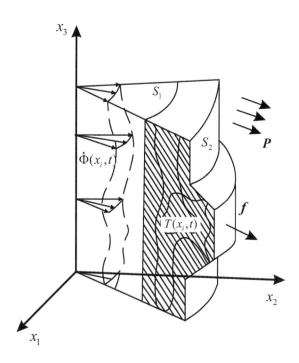

Fig. 5.1 Solid in thermal-force and radiation fields.

5 Creep and Irradiation Effects in Reactor Vessel Internals

$$P_i = P_i^0 + P_i^a \sin \frac{2\pi}{T_c}$$

In the assumptions formulated above, the mathematical formulation of the boundary - initial value problem of the creep of solids under the action of periodically varying loads can be represented by the following system of equations:

$$\sigma_{ij,j} + f_i = \rho \ddot{u}_i, \quad \sigma_{ij} n_j = P_i, \quad x_i \in S_2, \qquad (5.1)$$

$$\varepsilon_{ij} = \frac{1}{2}\left(u_{i,j} + u_{j,i} + u_{k,i} u_{k,j}\right), \quad x_i \in V, \quad u_i|_{S_1} = \bar{u}_i, x_i \in S_1,$$

$$\varepsilon_{ij} = e_{ij} + c_{ij} + c_{ij}^r + \varepsilon_{ij}^{sw}, \quad e_{ij} = e_{ij}^e + \varepsilon_{ij}^T,$$

$$\sigma_{ij} = D_{ijkl}\left(e_{kl} - c_{kl} - c_{kl}^r - \varepsilon_{kl}^{sw}\right),$$

$$u_i(x,0) = c_{ij}(x,0) = c_{ij}^r(x,0) = \varepsilon_{ij}^{sw}(x,0) = 0.$$

where ρ is the mass density; **n** is unit normal to the solid boundary; D_{ijkl} are the components of the elastic properties tensor, $i = 1,2,3, \bar{u}_i$ are the known displacement values in the surface part S_1, which do not vary in time.

To determine the limits of the study, we will introduce several assumptions. Due to the fact that we are considering structural elements in which large displacements and strains are prohibited by their purpose, we will limit ourselves to the Lagrange approach. We will consider quasi-static load processes, which is due to the fact that during design, a search is made for the eigen frequencies of the system and it is verified that the frequencies of its forced oscillations are far from them. In connection with this, we will neglect the term corresponding to inertial forces in the equation of motion, system (5.1). The hypothesis of strains additivity is applied. In the case when processes characterized by a coupling between irradiation creep and swelling strains take place, it is possible to postulate the existence of the hypothesis of additivity over an infinitesimally small interval of time. But such processes will not be considered.

When considering the processes of cyclic varying the stresses due to the fact that such problems for non-isotropic solids are just beginning to be investigated and simple qualitative conclusions are needed, we will limit ourselves to the case of stress varying according to a harmonic law with a low frequency. Such a case corresponds to a varying of a pressure from a gas flow and is considered in [10] for the isotropic properties of the material. To solve the problem, we will use the approaches of the method of many time scales and averaging over the period of varying stresses. At the same time, two systems of equations are obtained, one of which describes the deformation of a solid under the influence of a thermal-force field and radiation exposure and the action of only time-invariant load components, and the other is analogous to the problem of quasi-static elastic loading in a cycle. The systems are connected by the constitutive equations, which will be given below, in the Sect. 5.3. The used approach of reducing the cyclic loading process to the equivalent quasi-static one is described in [20–22] and also presented by the authors in another paper of this symposium. The approaches developed by the authors use FEM for solving boundary-value problems, and finite difference approaches for solving initial problems in time. To describe the development of a macrodefect (crack), which occurs after the completion of hidden damage accumulation, a process based on a series of reformulated boundary-initial value problems is used. In them, the initial conditions are determined by the distributions of components of the stress-strain state and damage parameter in finite elements acquired before the completion of the next hidden damage accumulation in next finite element, and the boundary conditions are determined by the current configuration of the considered structural element, taking into account its destroyed parts. A description of the method and corresponding algorithms can be found in [23, 24]. In addition to Solid Mechanics problems, the ability to determine temperature fields in stationary and non-stationary heat conduction problems is also implemented. A description of the method and software can be found in [25, 26]. During the last decades, experimental verification of the method, constitutive equations, and implemented software under uniaxial and complex stress states was performed, some results and references are presented in [21, 22, 24, 26].

5.3 Constitutive Equations

To demonstrate the capabilities of the calculation method, we will consider two versions of the constitutive equations: first for describing the behavior of materials with isotropic properties as well as second one for non-isotropic (in this case, transversal isotropic) properties of deformation and accumulation of hidden damage.

5.3.1 Materials with Isotropy of Properties

Here we use the equations only for static loading and irradiation effects. Cyclic behaviour of structures made from materials with isotropic properties is described by appropriate equations and analyzed in [21, 22, 24, 26].

The description of thermal creep and the damage accumulation associated with it will be performed using the following equations:

$$\dot{c} = \frac{3}{2} b \frac{\sigma_{vM}^{n-1}}{(1-\omega)^l} \left[\tilde{B}\right] s \tag{5.2}$$

$$\dot{\omega} = d \frac{\sigma_e^r}{(1-\omega)^l} \tag{5.3}$$

where $s = s(x_i, t)$ is the deviator of stress tensor, ω is scalar damage parameter introduced by Kachanov-Rabotnov approach, σ_{vM}, σ_e are von Mises equivalent stress and equivalent stress is obtained by use of fracture criterion which appropriate for considered material [6]; b, n, d, r, l are the material constants obtained only from experiments on creep and long-term strength under static load and constant temperature values on standard samples under tension. Here we will limit ourselves to cases of either constant temperature over the entire volume of the structural element, or moderate temperature changes, when thermo-physical and long-term physical and mechanical properties can be considered unchanged.

The irradiation swelling strain $\varepsilon_{ij}^{\text{sw}}$ is volumetric, so $\varepsilon_{ij}^{\text{sw}} = 0$ at $i \neq j$. In general case it describes by function S_Φ, which depends upon neutron fluency Φ, time t and temperature T:

$$\varepsilon_{ij}^{\text{sw}} = \frac{1}{3} \dot{S}_\Phi (\Phi, t, T) \delta_{ij} \tag{5.4}$$

As noted, the expression for swelling function S_Φ is often determined numerically using approximation procedures. For example, for steel type 0.08 % C, 16 % Cr, 11 % Ni, Mn 3%, 0,2-0.4 % Nb for the neutron fluence range $\dot{\Phi} = 4..6 \cdot 10 \cdot 19$ neutron/m²s and temperature values 458-790 K it was determined as the following dependence [10]:

$$\dot{S}_\Phi = A_1 \beta_1 (\alpha_1 \dot{\Phi})^{\beta_1} t^{\beta_1 - 1} \exp\left(0.0235T - \frac{83.5}{T - 630} - \frac{1782}{980 - T}\right), \tag{5.5}$$

where $A_1 = 5.33 \cdot 10^{-9}, \beta_1 = 0.19 + 1.63 \cdot 10^{-3} T, \alpha_1 \dot{\Phi} = 9.37 \cdot 10^{-3}$ dpa/h. Irradiation creep strains will be calculated using the linear dependence of the strain rate on stresses [4, 5]:

$$\dot{c} = \frac{3}{2} b_{\text{rc}} \sigma_{\text{vM}} [\tilde{B}] s \tag{5.6}$$

where b_{rc} is the material constant for considered temperature range.

5.3.2 Materials with Transversal Isotropy of Properties

As is known [6], at elevated temperatures, even materials that are isotropic during elastic deformation, can exhibit non-isotropic properties of creep and hidden damage. Let us consider the constitutive equations proposed by Morachkovsky for materials with transversely isotropic properties of creep and damage [27] and modified in [28] for the case of cyclic loading. We consider the case of a quasi-static harmonic load with frequencies in the range of 1...3 Hz (that is, forced oscillations that occur far from the eigen frequencies of the structural element during gas flow oscillations. At the same time, the contribution of inertial forces is neglected). The used equations are as follows:

$$\dot{c} = \tilde{B} H(A) \frac{\sigma_v^{m-1}}{(1-\eta)^m} [\tilde{B}] \sigma, \quad A = \frac{\sigma_v^a}{\sigma_v} \tag{5.7}$$

$$\dot{\omega} = d_{1111}^{p/2} K(A_D) \frac{\sigma_D^{p-2}}{(1-\eta)^{p+s-1}} [\tilde{D}] \sigma, \quad \dot{\eta} = d_{1111}^{p/2} K(A_D) \frac{\sigma_D^p}{(1-\eta)^{p+s}}, \tag{5.8}$$

$$\eta(0) = 0, \quad \eta(t_*) = 1, \quad A_d = \frac{\sigma_D^a}{\sigma_D}$$

where ω is the damage tensor. $\sigma_v = \sigma^T[\bar{B}]\sigma$, $\sigma_D = \sigma^T[\bar{D}]\sigma$ are the equivalent stresses which are the joint invariants of stress tensors and tensors of material constants; $\eta = \eta(t)t$ is scalar damage measure. The index a denotes the equivalent stresses calculated by use the amplitude components of the stress tensor under cyclic loading. At the time t_* of the finishing the hidden damage accumulation process $\eta(t_*)=1$. Matrix $[\bar{B}]$ contains the components of the material creep properties tensor b_{ijkl}:

$$[\bar{B}] = \begin{bmatrix} 1 & \beta_{12} & 0 \\ \beta_{12} & \beta_2 & 0 \\ 0 & 0 & 4\beta \end{bmatrix}, \quad \beta_{12} = -\frac{1}{2}b_{1111}, \quad \beta_{22} = \frac{b_{2222}}{b_{1111}}, \quad 4\beta = \frac{b_{1212}}{b_{1111}} \qquad (5.9)$$

Matrix $[\bar{D}]$ contains the components of the material creep damage properties tensor d_{ijkl}:

$$[\bar{D}] = \begin{bmatrix} 1 & \delta_{12} & 0 \\ \delta_{12} & \delta_2 & 0 \\ 0 & 0 & 4\delta \end{bmatrix}, \quad \delta_{12} = -\frac{1}{2}d_{1111}, \quad \delta_{22} = \frac{d_{2222}}{d_{1111}}, \quad 4\delta = \frac{d_{1212}}{d_{1111}} \qquad (5.10)$$

The functions $H(A)$ and $K(A_D)$ are obtained after expanding the functions of the creep strain and the damage measure into an asymptotic series on a small parameter

$$\mu = \frac{T_c}{t}$$

in two time scales (slow t and fast $\xi = \tau/T$, $\tau = t/\mu$):

$$c \cong c^0(t) + \mu c^1(\xi), \qquad \eta \cong \eta^0(t) + \mu \eta^1(\xi), \qquad (5.11)$$

where T_c is the period of cyclic load; $c_0(t)$, $\eta_0(t)$t, $c_1(t)$, $\eta_1(t)$ are the functions corresponding to the main creep and damage processes in slow (0) and fast (1) time scales. The expressions of these functions can be found in [20, 22, 26]. Then, with the help of the averaging procedure over the period T_c, constitutive equations (5.7) - (5.8) are obtained. At the same time, the main system of equations (5.1) remains unchanged, but now it describes the deformation motion of its points already in the main, slow time. The derivation of the equations is described in more detail in [20, 22, 26].

The components of the tensors b_{ijkl} and d_{ijkl} are determined by the results of experiments on the creep and long term strength of samples cut from the material in three directions. In the case of sheet materials produced by rolling, these are the directions along, across the rolling and at an angle of $45°$ to them. $\tilde{B}, m, p, s$ are the constants determined experimentally.

Next, we will consider the equation for determining the components of the irradiation swelling strain tensor. In the case of anisotropy of the swelling properties, we write

$$\dot{\varepsilon}_{ij}^{sw} = \frac{1}{3}\beta_{ij}\dot{S}_\Phi\left(\Phi,t,T\right)\delta_{ij} \qquad (5.12)$$

where β_{ij} are the coefficients reflecting the effect of anisotropy on the swelling process. Due to the information about the isotropic nature of the irradiation creep [29], we will describe it by Eq. (5.6).

5.4 Deformation, Damage Accumulation and Fracture in RVI

Let us consider examples of stress-strain state, damage accumulation and fracture modeling, obtained by use of 2D and 3D models of RVI structural elements.

5.4.1 Creep of T-joint of Tubes

As a first example, let's consider the results of numerical modeling of the processes of thermal creep and irradiation swelling of the T-joint of the cooling tubes from system of the reactor core (Fig. 5.2) [26]. The following dimensions are used: inner radius 50 mm, outer radius 100 mm. For the operational temperature range, the following data are accepted: Young's modulus $E = 1.55 \cdot 10^5$ MPa, Poisson's ratio $\nu = 0.3$, coefficient of thermal expansion $\alpha = 14.7 \cdot 10^{-6}(C°)^{-1}$, yield limit $\sigma_y = 650$ MPa, ultimate strength $\sigma_u = 980$ MPa.

Let us use the isotropic material model (5.2) without taking into account the damage of the material. The creep constants have the following values:

$$b = 3.6 \cdot 10^{-19.9} \text{MPa}^{-n}/\text{h}, \qquad n = 4.9.$$

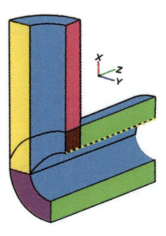

Fig. 5.2 Model of T-joint.

The T-joint is loaded with a uniformly distributed internal pressure of 120 MPa, the temperature field has a steady character and is characterized by a logarithmic distribution of temperatures in the radial direction, at the outer and inner radii temperature values of 740 K and 752 K are maintained, respectively. The rate of the integral flux function is given in the form of the experimentally obtained Eq. (5.5).

The calculations were carried out in the general 3D statement, due to the joint's symmetry, one half of it was considered as a FE model (Fig. 5.2). In this figure surfaces for which different formulations of boundary conditions symmetry were used are marked with different colors. Calculations were made for a time of 10,000 h. Figure 5.3 shows the varying of the maximum von Mises stress, which obtained in the stress concentrator (situated at the transition zone between the tubes). Curve 1 is built only for the creep problem as well as curve 2 presents the result of combined problem of creep and irradiation swelling. Figure 5.4 shows the distributions of von Mises stress in the cross-section of the T-joint also when solving the separate problem of thermal creep (a) and the combined problem of creep and irradiation swelling (b).

Thermal creep of tubes is well studied, it is characterized by significant redistribution of stresses with their general relaxation [21, 30]. Curve 1 of Fig. 5.3 also confirms this property of creep under the action of internal pressure. But adding to the analysis the effect of radiation exposure, which leads to the development of swelling, qualitatively changes the nature of the change in the stress state. Comparing Fig. 5.4 a) and b) we can see that the stress levels differ from 150 MPa for the maximum and 80 MPa for the minimum values. Due to the effect of swelling strains, the stress level increases, and due to stress relaxation during creep, it mainly decreases. Obviously, it is possible to find such a value of internal pressure that will compensate the effect of swelling. A more detailed description of the results can be found in [26].

5.4.2 Damage Accumulation and Fracture of Reactor Fuel Element

It is known [31] that the facts of the fracture of fuel elements claddings are the most dangerous factor from the point of view of ensuring their durability. At the same time,

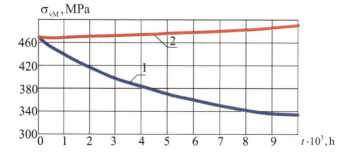

Fig. 5.3: Evolution of maximum von Mises stress in T-joint.

5 Creep and Irradiation Effects in Reactor Vessel Internals

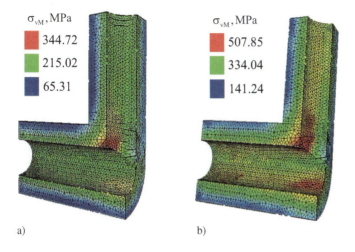

Fig. 5.4: Distribution of the von Mises stresses in the T-joint. a) thermal creep, b) thermal creep and irradiation swelling.

the fracture can go together - not only in the cladding, but also in the nuclear fuel. The resulting crack causes rapid loss the operability in both elements. As an example, Fig. 5.5 [31] shows the appearance of a crack developing in the fuel cladding and in the fuel itself. During the operation of fuel rods, their stress-strain state is determined by two main factors - the internal pressure of the gas gap and the mechanical load that occurs as a result of bending in the spaces between the tube boards [4].

Let us analyze the process of fuel rod fracture. It can take place under the condition that the gas gap is already absent due to the irreversible deformation of the fuel. In this regard, a calculation scheme without a gap between the fuel and the fuel cladding is involved in the simulation.

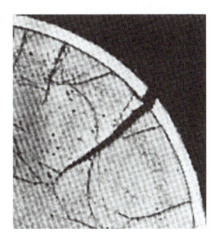

Fig. 5.5 Fracture of nuclear fuel together with the fuel cladding [B.R.T. Frost. Nuclear Fuel Elements. Oxford: Pergamon Press, 1982]

To simulate the bending of the fuel rod in its cross-section, the following calculation model was built. At the first stage, fuel rod is considered as a hinged beam during bending. According to the approaches of the theory of beam bending, the stress state was determined. The mid-length cross-section of the fuel rod is considered. The determined stresses were applied at the points of the cross-section in 2D model (plane strain scheme) of the fuel rod with a maximum value of static load on the surfaces of 25 MPa. Due to the symmetry of the stress state, a model for one half of the section was used.

In the calculations, it is assumed that the cladding is made of IN100 alloy. The creep-damage constants of this material were obtained:

$$b = 3.43 \cdot 10^{-29} (\text{MPa})^{-n}/\text{h}, n = 9.7, d = 7.5 \cdot 10^{-15} (\text{MPa})^{-r}/\text{h}, r = 5.2, l = 15.$$

Material parameters for the fuel material:

$$b = 1.25 \cdot 10^{-7} (\text{MPa})^{-n}/\text{h}, n = l = r = 3, d = 3.125 \cdot 10^{-7} (\text{MPa})^{-r}/\text{h}$$

were obtained by use experimental data [26].

We present the results of numerical simulation of creep, damage accumulation and subsequent fracture development for a cross-section of fuel rod under constant bending by the load linearly increasing up to 25 MPa on the surface of the cladding. According to the simulation data, it was established that after 318,443 h of creep, accompanied by hidden damage accumulation, a macrodefect appears on the outer surface of the fuel cladding (Fig. 5.6). Its material is destroyed simultaneously in several finite elements. Further, the development of the macrodefect proceeds quite quickly, within 0.001 h (3 s) it spreads to the fuel area (Fig. 5.7). After a few more seconds, the complete fracture of the fuel in the cross-section occurs [26].

Analysis of the obtained numerical results shows that the existence of a macrodefect in the fuel rod under consideration is limited to a few seconds. This means that for the considered material composition the assessment of its long-term strength can be performed only by calculating the processes of damage accumulation, without the involvement of a specialized software tool for simulation of fracture.

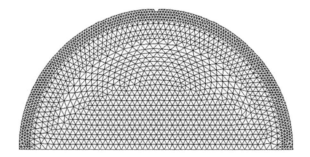

Fig. 5.6 View of the fuel rod cross-section model at the end of the hidden damage accumulation, $t = 318,443$ h

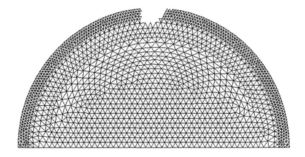

Fig. 5.7 View of the fuel rod cross-section model during fracture process at the time $t = 318,444$ h.

5.4.3 Transversal-Isotropic Creep-Damage Behaviour of Aluminium Notched Plate

The next example is related to the demonstration of the behavior of a representative of light alloys, which present class of materials exhibiting anisotropy of properties [15]. Consider the processes of thermal and irradiation creep, accompanied by the accumulation of hidden damage and the growth of irradiation swelling strains in a plate with side circular notches. Plate material is aluminum alloy type 2024. This problem is classic for the creep theory, but it is also of practical importance for some elements of RVI fasteners.

The plate is loaded by traction of 1 MPa, it is in a non-uniform temperature field $(230-280°C$ and under the action of radiation exposure. The material of the plate is characterized by significant anisotropy of creep properties and accumulation of hidden damage. The constants for the constitutive equations (5.7), (5.8) for this alloy are given in [32]. They are equal to:

$$b_{1111} = 6.669 \cdot 10^{-5}, b_{1122} = -3.334 \cdot 10^{-5}, (\text{MPa})^{(-2n/(n+1))}/\text{h}^{(2/(n+1))},$$

$$b_{2222} = 7.653 \cdot 10^{-5}, b_{1212} = 5.332 \cdot 10^{-5}, (\text{MPa})^{(-2n/(n+1))}/\text{h}^{(2/(n+1))},$$

$$d_{1111} = 1.159 \cdot 10^{-5}, d_{1122} = -5.794 \cdot 10^{-6}, (\text{MPa})^{-2}/\text{h}^{2/p},$$

$$d_{2222} = 1.385 \cdot 10^{-5}, d_{1212} = 9.437 \cdot 10^{-6}, (\text{MPa})^{-2}/\text{h}^{2/p},$$

$$m = p = 3.4, s = 0.$$

As noted in the Introduction, the description of irradiation swelling processes due to the complexity of experimental measurements during radiation exposure is a difficult task. Complete data for aluminum alloys, from which it is possible to construct an equation of state, are very scarce. In this regard, we will use the data given in various publications [15–17, 29] in order to build a qualitatively reliable model of type (5.4). To construct an equation for describing the strains of irradiation swelling, which are dependent on the values of temperature T and neutron fluency F, we apply exponential dependencies:

$$\dot{\varepsilon}_{ij}^{sw} = \frac{1}{3}\beta_{ij}\exp\left(-\frac{Q_T}{T}\right)\exp\left(-\frac{F_F}{F_1+10F}\right)\delta_{ij}, \qquad Q_T = \frac{U_T}{R}, \qquad (5.13)$$

where U_T is the value of activation energy for irradiation swelling process,

$$Q_T = 6968.8 K, F_F = -5.92 \cdot 10^{26} \text{neutron}/m^2, F_1 = 7 \cdot 10^{26} \text{neutron}/m^2.$$

The value of the threshold time for the beginning of the development of the irradiation swelling process is 15,000 h. Using data from [29, 33] for the temperature range under consideration, the value of the radiation creep constant was obtained:

$$b_{rc} = 3 \cdot 10^{-6} \text{MPa}^{-1}/h.$$

Due to the symmetry of the plate with two side notches, calculations were made for its fourth part. The results of the calculations in the form of distributions of temperatures (Fig. 5.8 a), damage measure and von Mises strains are presented in Figs. 5.8 - 5.12.

The cycle of calculation studies included the following variants of modeling: only thermal creep and damage of the plate without the influence of radiation exposure (variant 0, Figs. 5.8 b) and 5.9 a); thermal and irradiation creep and damage of the plate without the influence of irradiation swelling (variant 1, Fig. 5.11 a); thermal and irradiation creep and damage, taking into account the effect of isotropy of irradiation swelling (variant 2, $\beta_{ij} = 1$, Figs. 5.9 b) and 5.11 b) and in cases of transversal isotropy (variant 21 ($\beta_{11} = 2, \beta_{22} = 0.5$, Figs. 5.10 a) and 5.12a) and variant 22 ($\beta_{11} = 0.5$, $\beta_{22} = 2$, Figs. 5.10 b) and 5.12 b).

We will present the obtained values of the time until the completion of hidden damage accumulation: variant 0: $t_* = 50175$ h; variant 1: $t_* = 37869$ h; variant 2: $t_* = 37728$ h; variant 21: $t_* = 50727$ h; variant 22: $t_* = 33864$ h. As can be seen, the

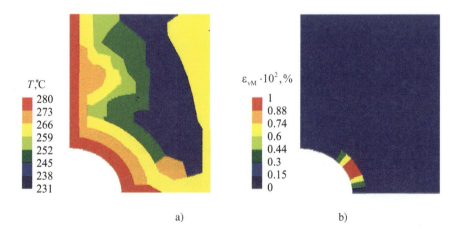

Fig. 5.8: Distribution of temperature a) and von Mises strains, b) in an aluminum plate with a notch, thermal creep, $t_*=50175$ h, static loading.

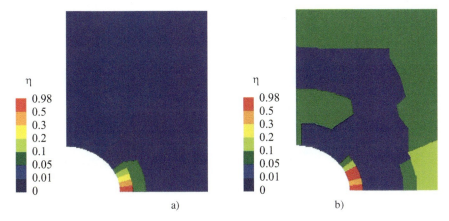

Fig. 5.9: Distribution of the damage measure in an aluminum plate with a notch a) thermal creep $t_* = 50175$ h, b) thermal, irradiation creep and irradiation swelling, $t_* = 37728$ h.

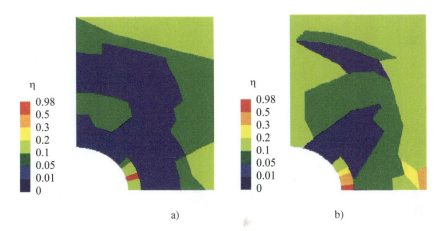

Fig. 5.10: Distribution of the damage measure in an aluminum plate with a notch. Thermal, irradiation creep and anisotropic irradiation swelling a) variant 21, $t_* = 50727$ h, b), variant 22, $t_* = 33864$ h.

addition of another process of creep (irradiation, variant 1) significantly intensifies the process of stress redistribution and, thanks to this, an increase in damage rate, which lead to more fast fracture. Adding isotropic irradiation swelling to the analysis (variant 2) leads to a further, but insignificant, reduction in the time to completion of hidden damage accumulation. Finally, the anisotropy of swelling properties can both significantly, by 4000 h, reduce the time t_* obtained in the isotropic analysis (variant 22), and significantly increase it, even in comparison with purely thermal processes (variant 21).

The analysis of the distribution of hidden damage in the case of not taking into account the influence of radiation exposure (variant 0, Fig. 5.9 a) shows that it is the

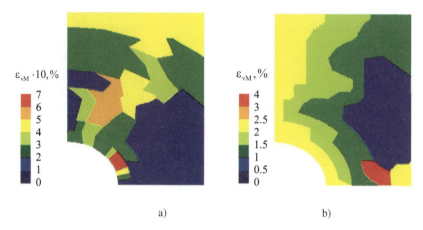

Fig. 5.11: Distribution of the von Mises strains in an aluminum plate with a notch a) thermal and irradiation creep $t_* = 37869$ h, b) thermal, irradiation creep and irradiation swelling, $t_* = 37728$ h.

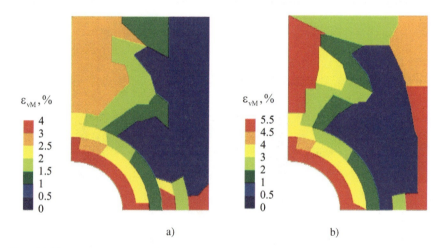

Fig. 5.12: Distribution of the von Mises strains in an aluminum plate with a notch. Thermal, irradiation creep and anisotropic irradiation swelling a) variant 21, $t_* = 50727$ h, b), variant 22, $t_* = 33864$ h.

same as in the classical analysis of creep with damage [6]: zones with significant damage are concentrated near the notch. Adding the influence of radiation taking into account irradiation creep and isotropy of swelling (variant 2, Fig. 5.9 b) significantly intensifies damage in the axial direction. If the swelling properties are considered anisotropic, then due to the higher level of strains and stresses in the axial direction (variant 22, Fig. 5.10 b), the damaged zone increases significantly. We also note that the place of completion of hidden damage accumulation may slightly shift along the edge of the notch.

The effect of radiation on the strain distribution is much greater than on the damage in the plate. This is obviously related to the addition of various mechanisms of deformation growth to the analysis. If, without taking into account radiation exposure, the strains do not exceed 0.01% (variant 0, Fig. 5.8 b), then the combined effect of thermal and irradiation creep (variant 1, Fig. 5.11 a) increases the maximum strain value to 0.7%. The main contribution to the strain level is made by irradiation swelling strains, their maximum is 4-5%.

The distribution along the plate of the maximum strain values also changes qualitatively – from a concentration around the notch in case of thermal creep (variant 0, Fig. 5.8 b) to a significant zone on the sides of the plate in case of radiation creep (variant 1, Fig. 5.11 a) and almost complete significant deformation under the action of irradiation swelling (Fig. 5.11, Fig. 5.12, variants 2, 21, 22). The anisotropy of irradiation swelling changes the location of the zones of maximum strains – now they are located on the sides of the plate, but in variant 22, when the deformation is more intense in the axial direction, their maximum level is greater (5.5%, Fig. 5.12 b). In variant 21, when the higher rate of swelling strains is in the direction perpendicular to the plate axis, due to the mutual influence of stresses, the time to failure is even longer than in case of purely thermal creep. The strain level at the edges of the plate increases significantly, by a factor of two, compared to the isotropic case (variant 2).

Now, we will present the results of the calculations of the deformed state and the distribution of the damage measure when adding a cyclic harmonic loading. We suppose that the stresses in the plate are caused by the cyclic temperature change according to the asymmetry parameter of the heating-cooling cycle H and the motion of the plate edge which is is run according to the harmonic law of traction with the cycle asymmetry parameter L:

$$P = P_0\left(1 + L\frac{2\pi}{T_c}\right), \quad T = T_0\left(1 + H\frac{2\pi}{T_c}\right) \tag{5.14}$$

For numerical modeling, we apply the constitutive equations (5.7) and (5.8). Let us take $L = H = 0.15$. The calculation results are given in Fig. 5.13, which shows the distribution of the damage measure a) and von Mises strains b). The time to the completion of hidden damage accumulation was shorter than in a case with constant stresses. It was equivalent to 28,914 h.

Analyzing the calculation results shown in Fig. 5.13, we come to the conclusion that due to the intensification of the process of damage growth, which ends in the area of the stress concentrator earlier than it was in the case of an unvarying stress field, in the vicinity of the plate far from the concentrator, the values of the damage measure even at $\eta = 0.05$ are absent (Fig. 5.13 a), as was in the results of variant 2 (Fig. 5.9 b). Due to the shorter time to failure, the overall strain level inside the plate also decreases (Fig. 5.13 b). At the same time, we stress that due to the fact that the strain rate is greater during cyclic loading, their overall level at the edges of the plate becomes the same, 2.5%, in a shorter time, as in purely static processes.

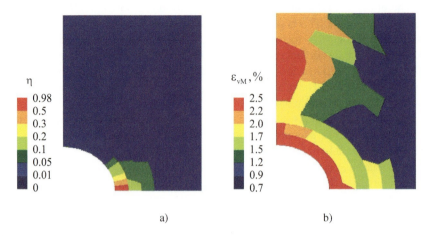

Fig. 5.13: Distribution of the damage measure a) and von Mises strains b) in an aluminum plate with a notch. Cyclic loading. Thermal, irradiation creep and irradiation swelling, $t_* = 28914$ h.

5.5 Conclusions

The approach and method of solving the problems of Creep theory with the application of Continuum Damage Mechanics in the case of action of radiation exposure are considered. The main difference from the classical problems of the Creep theory is the consideration in the analysis of two additional mechanisms of deformation - irradiation creep and irradiation swelling. If the consideration of irradiation creep, which is modeled by a power-law, or quite often even a linear dependence of the strain rate on stresses, does not present difficulties and is easily implemented into the general calculation scheme, then the modeling of irradiation swelling, starting from the construction of the constitutive equation to numerical procedures, is not a standard task. It is not implemented in most software complexes of engineering analysis.

The main problem is the construction of an adequate constitutive equation that reflects the dependence of irradiation swelling deformation on time. It is clear from the point of view of Continuum Mechanics approaches, that the formulation of the equation for the strain rate is better. If such an equation is built and verified, then with the possibility of implementing an additional unit to the FE software tool, the implementation of the calculation method of analysis taking into account the development of volume strains of irradiation swelling will not be a difficult task.

Due to the fact that the evaluation of the processes of joint development in time of thermal and irradiation creep strains, which are also accompanied by the accumulation of hidden damage, together with irradiation swelling strains, an adequate analytical modeling is not possible for the case of a complex stress state with a complex shape of a structural element and boundary conditions. For this, our approach uses a developed calculation method based on a combination of FEM and difference methods of time integration. A demonstration of the method's capabilities for various cases of deformation and damage of RVI elements is provided for 3D and 2D problems. The possibility of modeling materials with isotropy and anisotropy of long-term material properties, using scalar and tensor parameters of damage, continuing the analysis of fracture after the completion of the accumulation of hidden damage at some point of the element is shown.

The above presented results of calculations show that the contribution of strains caused by radiation exposure of the material in most cases leads to an increase in the overall level of strains and a reduction in the time until the completion of hidden damage. But for the case of materials even with transversal isotropy of properties, due to the contribution of irradiation swelling strains and different nature of stress redistribution in this case, the time to completion of hidden failure may even increase. The addition of a cyclic component of the thermal-force loading in most cases leads to an increase in the overall level of strains and a decrease in the values of time to completion of hidden damage accumulation.

Even with the isotropy of long-term properties, due to the combination of stress relaxation processes during creep and their growth during the growth of irradiation swelling strains with time, it is possible to obtain different dependences of the resulting stress on time, which will affect the time until the completion of hidden failure.

All presented examples show that an adequate analysis of long-term deformation and strength of structural elements exposed to thermal - force and radiation fields is possible only with the use of appropriate calculation tools. Due to the essential nonlinearity of the problems, even a moderate deviation of the parameters can lead to qualitatively different distributions of the components of the stress-strain state. Therefore, the basis of the methods should be adequate and verified constitutive equations, primarily for describing the irradiation swelling strains in time, understanding their isotropic or anisotropic nature.

Acknowledgements Dmytro Breslavsky acknowledges the support by the Volkswagen Foundation within the programme "Visiting research program for refugee Ukrainian scientists" (Az. 9C184).

References

[1] Brumovsky M (2018) Evaluation of Reactor Internals Integrity and Lifetime According to the NTD ASI. In: Proceedings of the ASME 2018 Pressure Vessels and Piping Conference, Pressure Vessels and Piping Conference, vol 1A: Codes and Standards
[2] Chopra OK (2010) Degradation of LWR core internal materials due to neutron irradiation. Argonne National Laboratory, NRC Job Code N6818, Office of Nuclear Regulatory Research
[3] Lee KH, Park JS, Ko HO, Jhung MJ (2013) Analysis for aging and operating experiences of reactor vessel internals. In: Transactions of the Korean Nuclear Society Autumn Meeting Gyeongju, KNS, Daejon

[4] Ma BM (1983) Nuclear Reactor Materials and Applications. Iowa State University
[5] Zinkle SJ, Tanigawa H, Wirth BD (2019) Chapter 5 - radiation and thermomechanical degradation effects in reactor structural alloys. In: Odette GR, Zinkle SJ (eds) Structural Alloys for Nuclear Energy Applications, Elsevier, Boston, pp 163–210
[6] Naumenko K, Altenbach H (2016) Modeling High Temperature Materials Behavior for Structural Analysis - Part I: Continuum Mechanics Foundations and Constitutive Models, Advanced Structured Materials, vol 28. Springer
[7] Ozaltun H, Herman Shen MH, Medvedev P (2011) Assessment of residual stresses on U10Mo alloy based monolithic mini-plates during hot isostatic pressing. Journal of Nuclear Materials **419**(1):76–84
[8] Pandit AM, Blom FJ, Baas PJ (2019) Stress assessment of baffle former bolt of PWR reactor for IASCC. In: 19th International Conference on Environmental Degradation of Materials in Nuclear Power Systems-Water Reactors, Boston, pp 1016–1023
[9] Sangjeung L, Eunho L, NoCheol P, Jongsung K, Youngin C, Jongbeom P (2019) Finite Element Analysis Methodology Reflecting Neutron Irradiation Effects For Structural Analysis. In: Transactions SMiRT-25 (Charlotte, NC, USA, August 4-9, 2019, Division III)
[10] Breslavsky D, Chuprynin A, Morachkovsky O, Tatarinova O, Pro W (2019) Deformation and damage of nuclear power station fuel elements under cyclic loading. The Journal of Strain Analysis for Engineering Design **54**(5-6):348–359
[11] Breslavsky D, Senko A, Tatarinova O, Voevodin V, Kalchenko A (2021) Stress–strain state of nuclear reactor core baffle under the action of thermal and irradiation fields. In: Altenbach H, Amabili M, Mikhlin YV (eds) Nonlinear Mechanics of Complex Structures: From Theory to Engineering Applications, Advanced Structured Materials, Springer, Cham, Advanced Structured Materials, vol 157, pp 279–293
[12] Tonks M, Gaston D, Permann C, Millett P, Hansen G, Wolf D (2010) A coupling methodology for mesoscale-informed nuclear fuel performance codes. Nuclear Engineering and Design **240**(10):2877–2883
[13] Garner FA (2020) Radiation-induced damage in austenitic structural steels used in nuclear reactors. In: Konings RJM, Stoller RE (eds) Comprehensive Nuclear Materials, vol 3, 2nd edn, Elsevier, pp 57–168
[14] Breslavsky D, Tatarinova O, Kalchenko O, Tolstolutska G (2022) Effect of irradiation and thermo-induced processes on reactor in vessel elements during long-term operation. Voprosy Atomnoj Nauki i Tekhniki (2 (138)):3–8
[15] Asundi MK (1961) The role of light metals in nuclear engineering. In: Symposium on Light Metal Industry In India, NML, Jamshedpur, pp 171–179
[16] Garric V, Colas K, Donnadieu P, Loyer-Prost M, Leprêtre F, Cloute-Cazalaa V, Kapusta B (2021) Impact of the microstructure on the swelling of aluminum alloys: Characterization and modelling bases. Journal of Nuclear Materials **557**:153,273

[17] McDonell WR (1972) Irradiation swelling and growth in uranium and other anisotropic metals. Trans Amer Nucl Soc **15**(1)
[18] Dubyk Y, Filonov V, Filonova Y (2019) Swelling of the WWER-1000 Reactor Core Baffle. In: Transactions SMiRT-25 (Charlotte, NC, USA, August 4-9, 2019, Division II)
[19] Yang JS, Lee JG, Oh SJ, Won SY (2015) Irradiation-induced degradation effects on baffle-former-barrel assembly of reactor vessel internal. In: Transactions of the Korean Nuclear Society Spring Meeting reactor vessel internal, KNS, Jeju
[20] Breslavskii DV, Morachkovskii OK (1998) Nonlinear creep and the collapse of flat bodies subjected to high-frequency cyclic loads (in Russ. International Applied Mechanics **34**(3):287–292
[21] Breslavs'kyi DV, Korytko YM, Morachkovs'kyi OK (2011) Cyclic thermal creep model for the bodies of revolution. Strength of Materials **43**(2):134–143
[22] Breslavsky D, Morachkovsky O, Tatarinova O (2014) Creep and damage in shells of revolution under cyclic loading and heating. International Journal of Non-Linear Mechanics **66**:87–95
[23] Breslavsky D, Kozlyuk A, Tatarinova O (2018) Numerical simulation of two-dimensional problems of creep crack growth with material damage consideration. Eastern-European Journal of Enterprise Technologies **2**(7 (92)):27–33
[24] Breslavsky D, Senko A, Tatarinova O (2021) Creep damage and fracture of notched specimens under static and fast periodic loading. International Journal of Damage Mechanics **30**(6):964–983
[25] Breslavsky DV, Korytko YN, Tatarinova OA (2017) Design and development of finite element method software (in Ukrain.). Pidruchnyk NTU KhPI
[26] Breslavsky DV (2020) Deformation and Long Term Strength of Structural Elements of Nuclear Reactors. Madrid, Kharkiv
[27] Morachkovsky OK, Pasynok MA (1997) Creep of isotropic and anisotropic plane bodies (in Russ.). In: Proceedings of International Scientific and Technical Conference "Information Technologies: Science, Engineering, Technology, Education, Health", vol 1, Kharkiv State Polytechnic University, Kharkiv, pp 127–131
[28] Altenbach H, Breslavsky D, Mietielov V, Tatarinova O (2020) Short term transversally isotropic creep of plates under static and periodic loading. In: Naumenko K, Krüger M (eds) Advances in Mechanics of High-Temperature Materials, Springer, Cham, Advanced Structured Materials, vol 117, pp 181–211
[29] Farrell K, King R, Jostsons A (1973) Examination of the irradiated 6061 aluminum hfir target holder. Oak Ridge National Lab., Tenn.
[30] Galishin AZ, Sklepus SN (2019) Prediction of the time to failure of axisymmetrically loaded hollow cylinders under conditions of creep. Journal of Mathematical Sciences **240**(2):194–207
[31] Frost BRT (2013) Nuclear Fuel Elements: Design, Fabrication and Performance. Elsevier
[32] Konkin VN, Morachkovskii OK (1987) Creep and long-term strength of light alloys with anisotropic properties. Strength of Materials **19**(5):626–631

[33] Aitkhozhin ES, Chumakov EV (1996) Radiation-induced creep of copper, aluminium and their alloys. Journal of Nuclear Materials **233-237**:537–541

Chapter 6
Analysis of Damage and Fracture in Anisotropic Sheet Metals Based on Biaxial Experiments

Michael Brünig, Sanjeev Koirala, and Steffen Gerke

Abstract The paper discusses the effect of stress state and of loading with respect to the rolling direction on damage and failure of anisotropic ductile sheet metals. For the investigated aluminum alloy EN AW-2017A experiments with uniaxially and biaxially loaded flat specimens have been performed to identify elastic-plastic material parameters. The focus is on numerical analysis on the micro-scale examining the deformation and damage behavior of differently loaded void-containing unit cells to detect damage and failure processes. Results of the finite element calculations show that the stress state and the loading direction with respect to the rolling direction have an effect on formation of damage mechanisms on the micro-level as well as on corresponding macroscopic damage strains.

6.1 Introduction

Analysis of inelastic deformations as well as of damage and fracture behavior in engineering structures must be based on accurate and efficient theoretical models. For example, on the micro-level damage and fracture processes in ductile metals are mainly related to nucleation, growth and coalescence of micro-defects which may lead to formation of macro-cracks resulting in final failure of structures. Besides experiments numerical analysis on the micro-level of the behavior of individual micro-defects in elastic-plastic materials will reveal detailed information for the formulation of appropriate constitutive frameworks [1, 2]. The results of these numerical investigations can be taken to develop continuum models which can be

---

Michael Brünig · Sanjeev Koirala · Steffen Gerke
Institut für Mechanik und Statik, Universität der Bundeswehr München, 85577 Neubiberg, Germany,
e-mail: michael.bruenig@unibw.de, sanjeev.koirala@unibw.de, steffen.gerke@unibw.de

used to predict inelastic deformations as well as damage and fracture behavior of ductile metals.

Different research groups (see, for example, [1, 3–8]) carried out three-dimensional finite element calculations to detect damage and fracture processes in ductile metals and to analyze the deformation behavior of micro-defects under different stress states. These numerical studies take into account isotropic elastic-plastic material behavior and clearly demonstrate that the current stress state has a remarkable influence on the microscopic damage and fracture mechanisms as well as on the macroscopic failure behavior. Based on the results of the finite element analyses with the void-containing representative volume elements damage evolution equations in phenomenological constitutive approaches can be proposed and validated. In addition, the numerical data can be taken to identify micro-mechanically motivated material parameters [1, 2].

Internal changes in the crystallographic structure of ductile metals occurring during manufacturing processes such as deep drawing, extrusion or rolling lead to anisotropies in material behavior. These deformation-induced anisotropies cannot be neglected in simulation of the deformation and fracture behavior of thin metal sheets and, therefore, have to be incorporated in material models. In the literature, various hydrostatic-stress-independent anisotropic yield conditions have been discussed using quadratic [9, 10], non-quadratic [11–13] or spline functions [14] of stresses. On the other hand, in order to take into account the hydrostatic stress dependence the Hoffman yield criterion [15] has been proposed.

In the present paper a continuum framework incorporating plastic anisotropy modeled by the Hoffman yield condition is presented. To analyze in detail the damage and failure mechanisms in the investigated aluminum alloy EN AW-2017A numerical simulations on the micro-scale are performed considering void-containing representative volume elements under various load combinations with respect to the principal axes of anisotropy. Macroscopic damage strains are numerically predicted which are affected by the stress state, the load ratio and the loading direction.

6.2 Constitutive Framework

The analysis is based on the continuum damage framework proposed by [1, 16, 17] and its generalization for anisotropic plasticity by [18–20]. The basic idea of the phenomenological model is the introduction of the damage strain tensor, $\boldsymbol{A}^{da}$, describing macroscopic inelastic strains related to damage and failure mechanisms on the micro-level. Additionally, the kinematic approach considers the additive decomposition of the strain rate tensor $\dot{\boldsymbol{H}}$ into elastic, $\dot{\boldsymbol{H}}^{el}$, effective plastic, $\dot{\bar{\boldsymbol{H}}}^{pl}$, and damage parts, $\dot{\boldsymbol{H}}^{da}$ [16].

For the investigated aluminum alloy EN AW-2017A the Hoffman yield condition [15]

$$f^{pl} = \boldsymbol{C} \cdot \bar{\boldsymbol{T}} + \sqrt{\frac{1}{2}\bar{\boldsymbol{T}} \cdot \mathcal{D}\bar{\boldsymbol{T}}} - c = 0 \tag{6.1}$$

6 Damage and Fracture in Anisotropic Sheet Metals

is used to model the anisotropic plastic behavior, where $\bar{T}$ represents the effective Kirchhoff stress tensor introduced in the fictitious undamaged configuration and c denotes the equivalent stress measure. Constitutive parameters modeling the plastic anisotropy appear in the tensor of coefficients

$$C = C^i_{.j} \, g_i \otimes g^j = C_{(i)} \, g_i \otimes g^i \tag{6.2}$$

with the components (in Voigt notation)

$$\left[C^i_{.j}\right] = [C_1 \ C_2 \ C_3 \ 0 \ 0 \ 0]^T \tag{6.3}$$

as well as in the tensor

$$\mathcal{D} = D^{i.k}_{.j.l} \, g_i \otimes g^j \otimes g_k \otimes g^l \tag{6.4}$$

with the respective coefficients given in matrix representation

$$\left[D^{i.k}_{.j.l}\right] = \begin{bmatrix} C_4+C_5 & -C_4 & -C_5 & 0 & 0 & 0 \\ -C_4 & C_4+C_6 & -C_6 & 0 & 0 & 0 \\ -C_5 & -C_6 & C_5+C_6 & 0 & 0 & 0 \\ 0 & 0 & 0 & C_7 & 0 & 0 \\ 0 & 0 & 0 & 0 & C_8 & 0 \\ 0 & 0 & 0 & 0 & 0 & C_9 \end{bmatrix}. \tag{6.5}$$

For the investigated aluminum alloy these parameters have been determined by testing uniaxially loaded dog-bone shaped specimens cut in different directions from thin sheets with respect to the rolling direction [20]. This procedure leads to the values given in Table 6.1.

The equivalent yield stress c of the undamaged metal is identified by testing unnotched flat specimens cut in rolling direction. For the investigated ductile anisotropic aluminum alloy EN AW-2017A the Voce law is used

$$c = c_o + R_o \epsilon^{pl} + R_\infty \left(1 - e^{-b \, \epsilon^{pl}}\right) \tag{6.6}$$

where c_o is the initial yield strength, R_o and R_∞ represent hardening moduli, b is the hardening exponent and ϵ^{pl} denotes the equivalent plastic strain measure. For the investigated ductile anisotropic aluminum alloy EN AW-2017A the parameters listed in Table 6.2 have been chosen leading to good agreement of experimental data and the numerical curve. In the analysis of anisotropic ductile metals generalized invariants of the effective Kirchhoff stress tensor $\bar{T}$ are introduced [20]. In particular,

Table 6.1: Anisotropy parameters.

C_1	C_2	C_3	C_4	C_5	C_6	C_7	C_8	C_9
-0.0424	-0.0102	0.0000	0.8123	1.3607	1.3103	3.7580	3.0000	3.0000

Table 6.2: Plastic material parameters.

	c_o [MPa]	R_o [MPa]	R_∞ [MPa]	b [-]
RD	333	488	142	19

the first Hoffman stress invariant is defined by

$$\bar{I}_1^H = \frac{1}{a}\mathbf{C}\cdot\bar{\mathbf{T}} \quad \text{with} \quad a = \frac{1}{3}\text{tr}\mathbf{C} \tag{6.7}$$

and the second and third deviatoric Hoffman stress invariants are taken to be

$$\bar{J}_2^H = \frac{1}{2}\bar{\mathbf{T}}\cdot\mathcal{D}\bar{\mathbf{T}} \tag{6.8}$$

and

$$\bar{J}_3^H = \det\left(\mathcal{D}\,\bar{\mathbf{T}}\right). \tag{6.9}$$

Using these definitions further generalized stress parameters are defined. For example, the Hoffman stress triaxiality

$$\bar{\eta}^H = \frac{\bar{I}_1^H}{3\sqrt{3\bar{J}_2^H}} \tag{6.10}$$

and the generalized Hoffman-Lode parameter

$$\bar{L}^H = \frac{-3\sqrt{3}\,\bar{J}_3^H}{2\,(\bar{J}_2^H)^{(3/2)}} \tag{6.11}$$

are used to characterize the influence of the current stress state on the behavior of anisotropic materials.

In addition, the flow rule

$$\dot{\bar{\mathbf{H}}}^{pl} = \dot{\gamma}\bar{\mathbf{N}} \tag{6.12}$$

models the evolution of the effective plastic strains in the undamaged configuration where $\dot{\gamma}$ represents the equivalent plastic strain rate characterizing the amount of the plastic strain rate and the normalized deviatoric effective stress tensor

$$\bar{\mathbf{N}} = \frac{\mathcal{D}\bar{\mathbf{T}}}{\|\mathcal{D}\bar{\mathbf{T}}\|} \tag{6.13}$$

defines its direction.

Furthermore, in the damaged configurations the damage criterion

$$f^{da} = \alpha \bar{I}_1^H + \beta\sqrt{\bar{J}_2^H} - \sigma = 0 \tag{6.14}$$

is assumed to model the onset and evolution of damage in plastically anisotropic ductile materials, where the generalized first and second deviatoric Hoffman invariants of the Kirchhoff stress tensor formulated with respect to the damaged configurations, I_1^H and J_2^H, characterize the current stress state and σ denotes the equivalent damage stress measure. The stress-state- and loading-direction-dependent parameters α and β have been determined by series of biaxial experiments performed with differently loaded specimens, see [20] for further details.

In addition, the damage rule

$$\dot{\boldsymbol{H}}^{da} = \dot{\mu}\left(\frac{1}{\sqrt{3}}\tilde{\alpha}\mathbf{1} + \tilde{\beta}\boldsymbol{N}\right) \qquad (6.15)$$

models formation of macroscopic irreversible strains caused by damage and failure processes on the micro-scale where the normalized deviatoric part of the Kirchhoff stress tensor

$$\boldsymbol{N} = \frac{\text{dev}\boldsymbol{T}}{\|\text{dev}\boldsymbol{T}\|} \qquad (6.16)$$

has been used and the parameters $\tilde{\alpha}$ and $\tilde{\beta}$ represent the stress and loading direction dependence of the damage strain rate tensor (6.15) which will be identified by further experiments as well as by numerical calculations on the micro-level discussed in the following section.

6.3 Numerical Simulations and Results

In the continuum damage model discussed above the macroscopic damage strain tensor $\boldsymbol{A}^{da}$ is used to model the evolution of damage caused by different damage and failure mechanisms on the micro-scale. The stress state and loading direction dependence of this tensor is taken into account by the parameters $\tilde{\alpha}$ and $\tilde{\beta}$ in the damage rule (6.15). To determine these parameters numerical simulations analyzing the deformation behavior of three-dimensionally loaded void-containing representative volume elements with initial porosity of 3% shown in Fig. 6.1 have been performed. Therefore, the finite element program ANSYS has been enhanced by a user-defined subroutine based on the proposed anisotropic continuum framework. In the numerical

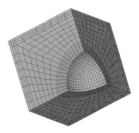

Fig. 6.1 Finite element mesh of one eighth of the unit cell.

analysis, eight-node elements of type SOLID185 and symmetry boundary conditions are used. The finite elements are elastically and plastically deformed leading to macroscopic elastic and plastic strain rates. The changes in size and shape of the initially spherical voids correspond to the macroscopic damage strains.

Taking into account the kinematic approach with additive decomposition of the strain rates into elastic, plastic and damage parts the components of the macroscopic strain rate tensor in the principal directions (i) can be written in the form

$$\dot{H}^{unit-cell}_{(i)} = \dot{H}^{el}_{(i)} + \dot{H}^{pl}_{(i)} + \dot{H}^{da}_{(i)}. \tag{6.17}$$

In the finite elements microscopic elastic and plastic strain rates, $\dot{\boldsymbol{h}}^{el}$ and $\dot{\boldsymbol{h}}^{pl}$, are computed during the loading process. They lead to the elastic-plastic macroscopic strain rates

$$\dot{\boldsymbol{H}}^{ep} = \dot{\boldsymbol{H}}^{el} + \dot{\boldsymbol{H}}^{pl} = \frac{1}{V} \int_{V_{matrix}} \left(\dot{\boldsymbol{h}}^{el} + \dot{\boldsymbol{h}}^{pl} \right) dv \tag{6.18}$$

of the unit-cell where V is the current volume of the representative volume element and V_{matrix} is the current volume of the matrix material (finite elements). Using Eqs. (6.17) and (6.18) the macroscopic damage strain rate tensor is given by

$$\dot{H}^{da}_{(i)} = \dot{H}^{unit-cell}_{(i)} - \dot{H}^{ep}_{(i)}. \tag{6.19}$$

This leads to the principal components of the damage strain tensor

$$A^{da}_{(i)} = \int \dot{H}^{da}_{(i)} dt. \tag{6.20}$$

Furthermore, the evolution of the void volume fraction f of the micro-defect containing representative volume element

$$\dot{f} = (1-f) \operatorname{tr} \dot{\boldsymbol{H}}^{da} \tag{6.21}$$

can be expressed in terms of the volumetric part of the damage strain rate tensor leading to

$$f = \int \dot{f} dt, \tag{6.22}$$

see [16] for further details. To visualize the evolution of macroscopic damage strain components in an adequate manner the equivalent strain rate

$$\dot{\epsilon}_{eq} = \sqrt{\frac{2}{3} \dot{\boldsymbol{H}} \cdot \dot{\boldsymbol{H}}} \tag{6.23}$$

and the equivalent strain

$$\epsilon_{eq} = \int \dot{\epsilon}_{eq} dt \tag{6.24}$$

are introduced characterizing the amount of strain rates and strains.

The influence of different load ratios $F_x/F_y/F_z$ on the deformation behavior of the void-containing representative volume element has been studied. In the present paper two cases are numerically investigated and the results are compared with experimental observations. For example, in [19] the H-specimen had been biaxially deformed with the load ratios $F_1/F_2 = 0/1$ and $1/+1$ leading to tensile and shear dominated stress states, respectively.

In the experiment with the H-specimen loaded by $F_1/F_2 = 0/1$ the Hoffman stress triaxiality $\eta^H = 0.75$ and the Hoffman Lode parameter $L^H = 0.3$ had been determined by corresponding numerical simulations [19]. For comparison, the micro-defect containing unit cell is deformed by the load ratio $F_x/F_y/F_z = 1.0/0.63/0.27$ leading to the similar stress parameters $\eta^H = 0.75$ and $L^H = 0.23$. The increase of the principal values of the damage strain tensor $A^{da}_{(i)}$ (6.20) with increasing equivalent strain measure (6.24) is shown in Fig. 6.2. The damage strain A^{da}_x shows an increase up to 0.065 where slightly larger increase can be seen for loading in transverse direction (90°) and slightly smaller one for loading in diagonal direction (45°). On the other hand, the damage strain components A^{da}_y and A^{da}_z show smaller increases up to 0.04 and 0.01, respectively, with the same dependence of the loading direction with respect to the rolling direction. In addition, the evolution of the void volume fraction f (6.22) with increasing loading is shown in Fig. 6.3. A remarkable increase with maximum values of 0.13 has been numerically predicted for loading in the rolling direction. Again, slightly larger increase of the void volume fraction can be seen for

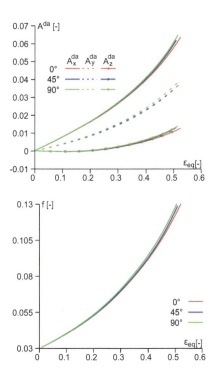

Fig. 6.2 Formation of principal components of the damage strain tensor for $\eta^H = 0.75$ and $L^H = 0.23$.

Fig. 6.3 Formation of the void volume fraction for $\eta^H = 0.75$ and $L^H = 0.23$.

loading in transverse direction (90°) and slightly smaller one for loading in diagonal direction (45°). This damage behavior was also visible in the pictures of scanning electron microscopy in [19] where large pores occur for loading in transverse (90°) and in rolling (0°) direction whereas less and smaller voids were visible after loading in diagonal direction (45°).

For the biaxial experiments with the load ratio $F_1/F_2 = 1/+1$ the Hoffman stress triaxiality $\eta^H = 0.4$ and the Hoffman-Lode parameter $L^H = 0.2$ had been predicted in corresponding numerical simulations [19]. Similar stress parameters $\eta^H = 0.4$ and $L^H = 0.25$ have also been achieved in the unit cell calculations with the load ratio $F_x/F_y/F_z = 1.0/0.37/-0.27$. The evolution of the principal values of the damage strain tensor $A_{(i)}^{da}$ (6.20) with increasing equivalent strain measure (6.24) is shown in Fig. 6.4. In this shear-dominated loading case the damage strain component A_x^{da} increases up to 0.022 and A_z^{da} shows a decrease up to -0.025 whereas the component A_y^{da} only reaches 0.005. This means that during shear loading the initially spherical void is deformed into an ellipsoid. In Fig. 6.4 nearly no effect of the loading direction on the macroscopic damage strain behavior can be seen. The void volume fraction f shown in Fig. 6.5 only shows a small increase up to 0.033 and a slightly larger increase in diagonal loading direction (45°) can be seen. This damage behavior had also been seen in the pictures of scanning electron microscopy published in [19]. The photos showed shear mechanisms on the micro-scale with superimposed increase of voids which were deformed in shear direction.

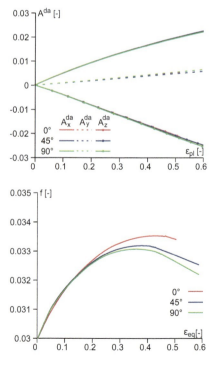

Fig. 6.4 Formation of principal components of the damage strain tensor for $\eta^H = 0.4$ and $L^H = 0.25$.

Fig. 6.5 Formation of the void volume fraction for $\eta^H = 0.4$ and $L^H = 0.25$.

6.4 Conclusions

In the present paper the influence of the stress state and the loading direction with respect to the rolling direction on damage and failure behavior of the aluminum alloy EN AW-2017A has been analyzed. Numerically predicted deformations of micro-defect containing unit cells for two load cases corresponding to biaxial loading scenarios of the H-specimen have been presented. For the investigated anisotropic ductile metal the stress state was characterized by the generalized stress triaxiality and the generalized Lode parameter based on stress invariants taken from the Hoffman yield condition. Elastic-plastic material parameters of the analyzed aluminum alloy were identified by experiments performed with uniaxially loaded flat specimens cut from sheets in different directions with respect to the rolling direction. These parameters were taken into account in the finite element analysis to study the deformation and damage behavior of micro-defect-containing representative volume elements. Various three-dimensional load ratios have been considered and numerically obtained evolutions of the principal components of the damage strain tensor and of the void volume fraction as well as the corresponding damage and failure processes on the micro-scale have been discussed. These numerically predicted results have been compared with photos from scanning electron microscopy of fracture surfaces of the tested H-specimen published in the literature. The numerical results for the macroscopic damage strains can be seen as quasi-experimental results and can be taken to develop laws for the damage strain rates used in sophisticated continuum damage models. Based on this theoretical framework numerical simulation of experiments can be performed to numerically predict the deformation and failure behavior of structural components built with anisotropic sheet metals.

Acknowledgements Financial support from the Deutsche Forschungsgemeinschaft (DFG, German Research Foundation) under project number 394286626 is gratefully acknowledged.

References

[1] Brünig M, Gerke S, Hagenbrock V (2013) Micro-mechanical studies on the effect of the stress triaxiality and the Lode parameter on ductile damage. International Journal of Plasticity **50**:49–65
[2] Brünig M, Hagenbrock V, Gerke S (2018) Macroscopic damage laws based on analysis of microscopic unit cells. ZAMM - Journal of Applied Mathematics and Mechanics / Zeitschrift für Angewandte Mathematik und Mechanik **98**(2):181–194
[3] Barsoum I, Faleskog J (2011) Micromechanical analysis on the influence of the Lode parameter on void growth and coalescence. International Journal of Solids and Structures **48**(6):925–938
[4] Gao X, Wang T, Kim J (2005) On ductile fracture initiation toughness: Effects of void volume fraction, void shape and void distribution. International Journal

of Solids and Structures **42**(18):5097–5117
[5] Gao X, Zhang G, Roe C (2010) A study on the effect of the stress state on ductile fracture. International Journal of Damage Mechanics **19**(1):75–94
[6] Kim J, Gao X, Srivatsan T (2003) Modeling of crack growth in ductile solids: a three-dimensional analysis. International Journal of Solids and Structures **40**(26):7357–7374
[7] Kuna M, Sun D (1996) Three-dimensional cell model analyses of void growth in ductile materials. International Journal of Fracture **81**(3):235–258
[8] Scheyvaerts F, Onck P, Tekoğlu C, Pardoen T (2011) The growth and coalescence of ellipsoidal voids in plane strain under combined shear and tension. Journal of the Mechanics and Physics of Solids **59**(2):373–397
[9] Hill R (1948) A theory of the yielding and plastic flow of anisotropic metals. Proceedings of the Royal Society of London Series A Mathematical and Physical Sciences **193**(1033):281–297
[10] Stoughton TB, Yoon JW (2009) Anisotropic hardening and non-associated flow in proportional loading of sheet metals. International Journal of Plasticity **25**(9):1777–1817
[11] Barlat F, Aretz H, Yoon J, Karabin M, Brem J, Dick R (2005) Linear transfomation-based anisotropic yield functions. International Journal of Plasticity **21**(5):1009–1039
[12] Ha J, Baral M, Korkolis YP (2018) Plastic anisotropy and ductile fracture of bake-hardened AA6013 aluminum sheet. International Journal of Solids and Structures **155**:123–139
[13] Hu Q, Yoon JW, Manopulo N, Hora P (2021) A coupled yield criterion for anisotropic hardening with analytical description under associated flow rule: Modeling and validation. International Journal of Plasticity **136**:102,882
[14] Tsutamori H, Amaishi T, Chorman RR, Eder M, Vitzthum S, Volk W (2020) Evaluation of prediction accuracy for anisotropic yield functions using cruciform hole expansion test. Journal of Manufacturing and Materials Processing **4**(2):43
[15] Hoffman O (1967) The brittle strength of orthotropic materials. Journal of Composite Materials **1**(2):200–206
[16] Brünig M (2003) An anisotropic ductile damage model based on irreversible thermodynamics. International Journal of Plasticity **19**(10):1679–1713
[17] Brünig M (2016) A thermodynamically consistent continuum damage model taking into account the ideas of CL Chow. International Journal of Damage Mechanics **25**(8):1130–1141
[18] Brünig M, Gerke S, Koirala S (2021) Biaxial experiments and numerical analysis on stress-state-dependent damage and failure behavior of the anisotropic aluminum alloy EN AW-2017A. Metals **11**(8):1214
[19] Brünig M, Koirala S, Gerke S (2022) Analysis of damage and failure in anisotropic ductile metals based on biaxial experiments with the H-specimen. Experimental Mechanics **62**(2):183–197
[20] Brünig M, Koirala S, Gerke S (2023) A stress-state-dependent damage criterion for metals with plastic anisotropy. International Journal of Damage Mechanics **0**(0):10567895231160,810

Chapter 7
Effect of Physical Aging on the Flexural Creep in 3D Printed Thermoplastic

Marcel Fischbach and Kerstin Weinberg

Abstract Extrusion-based 3D printing has become one of the most common additive manufacturing methods and is widely used in engineering. This contribution presents the results of flexural creep experiments on 3D printed PLA specimens, focusing on changes in creep behavior due to physical aging. It is shown experimentally that the creep curves obtained on aged specimens are shifted to each other on the logarithmic time scale in a way that the theory of physical aging can explain. The reason for the physical aging of 3D printed thermoplastics is assumed to be the special heat treatment that the polymer undergoes during extrusion. Additionally, results of a long-term flexural creep experiment are shown, demonstrating that non-negligible creep over long periods can be observed even at temperatures well below the glass transition temperature. Such creep effects should be considered for designing components made of 3D printed thermoplastics.

7.1 Introduction

Over the past decade, 3D printing technologies have become essential to modern manufacturing processes. Previously used primarily for rapid prototyping, 3D printing technologies are now employed to produce end-use parts in various applications, particularly for complex geometries [1]. Since it is often necessary to prove that these printed parts can withstand the prevailing loads, knowledge of their short- and long-term material properties is required.

One of the most widely used 3D printing processes is fused filament deposition, in which plastic parts are built up layer by layer from a thermoplastic polymer filament. To generate the geometry of a single layer, the filament is heated above its melting temperature in an extruder and deposited through a nozzle on a build plate or a

---

Marcel Fischbach · Kerstin Weinberg
Chair of Solid Mechanics, University of Siegen, Paul-Bonatz-Straße 9-11, 57076 Siegen, Germany,
e-mail: `marcel.fischbach@uni-siegen.de,kerstin.weinberg@uni-siegen.de`

previously printed layer, Fig. 7.1. When the term 3D printing is used in the following, we refer to this extrusion-based process. The most commonly used filament materials for this type of 3D printing are polylactide and acrylonitrile butadiene styrene [1], abbreviated hereafter as PLA and ABS, respectively.

Depending on the chosen print settings, e.g. nozzle and platform temperature, print speed, infill density and infill pattern, to name only a few, printed parts have specific mechanical properties. Numerous studies have been conducted to quantify the relationship between chosen print parameters and the properties of 3D printed parts. For example, Ahn *et al.* [2] investigated the effect of selected print parameters on the tensile strength of 3D printed ABS using a two-level experimental design and analyzed the effect of print raster orientation on the tensile and compressive strength of printed compared to injection molded test specimens. Wittbrodt and Pearce [3] investigated the correlations between PLA filament color, nozzle temperature, degree of crystallinity, and yield strength of 3D printed tensile specimens. Fernandez-Vicente *et al.* [4] and Rismalia *et al.* [5] analyzed the influence of common infill patterns and infill density on the tensile strength and tensile modulus of 3D printed specimens made of ABS and PLA, respectively. Reppel and Weinberg [6] investigated the qualitative rupture behavior of printed tensile specimens made of thermoplastic polyurethane and modeled their deformation behavior using hyperelastic material models. Akhoundi *et al.* [7] studied the effects of nozzle temperature and heat treatment (annealing) on the crystallinity, the interlayer and intralayer adhesion, cf. [1], and the mechanical properties of 3D printed tensile specimens of high-temperature PLA (HTPLA). Khosravani *et al.* [8] analyzed the stress-strain behavior, the elastic modulus and tensile strength of 3D printed PLA specimens as a function of print raster orientation and two different print speeds.

In the field of time-dependent, i.e. dynamic, viscoelastic and viscoplastic properties of printed thermoplastics, numerous contributions can also be found in literature, e.g. [9–13]. However, to our knowledge, there are none on the influence of process-induced physical aging on the creep behavior of 3D printed thermoplastics under usual production conditions.

Physical aging is a specific type of polymer aging that occurs in both amorphous and semi-crystalline thermoplastics, which are rapidly cooled below their glass transition temperature [14]. In the resulting non-equilibrium state of low but not vanishing

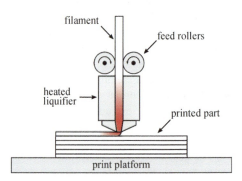

Fig. 7.1 Schematic of extrusion-based 3D printing.

molecular mobility, the polymer gradually evolves toward thermodynamic equilibrium. As a result, this slow, time-dependent, and asymptotic process leads to a measurable stiffening of the polymer, which is particularly noticeable in the creep behavior, cf. [15–18]. Additionally, the polymer's density and yield stress increase during aging while its impact strength decreases [16].

As detailed in [15], rapid cooling of polymers below their glass transition temperature can be artificially induced to study the effects of physical aging in a controlled manner. However, in addition to this academic approach, Struik [15] argued that similar cooling conditions could also prevail during the processing of plastics by extrusion or injection molding. Struik demonstrated from creep experiments on injection molded PVC samples that those can initiate significant physical aging. Thus, concerning the special heat treatment to which the polymer filament is subjected during extrusion in 3D printing, the question arises as to whether and to what extent 3D printed components exhibit physical aging. Therefore, we address here the question of physical aging experimentally. Using a three-point bending setup, we investigate the flexural creep of 3D printed specimens made of PLA, a thermoplastic whose mechanical properties are affected by physical aging at moderate temperatures [19].

In this contribution, we briefly introduce the phenomena of physical aging and its relation to the viscoelasticity of thermoplastic polymers in Sect. 7.2. Then, in Sect. 7.3, the preparation of the specimens and the conducted experiments are explained. After the presentation of the experimental results in Sect. 7.4, we discuss the measured results and critically evaluate the experimental procedure in Sect. 7.5.

7.2 Theoretical Background

In this section, the essence of physical aging and its linkage to the creep of thermoplastics is presented. In plastics, the term *creep* refers to a time- and temperature-dependent increase in deformation under constant loading which is typically described by the viscoelastic compliance, i.e. the reciprocal of the material's stiffness.

7.2.1 Viscoelasticity of Thermoplastics

Viscoelasticity describes the time-dependent mechanical properties of polymers at temperatures below melting temperature. The macroscopic material behavior is typically symbolized by a general Maxwell model, i.e. a number of N spring-damper elements in parallel to an elastic spring. For linear elastic and viscous relations and for $N \to \infty$, the dependence of the material's compliance J from time $t \in [0, \infty)$ is given by the creep curve

$$J(t) = J_0 + \int_0^\infty f(\tau) \left(1 - \exp\left(-\frac{t}{\tau}\right)\right) d\tau. \tag{7.1}$$

Here $f(\tau)$ denotes the (measured or modeled) retardation spectrum and J_0 is the initial compliance.

Experimentally, such creep curves are obtained by measuring the history of strain $\varepsilon(t)$ related to the loading with stress $\sigma(t)$,

$$J(t) = \frac{\varepsilon(t)}{\sigma(t)}. \tag{7.2}$$

Because of the exponential relation (7.1), the creep curve is typically plotted over a logarithmic time axis, $\log(t)$. Since the degree of molecular mobility essentially determines the viscosity of microscopic polymeric chain networks, the macroscopic behavior of a polymer is characterized by a strong temperature dependence. For thermo-rheologically simple materials, however, it is assumed that temperature only affects the velocity of molecular movements but not their type and number. Therefore, the shape of experimentally obtained creep curves remains the same at different temperatures and only their position on the time scale is different.

This implies that from creep curves measured at different temperatures but within the same time interval a master curve at a reference temperature T_0 can be obtained by applying a time shift $\log(a_T)$ as illustrated in Fig. 7.2. Depending on the reference temperature, the shift factor a_T can be expressed as a function of temperature T by the well-known Williams-Landel-Ferry (WLF) equation or an Arrhenius approach [20]. This method, known as *time-temperature superposition principle* (TTSP) is commonly used to determine the rapid and long-term creep behavior of viscoelastic solids. Instead of conducting isothermal creep tests over inconvenient time periods, short-term tests are carried out at different temperatures and their results are combined. We remark that for thermo-rheologically complex materials a TTSP master curve does not result from horizontal shifting alone. Here, additional vertical shifts may be required [18].

The common rheological models with spring-damper arrangements can cover the full range of the material's viscoelastic behavior according to Eq. (7.1). A simplification, which corresponds to a linearization over short logarithmic time spans, gives a model which was proposed as early as 1863 by Kohlrausch [21]

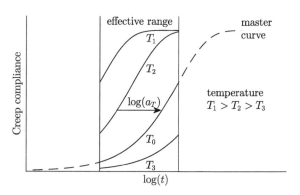

Fig. 7.2 Master curve generation by means of time-temperature superposition (TTSP).

$$J(t) = J_0 \exp\left(\left(\frac{t}{\tau}\right)^m\right). \tag{7.3}$$

Again, J_0 describes the instantaneous creep response, while τ is a characteristic retardation time associated with the active creep mechanism. This decay time increases with isothermal aging, which leads to a stretching of the curve. Exponent m is the Kohlrausch coefficient, $0 < m \leq 1$, which is associated with the dispersion of retardation times. It is about $1/3$ for polymers [15].

For obvious reasons, Eq. (7.3) cannot represent the long-term creep behavior of glassy polymers over the entire retardation spectrum. Nonetheless, it is well suited for the representation of short-time creep and, in particular, for the practical determination of horizontal and vertical shifts of experimentally obtained creep curves. Therefore, it has been employed by various experimentalists, cf. [15, 16, 18].

7.2.2 Physical Aging

Physical aging explains, why in the long-term creep of polymers, the observed creep compliance gradually deviates from the corresponding TTSP master curve. In his extensive studies on a number of synthetic polymers, Struik [15] showed that the aging time t_e, which elapses between quenching the polymer below its glass transition temperature and material testing, has a significant effect on the mechanical properties. This phenomenon is visible in the creep behavior of amorphous and semi-crystalline polymers and can be explained by the free volume theory of molecular motion. The theory roughly says that the molecular mobility of amorphous polymers at temperatures above glass transition, $T > T_g$, is sufficient to immediately reach thermodynamic equilibrium, thanks to enough free volume. For $T < T_g$, the molecular mobility becomes so small that there is a difference between the actual and the equilibrium free volume. This difference is larger the faster the polymer is cooled and acts as a driving force to reach the equilibrium state with the remaining molecular mobility. The required molecular rearrangements are slow, self-delaying, lead to time-dependent changes in mechanical properties, and come to a standstill at very low temperatures.

Since molecular mobility depends directly on the available free volume, it decreases with aging time. This in turn increases the relaxation times, which is mapped by a certain shift factor a_e. As a consequence, isothermal creep curves obtained at increasing aging times t_e are shifted to the right on the logarithmic time scale, Fig. 7.3. This inverse proportionality between molecular mobility and aging time was confirmed by experimental results of isothermal short-term creep tests at small strains in [15]. On double-logarithmic representations of the shift factors a_e over the corresponding aging times t_e, a linear relation was found, see inset of Fig. 7.3. The slope of this relationship is denoted as the aging shift rate

$$\mu = -\frac{\Delta \log(a_e)}{\Delta \log(t_e)}. \tag{7.4}$$

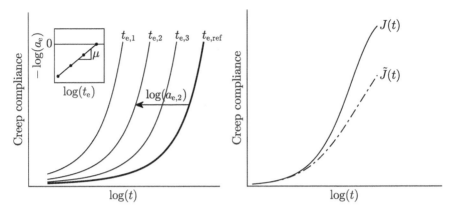

Fig. 7.3: Isothermal creep curves at different aging times t_e; the inset shows the curve shifts on the logarithmic time scale with respect to the reference curve with aging time $t_{e,\text{ref}}$.

Fig. 7.4: Prediction of long-term creep curve $\tilde{J}(t)$ obtained by correction of the momentary master curve $J(t)$ by means of the effective time theory.

With the sign convention according to literature, $-\log(a_e)$ denotes a horizontal curve shift to the right with respect to the reference curve. The shift rate is constant for a given temperature and has values of about one for a wide range of polymers.

To incorporate the physical aging phenomena in the macroscopic creep function (7.1) or its simplified form (7.3), an effective time approach according to Struik [15] will be utilized. In the effective time theory, the shift factor a_e is used to define a quasi-time function $\lambda(t)$ with

$$d\lambda = a_e(t, t_e) \, dt. \tag{7.5}$$

To integrate (7.5), the shift factor for a total aging time of $t_e + t$ with respect to a reference elapsed aging time t_e needs to be specified. Here the aging shift rate (7.4) is used to deduce

$$a_e(t, t_e) = \left(\frac{t_e}{t_e + t}\right)^\mu. \tag{7.6}$$

This model implies that all relaxation processes in the interval dt are slower by a factor of $1/a_e$ at time $t > 0$ than at time $t = 0$. Thus, integration of (7.5) gives

$$\lambda(t) = \int_0^t a_e(\xi, t_e) \, d\xi \stackrel{\mu \neq 1}{=} \frac{t_e}{1-\mu}\left[\left(1+\frac{t}{t_e}\right)^{1-\mu} - 1\right]. \tag{7.7}$$

The effective time function $\lambda(t)$ allows the conversion from the *momentary* master curve $J(t)$ to the corrected long-term creep compliance $\tilde{J}(t)$ by the relation

$$\tilde{J}(t) = J(\lambda), \tag{7.8}$$

see Fig. 7.4. When no aging takes places, i.e. $\mu = 0$, Eq. (7.7) corresponds to the momentary master curve creep time t.

To conclude this very brief introduction on the basic aspects of physical aging, it should be noted that these were initially developed for fully amorphus polymers. Here, the shift rate must be in the range $0.7 < \mu < 1$ to obtain reliable creep predictions over very long times [15]. However, Struik [15, 22] found that semi-crystalline polymers and filled rubbers are also affected by physical aging mechanisms at temperatures below the glass transition temperature of their amorphous phase, and the proposed effective time approach can be used as well.

7.3 Material and Methods

In what follows, an overview of the specimen fabrication and their geometry are given. Then, the loading regimes of the three-point bending tests conducted are explained in detail.

7.3.1 Test Specimens

The geometry of the test specimens used in the flexural creep tests follows the ISO 178 standard and is shown in Fig. 7.5. The specimens were printed in a semi-professional desktop 3D printer using PLA filament with the brand name Ultrafuse PLA from BASF 3D Printing Solutions and a diameter of 1.75 mm. The used print settings are summarized in Table 7.1.

To improve adhesion between the printing platform and the printed object, it is recommended in practice to set the platform temperature in the range of the glass transition temperature of the material. This increases the temperature in the first few layers of the object. In the layers atop, the effect of this heat source is not present, so the filament deposited here cools very quickly to the prevailing ambient temperature. For example, by means of a simplified heat transfer simulation of the cooling behavior of a deposited elliptical ABS fiber in a 3D printing process, Rodriguez et al. [23]

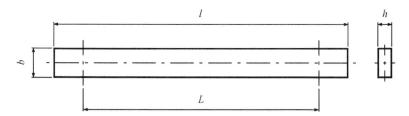

Fig. 7.5: Specimen dimensions according to ISO 178 with nominal length $l = 100$ mm, width $b = 10$ mm, height $h = 5$ mm and support span $L = 80$ mm.

Table 7.1: 3D print settings used for sample fabrication.

print setting	value
nozzle size	0.4 mm
extruder temperature	215°C
platform temperature	30°C
layer height	0.2 mm
print speed	60 mm/s
shell count	1
infill pattern	45°-90° line pattern
infill density	100 %
extrusion multiplier	102 %
cooling fan	always on

found that it takes less than one second for its core temperature to fall below the glass transition temperature of 94°C, assuming two-sided contact (e.g. left and bottom) to already printed material with a temperature of 55°C and a nozzle temperature of 270°C. Since the test specimens used in our study are low in height and consist of only 25 layers, their cooling behavior during printing would be significantly affected by elevated platform temperatures. To minimize this effect and to achieve high and uniform cooling rates, the temperature of the platform was set to a low value of 30°C, which is well below the glass transition temperature $T_g = 61°C$[1] of the PLA material used. The bond between the platform and the printed sample was improved by using a water-soluble adhesive.

After a specimen was finished printing, it was left in the printer for about 10 minutes until the printing platform reached the ambient temperature of about 20°C. The printer was located in the same room where the creep tests were performed.

7.3.2 Sequential Creep Tests

To study the creep behavior of the test specimens, three-point flexural creep tests were conducted in accordance with ISO 899-2 and under the conditions presented in the following. In our experiments, we use a three-point bending apparatus connected to a universal testing machine as shown in Fig. 7.6. Prior to testing, the specimens were measured with a caliper gauge in width and height with an accuracy of 0.05 mm.

In the sequential creep tests, the specimens were subjected to a constant load F for several short periods of time at aging times of 1.25, 2.5, 5, 10 and 20 hours, see Fig. 7.7. This sequential procedure was also used by Struik [15] and others, where the duration T_i of the i-th loading sequence is small compared to the previous aging time $t_{e,i}$. Thus, almost no additional aging occurs during the loading sequences and

---

[1] Ultrafuse PLA, Technical Data Sheet v4.4, BASF 3D Printing Solutions GmbH

Fig. 7.6 Three-point bending apparatus connected to a universal testing machine.

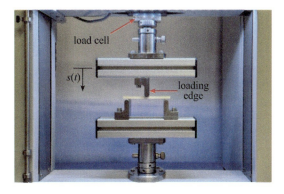

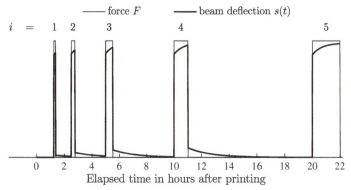

Fig. 7.7: Sequential creep test procedure.

momentary creep curves are obtained. In our study, the durations T_i were 10 % of the corresponding aging times $t_{e,i}$.

In order to analyze the individual creep sequences unambiguously in terms of the elapsed aging time, it must be ensured that the material properties do not change significantly as a result of the sequential loading. Therefore, the tests were carried out at small strains, where the assumption of linear viscoelasticity and Boltzmann's superposition principle are valid. In our tests, the specimens were loaded with a nominal force of 10 N to obtain small flexural strains well below 1 %. The force was measured with a load cell and kept nearly constant in a narrow range around the nominal value. The tests were performed in absence of any UV radiation at a constant temperature of about 20°C and relative air humidity between 30 and 50 %, which was confirmed by measurements during the tests.

Using this test setup in our experiments, the flexural creep strain $\varepsilon(t)$ of a loaded specimen was determined by the relation

$$\varepsilon(t) = \frac{6h}{L^2}(s(t) - s_0). \tag{7.9}$$

Here L denotes the span between the specimen supports, which must be a multiple of h for Euler-Bernoulli beam theory to be applicable; $s(t)$ is the crosshead travel distance and $s_0 < s(t = 0)$ its absolute value at which the loading edge (see Fig. 7.6) has initial contact to the specimen. In practice, the value of s_0 was determined at the time when the measured force signal begins to differ from zero.

As a measure of the creep behavior of the specimens considered, their flexural creep compliance $J(t)$ was calculated as a function of time by Eq. (7.2), where $\varepsilon(t)$ is the strain determined by (7.9) and $\sigma(t)$ is the nearly constant flexural stress obtained by

$$\sigma(t) = \frac{3L}{2bh^2} F(t). \tag{7.10}$$

Both strain and stress were calculated using the individual measured specimen dimensions h and b, which differed slightly from the nominal values in Fig. 7.5. The time at which a specimen is fully loaded is declared as $t_i = 0$ and defines the initial creep compliance, $J_{i,0} = J_i(t_i = 0)$, of the i-th creep sequence.

7.3.3 Long Term Creep Test

In addition to the sequential creep tests, a long-term creep experiment was conducted over a loading period of one week. Here the specimen was subjected to a nominal force of 10 N and held constant over 170 h. The test was performed to compare its result in the short-term range with the short-term creep of specimens of the same age that had already been subjected to several loading sequences. Therefore, the long-term test was started after an elapsed aging time of five hours, which corresponds to the specimen age at the beginning of the third sequence in the sequential tests according to Sect. 7.3.2.

7.4 Experimental Results

In the following, the steps of post-processing of the experimental data obtained with the methods described in 7.3.2–7.3.3 are detailed. A qualitative discussion of the final results presented can be found in Sect. 7.5.

7.4.1 Sequential Creep Tests

In total five specimens were printed and tested according to the procedures explained in Subsects. 7.3.1–7.3.2. Figure 7.8a shows the averaged creep compliance curves obtained by sequential creep tests for different aging times. The means of the initial creep compliances, $\bar{J}_{i,0}$, were computed for each set, $i = 1, \ldots, 5$, and all curves were

7 Effect of Physical Aging on the Flexural Creep in 3D Printed Thermoplastic

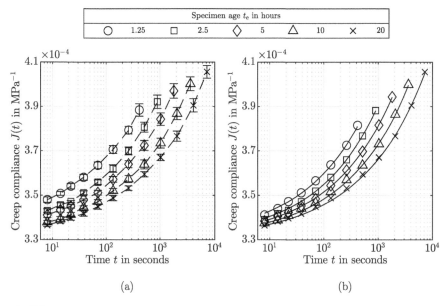

Fig. 7.8: a) Calibrated creep curves averaged over five specimens; b) creep curves shifted vertically to the reference initial compliance $\bar{J}_{5,0}$ and fitted individually to the creep model (7.3).

calibrated accordingly. The error bars in Fig. 7.8a refer to the remaining scatter of the calibrated curves, given as standard deviation.

Plotted in a semi-logarithmic diagram, the averaged creep curves show a similar shape and a nearly equidistant distribution with respect to the logarithmic time scale. The higher the specimen age, the more they are shifted to the right, which can be explained by an increase in relaxation times. In view of the explanations in Sect. 7.2.2, it is therefore plausible to assume physical aging as the cause of the observed creep behavior.

The 20 h aged curve is chosen as reference curve, $t_{e,\text{ref}} = t_{e,5} = 20\,\text{h}$. To quantify the curve shifts on the logarithmic time scale as a function of aging time t_e, the averaged creep curves were first arranged by vertical shifts $\Delta \bar{J}_i$ so that their initial compliances $\bar{J}_{i,0}$ coincide with the initial compliance $\bar{J}_{5,0}$ of the reference curve. Next, the experimental data calibrated this way, represented by symbols in Fig. 7.8b, were individually fitted to the creep model (7.3) using least squares minimization with fitting parameters τ and m. The resulting fit curves are shown as continuous lines in Fig. 7.8b. The corresponding fits of parameter m are listed in Table 7.2. Since these values differ only slightly from each other, it can be verified that all creep curves are characterized by the same shape and differ only by retardation time τ_i. Thus, parameter m is assumed to be a material constant and set to the mean value of $m = 0.35$ for subsequent analysis.

Given a constant value for m, the shift $\log(a_i)$ of a creep curve $J_i(t)$ relative to a reference curve $J_{\text{ref}}(t)$, each represented by (7.3), can be calculated by the relation

Table 7.2: Shift and fit results from data analysis.

creep sequence	i	1	2	3	4	5	unit
specimen age	$t_{e,i}$	1.25	2.5	5	10	20	h
vertical shift to $\bar{J}_{5,0}$	$\Delta \bar{J}_i$	-0.69	-0.39	-0.27	-0.04	0	10^{-5} MPa^{-1}
model parameter (1st fit)	m_i	0.35	0.36	0.35	0.35	0.34	-
retardation time (2nd fit)	τ_i	1.09	1.40	2.38	4.19	7.58	10^5 s

$$-\log(a_i) = \log\left(\frac{\tau_i}{\tau_{\text{ref}}}\right). \tag{7.11}$$

To calculate the shifts $\log(a_i)$ of the considered creep curves with respect to the chosen reference curve $\bar{J}_5(t)$, they were again fitted to the creep model (7.3) with fitting parameters τ_i and the now constant model parameter $m = 0.35$. The fitted values of $\tau_i, i = 1,...,5$ are listed in Table 7.2; they were used to calculate $\log(a_i)$, $i = 1,...,5$ by Eq. (7.11).

The results are shown in Fig. 7.9 over the aging time t_e in a double-logarithmic plot. The data points follow, to a good approximation, a linear relationship with slope $\mu = 0.72$ representing the shift rate (7.4) calculated by linear regression. Figure 7.10 shows the experimental creep curves each correspondingly shifted by $\log(a_i), i = 1,...,5$ to the reference aging time $t_{e,\text{ref}} = 20$ h. The solid line represents the fit result for the reference creep curve $\bar{J}_5(t)$.

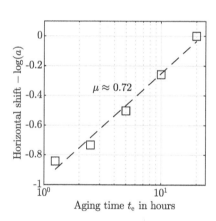

Fig. 7.9: Calculated time shifts as a function of specimen age t_e.

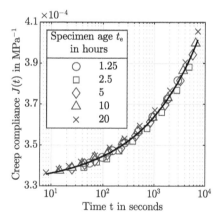

Fig. 7.10: Experimental creep curves shifted to the reference aging time $t_{e,\text{ref}} = 20$ h.

7.4.2 Long Term Creep Test

The creep curve obtained from the long-term experiment is shown as a solid line in Fig. 7.11. The diamond-shaped data points refer to the averaged short-term creep curve $\bar{J}_3(t)$ calculated as described in Sect. 7.4.1. To compare both curves with respect to their position on the logarithmic time scale, the short-term curve was shifted slightly vertically so that their initial compliances coincide. It can be clearly seen that the curves are not shifted against each other on the logarithmic time scale and have an identical shape.

The long-term creep curve shows that even after a one-week load, a state of equilibrium has not yet been reached and the material continues to creep. At the end of the recorded period, a creep modulus of 1040 MPa is calculated, which corresponds to only about 56 % of the quasi-static flexural elastic modulus of 1860 MPa given in the material data sheet.

7.5 Discussion

Sequential creep tests were performed to analyze the short-term flexural creep behavior of 3D printed, increasingly aged specimens, the results of which are presented in Sect. 7.4. Plotted on a logarithmic time scale, the obtained short-term creep curves show similar shapes but are horizontally shifted as a function of aging time. Their relative shifts to a reference curve were found to satisfy equation (7.4) with a calculated shift rate $\mu \approx 0.72$ to a good approximation.

By comparing two creep curves obtained on identically aged specimens with different preloading histories, see Sect. 7.4.2, it was shown that sequential loading of

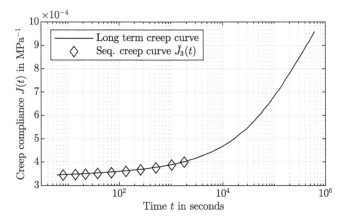

Fig. 7.11: Comparison of the averaged short-term creep curve $\bar{J}_3(t)$ with a long-term creep curve obtained using a single specimen with initial aging time $t_e = 5$ h.

specimens with small forces as described in Sect. 7.3.2 does not significantly affect their creep behavior. Since all creep tests were performed at constant temperatures and under exclusion of UV radiation, we assume that physical aging is the main reason for the observed increase in retardation times as a function of specimen age. Like the influence of internal stresses, cf. [15], the influence of post-crystallization was excluded from the considerations, since such crystallization is to be expected only for semi-crystalline polymers exposed to temperatures above their glass transition temperature for extended periods of time [24].

Struik [15] argued and showed experimentally that physical aging occurs in a certain temperature range below the glass transition temperature and disappears at both very low and high temperatures. Here the shift rate tends to zero and reaches a maximum at a temperature between these boundary states. From the test results obtained at an ambient temperature of 20°C we calculated a shift rate of $\mu \approx 0.72$, which is far from zero, but not very close to one. Since the testing temperature was about 40°C lower than the glass transition temperature of the PLA material used, we thus assume that the maximum possible shift rate is larger than the value we calculated and is reached at higher temperatures.

Motivated by Struik's argument that physical aging is not limited to individual polymers but is a more general phenomenon, we further assume that other 3D printed thermoplastics are affected by the same physical aging shown here using PLA as an example material. The strength of this aging influence depends on the glass transition temperature of the material printed, the ambient temperature during the creep test, and the cooling rate achieved during extrusion. It should be noted that the cooling process in 3D printing differs significantly from that in conventional extrusion processes, such as injection molding. While inhomogeneous temperature gradients can lead to locally different cooling conditions in the latter case, these can be assumed to be approximately the same over the entire component in 3D printing. The creep and aging properties determined on small 3D printed test specimens can therefore be directly transferred to large 3D printed components under otherwise identical conditions.

Since physical aging is temperature-dependent, special temperature treatments in advanced 3D printing, e.g. the use of heated print chambers or subsequent annealing, c.f. [7], would have a significant influence on the long-term creep behavior of printed parts. In our prints, the cooling rate was just large enough to achieve a shift rate $\mu > 0.7$, which according to [15] allows valid prediction of aging-influenced long-term creep behavior based on momentary TTSP master creep curves as described in Subsect. 7.2.2. However, further research is needed to investigate in detail the extent to which advanced heat treatments affect physical aging and thus the long-term creep of 3D printed thermoplastics.

To demonstrate that 3D printed thermoplastics creep significantly over long time periods even at temperatures well below their glass transition temperature, a long-term creep experiment was conducted over a week at a constant temperature. The corresponding results are presented in Subsect. 7.4.2 and are essential to consider when designing load-bearing structures made of 3D printed thermoplastics. We refer in particular to those applications where precise knowledge of the viscoelastic material properties is important, e.g. in the design of interference fits for shafts or ball bearings.

Using the test setup described in Subsect. 7.3.2, creep curves were obtained by calculating the creep compliance as a function of the load the specimens were subjected to and the time-dependent change in the crosshead travel distance of the testing machine. Therefore, no additional strain gauges or extensometers were used. However, the creep curves obtained by this method exhibited considerable scatter in the calculated initial compliances, even when recorded at identical aging times. Thus, in order to use them for a quantitative interpretation of the specimens' creep behavior, some corrections had to be made as described in Subsect. 7.4.1. The reasons for the observed scatter in the initial compliances are, in our opinion, related to the measurement of the reference crosshead travel distance s_0, see Eq. (7.9), which is affected by changing contact conditions between the loading edge and the specimen and by possibly inaccurate force measurements in the range of very small loads. For an accurate determination of the initial compliances of sequentially loaded creep specimens, our proposed test setup can therefore only be recommended to a limited extent. To improve the measurement results and to reduce the post-processing effort, we recommend the use of high-resolution load cells for the accurate measurement of small forces or direct strain measurements with strain gauges or extensometers, cf. [16, 18].

References

[1] Tan LJ, Zhu W, Zhou K (2020) Recent Progress on Polymer Materials for Additive Manufacturing. Adv Funct Mater **30**(43):2003,062

[2] Ahn SH, Montero M, Odell D, Roundy S, Wright PK (2002) Anisotropic material properties of fused deposition modeling ABS. Rapid Prototyp J **8**(4):248–257

[3] Wittbrodt B, Pearce JM (2015) The effects of PLA color on material properties of 3-D printed components. Addit Manuf **8**:110–116

[4] Fernandez-Vicente M, Calle W, Ferrandiz S, Conejero A (2016) Effect of Infill Parameters on Tensile Mechanical Behavior in Desktop 3D Printing. 3D Print Addit Manuf **3**(3):183–192

[5] Rismalia M, Hidajat SC, Permana IGR, Hadisujoto B, Muslimin M, Triawan F (2019) Infill pattern and density effects on the tensile properties of 3D printed PLA material. J Phys Conf Ser **1402**(4):044,041

[6] Reppel T, Weinberg K (2018) Experimental Determination of Elastic and Rupture Properties of Printed Ninjaflex. Tech Mech **38**(1):104–112

[7] Akhoundi B, Nabipour M, Hajami F, Shakoori D (2020) An Experimental Study of Nozzle Temperature and Heat Treatment (Annealing) Effects on Mechanical Properties of High-Temperature Polylactic Acid in Fused Deposition Modeling. Polym Eng Sci **60**(5):979–987

[8] Khosravani MR, Berto F, Ayatollahi MR, Reinicke T (2022) Characterization of 3D-printed PLA parts with different raster orientations and printing speeds. Sci Rep **12**(1):1016
[9] Zhang H, Cai L, Golub M, Zhang Y, Yang X, Schlarman K, Zhang J (2018) Tensile, Creep, and Fatigue Behaviors of 3D-Printed Acrylonitrile Butadiene Styrene. J Mater Eng Perform **27**(1):57–62
[10] Mohammadizadeh M, Fidan I, Allen M, Imeri A (2018) Creep behavior analysis of additively manufactured fiber-reinforced components. Int J Adv Manuf Technol **99**(5):1225–1234
[11] Tezel T, Kovan V, Topal ES (2019) Effects of the printing parameters on short-term creep behaviors of three-dimensional printed polymers. J Appl Polym Sci **136**(21):47,564
[12] Aghayan S, Bieler S, Weinberg K (2021) Determination of the high-strain rate elastic modulus of printing resins using two different split Hopkinson pressure bars. Mech Time-Depend Mater **26**:761–773
[13] Ye J, Yao T, Deng Z, Zhang K, Dai S, Xiangbing L (2021) A modified creep model of polylactic acid (PLA-max) materials with different printing angles processed by fused filament fabrication. J Appl Polym Sci **138**(17):50,270
[14] White JR (2006) Polymer ageing: physics, chemistry or engineering? Time to reflect. C R Chim **9**(11):1396–1408
[15] Struik LCE (1977) Physical aging in amorphous polymers and other materials. PhD thesis, Delft University of Technology
[16] Hastie RL (1991) The effect of physical aging on the creep response of a thermoplastic composite. PhD thesis, Virginia Polytechnic Institute and State University
[17] Sullivan JL, Blais EJ, Houston D (1993) Physical aging in the creep behavior of thermosetting and thermoplastic composites. Compos Sci Technol **47**(4):389–403
[18] Pierik ER, Grouve WJB, van Drongelen M, Akkerman R (2020) The influence of physical ageing on the in-plane shear creep compliance of 5HS C/PPS. Mech Time-Depend Mater **24**(2):197–220
[19] Pan P, Zhu B, Inoue Y (2007) Enthalpy Relaxation and Embrittlement of Poly(L-lactide) during Physical Aging. Macromolecules **40**(26):9664–9671
[20] Grellmann W, Seidler S (2013) Polymer Testing, 2nd edn, Hanser, Munich, pp 86–87
[21] Kohlrausch F (1863) Ueber die elastische Nachwirkung bei der Torsion. Ann Phys **195**(7):337–368
[22] Struik LCE (1987) The mechanical and physical ageing of semicrystalline polymers: 1. Polymer **28**(9):1521–1533
[23] Rodriguez JF, Thomas JP, Renaud JE (2000) Characterization of the mesostructure of fused-deposition acrylonitrile butadiene-styrene materials. Rapid Prototyp J **6**(3):175–185
[24] Kalinka G (1995) Ein Beitrag zur Kristallisation gefüllter und ungefüllter Thermoplaste. BAM Forschungsberichtreihe - 208

Chapter 8
Development of a Microstructure-Based Finite Element Model of Thermomechanical Response of a Fully Metallic Composite Phase Change Material

Elisabetta Gariboldi, Matteo Molteni, Diego André Vargas Vargas, and Konstantin Naumenko

Abstract Form-stable composite Phase Change Materials (C-PCMs), among which Al-Sn alloys, store/release heat during the transformation of one of the phases they are made of, without significant dimensional and shape changes. A microstructural based model of the thermomechanical behaviour of these materials can help to check their form-stability under various potential service conditions. The volume changes induced by melting/solidification of the low-melting Sn phase and different thermal expansion of solid Al and Sn can actually induce strains and stress fields during thermal cycles. A numerical simulation of a modelled Al-Sn C-PCM dilatometric test, a technique generally addressed toward the estimation of material thermal expansion, has been run and validated. The model describes the complex evolution of stress-strain fields and local plastic strains during the thermal cycle. The most critical conditions arise during cooling, after the completion of Sn solidification, at the highly stressed interface between Sn and Al phase. While at the end of a thermal cycle the overall shrinkage is minimal, the microscopic plastic strain remain locally. The phases properties and the representative microstructure are thus critical features for the development of a reliable model. The model could be used to consider repeated cycles and thermomechanical behaviour of C-PCMs.

Elisabetta Gariboldi · Matteo Molteni · Diego André Vargas Vargas
Politecnico di Milano, Dipartimento di Meccanica, Via La Masa 1, 20156 Milano, Italy,
e-mail: elisabetta.gariboldi@polimi.it, matteo1.molteni@polimi.it, diegoandre.vargas@mail.polimi.it

Konstantin Naumenko
Lehrstuhl für Technische Mechanik, Institut für Mechanik, Fakultät für Maschinenbau, Otto-von-Guericke-Universität Magdeburg, Universitätsplatz 2, 39106 Magdeburg, Germany,
e-mail: konstantin.naumenko@ovgu.de

8.1 Introduction

The development of renewable but intermittent power sources concurrently pushes the research toward the development of reliable energy storage systems, among which Phase Change Materials (PCMs) [1]. In particular, their capability of storing/releasing thermal energy as latent heat of transformation, at constant or modulated temperature, makes them appealing for Thermal Energy Storage (TES) and Thermal Management (TEM) purposes [2]. The solid-liquid phase change is the most exploited one, which allows high energy storability and low volume expansion. In particular, a lack of control on this latter, associated to the leakage of the liquid phase and, consequently, a reduction in the storage potential, threaten the performances of the PCM itself. Different strategies have been attempted in order to tackle the issue of leakage of PCM in the molten state either adopting containers [3, 4] or capsules from millimetric [5, 6] to nanometric size [7, 8]. More recently, the need of a confinement material has been overcome by the design of form-stable PCMs. These latter experience minimal volume changes induced by phase transition and original form recovery after a charge/discharge cycle [9]. Both chemical- and physical-based approaches are explored in the design of form-stable material. Among the former, mainly adopted for organic PCMs, the formation of covalent [10], hydrogen [11] or van der Waals bonds [12] between them and thermally stable materials are explored. The physical approaches mainly exploit capillarity [11, 13] for the achievement of form-stability. In metallic materials, the interaction of the liquid metal with surrounding environment is particularly critical and has to be considered together with form-stability. The use of immiscible alloys, such as Al-Sn, as proposed by [14] manufactured by powder metallurgy [14, 15] or by casting followed by medium-high cooling rates [16, 17] demonstrated to be a smart solution. Due their immiscibility mature, below the temperature of incipient melting, i.e., in the fully solid state, an Al-Sn alloy consists of two phases whose composition roughly corresponds to the starting pure elements. When produced by the abovementioned suitable processes, the microstructure of Al-Sn alloy resembles the one of composite materials, with Al as matrix and the low-melting Sn as inclusion. The melting starts close to the melting temperature of pure Sn and leading to a liquid which remains extremely rich of Sn up to about 30°C above the onset of melting. The binary Al-Sn system allows to tailor the thermal storage potential during thermal cycles by adjusting the Sn content. Dealing with this material as a composite where the only Sn-rich phase melts/solidifies, is thus reasonable. Under service, as for other composite materials, strains and thus stresses of thermal origin can arise. For the case of a composite PCM (C-PCM), i.e., Al-Sn in this study, thermal strains arise due not only to different coefficient of thermal expansions of the two phases [18, 19], but also to the volume change associated to the phase transition in the low-melting phase. These effects should be controlled during service of C-PCM material, characterized by the partial or the complete melting of C-PCM. Plastic deformation of the high and low melting phases can be triggered, which in the latter case are cancelled by melting. The result on the form stability purposes of the C-PCM during one or more repeated cycles is affected by several factors such as the alloy composition, the phase arrangement and coarseness.

Dilatometric tests represent a suitable tool to characterize the form-stability of the C-PCM during one or repeated thermal cycles. However, time-consuming experimental campaigns are required for a comprehensive characterization of the material. Nevertheless, the quantification of thermal strains and the triggering of the plastic deformation of each phase during thermal cycle is not possible. Numerical simulations based on a physically representative material model can easily foresee the thermomechanical response of the abovementioned phases, as well as the one of the whole C-PCM. Hence, the study proposes a first attempt in evaluating the stresses which develop between the two phases during a simulated thermal cycle. Stresses influence on the form stability during the phase change is taken into account as well. The aim of the study is achieved through the setting-up of a microstructure-based thermo-mechanical model for the description of the dilatometric behaviour of a Al-Sn composite PCMs, to be validated by means experimental dilatometric curves.

8.2 Microstructure-Based FE model of a Al-Sn C-PCM with Free Expansion

A reference Al-Sn alloy, characterized by the 40%mass Sn (roughly correspond to 20% in volume) was selected. The alloy with the same composition has been previously investigated both obtained by compressing mixed Al and Sn powders [14, 15] and produced via liquid metallurgy route with moderate-high cooling rate [16, 17]. Either in their as produced or thermally cycled condition, the microstructure consists of Sn inclusions surrounded by Al matrix, as shown by the scanning electron micrographs in Fig. 8.1. In both cases the microstructure consists of an Al matrix (darker region in the SEM micrographs), with minimal Sn in solid solution, and of Sn-rich inclusions (brighter areas in Fig. 8.1), with minimal Al content. These latter can be reasonably approximated to spheres, whose diameters vary in different ranges according to the material processing conditions.

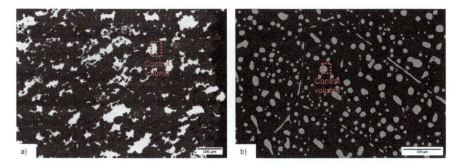

Fig. 8.1: Microstructure of Al-40%mass Sn alloy produced by powder metallurgy (a) and by casting followed by medium-high cooling rate (b) after thermal cycles.

In a first attempt to develop a microstructurally based FE model of the material, a representative control volume (as shown in Fig. 8.2a) modelled on the bases of the stable Al-Sn microstructure (Fig. 8.1), was selected as the reference geometry for the numerical simulations. The control volume consists of a single Sn sphere (radius $R = 42$ μm, Fig. 8.2a), surrounded by an Al cubic matrix (cell size $L_0 = 116$ μm). The volume of the sphere corresponds to the 20% of the whole geometry. The cell dimensions were chosen on the bases of the Sn areas, whose extension strictly depends on the production process selected (Fig. 8.1). No porosity nor interface detachment, i.e., perfect bonding, were considered (Fig. 8.2a). Considering the possibility of having asymmetric thermal loadings, the simulations were performed only on 1/4 of the reference model The numerical analyses, which simulate a dilatometric test on the control volume, were performed with the commercially available software Comsol Multiphysics, version 6.1. Pure Al and pure Sn have been considered as the reference materials of the control volume (Fig. 8.2a), taking into account for this latter solid-liquid phase transition. Heat transfer and solid mechanics physics were adopted for the simulations.

The Laplace equation was adopted for the description of the thermal problem. Heat conduction was considered as the only heat transfer mechanism for all the phases. Indeed, preliminary calculations demonstrate the absence of convection in the molten Sn sphere of the considered size. Temperature-dependent thermophysical properties of interest, i.e., thermal conductivity, specific heat at constant pressure, density and thermal expansion, were directly gathered from literature and assessed for both pure Al [19–22] and Sn [21, 23–25]. Temperature-dependent properties of Sn, undergoing phase transition, were estimated in its solid, as well as its liquid state. For all the listed material properties, solid Al, solid Sn and liquid Sn data were interpolated with a piecewise cubic function. Sn phase transformation was considered to occur in both heating and cooling over a temperature interval of 4°C, i.e., from 228°C to 232°C. The former corresponds to the horizontal eutectic temperature, whereas the latter to

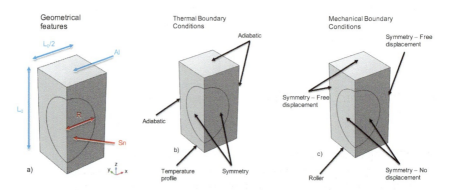

Fig. 8.2: a) Al-Sn control volume selected for the numerical simulations, b) Thermal boundary conditions applied to the control volume for the heat transfer physics, c) Mechanical boundary conditions for the solid mechanics physics.

the melting of pure Sn. Within this range, Sn was considered as composite partly made of liquid and solid, whose fractions vary with temperature. Sn liquid fraction changes linearly and with it the corresponding properties, apart from C_p. A peculiar trend was considered for C_p of Sn, whose temperature profile includes the latent heat of melting during phase transition (60000 J/kg [26]).

An elastoplastic description of the mechanical behaviour for both Al and Sn was chosen. The elastic behaviour of the former was modelled on the bases of temperature-dependent Young's modulus [27, 28] and Poisson's ratio [29], whereas temperature-dependent bulk modulus [30] and shear modulus [31] were selected for the description of Sn elastic response. Ludwik isotropic hardening model [32] was chosen instead for simulating the plasticity of both materials (8.1).

$$\sigma = \sigma_{ys} + K\varepsilon^n, \tag{8.1}$$

where σ_{ys} refers to the material yield strength and ε represents the plastic strain. K and n stand for strength coefficient and hardening exponent, respectively. These latter data were derived from the tensile curves proposed in [33, 34] for Al and Sn, respectively. As far as Sn mechanical properties are concerned, they decay close to 0 when the metal undergoes solid-liquid transformations. Once again, the Sn domain in its phase transition can be considered as a combination of solid and liquid Sn, whose volume fraction vary linearly with temperature. Thus, linear trends were adopted for the description of the mechanical properties in the phase change between solid and liquid. As aforementioned, the materials were simply modelled as elastoplastic and the effect of creep was not taken into account.

In the present preliminary numerical study, the model has been applied to the prediction of the free expansion of material which can be experimentally evaluated by means of dilatometric tests, where the length of cylindrical specimen prevails over its diameter. In this view, the boundary conditions applied to the reference volume allow the evaluation of its thermal expansion without any external mechanical loading during the thermal cycle selected. Thus, the following boundary conditions, displayed in Fig. 8.2, have been selected for the control volume. From the heat transfer physics point of view, thermal load history was applied at the bottom surface of the reference model. In particular, a relatively low cycle, leading to homogeneous temperature profiles, was selected. These conditions well simulate the possible service conditions of the C-PCM. Symmetry boundary conditions were chosen for surfaces where Al and Sn coexist. Adiabatic boundary conditions instead, were assigned to the remaining faces of the reference volume (Fig. 8.2b). As far as the mechanical boundary conditions are concerned (Fig. 8.2c), roller constraint was applied at the bottom face of the control volume. Symmetry boundary conditions were selected for all the remaining faces. No displacement was chosen for the faces where the Al-Sn interface is exposed, whereas free displacement was selected for the remaining surfaces.

The control volume was automatically meshed by the software on the bases of the type of physics selected. The mesh consists of tetrahedral elements with a maximum element size of approximately 6.4 μm. A triangular type thermal cycle from room temperature to 270°C at 2°C/min was set at the bottom surface of the control volume. A multistep transient study was performed for the investigation of the thermomechanical response of the selected material. A 10 s time step was chosen for exploring Al-Sn behavior out of the solid-liquid Sn transition, whereas it refines to 1 s when low-melting metal undergoes phase change.

8.3 Results and Discussion

The change of length ($\Delta L = L - L_0$) in vertical z direction, and the corresponding thermal strain ($\Delta L / L_0$) and its temperature derivative

$$\frac{d\Delta L/L_0}{dT}$$

have been computed during the set thermal cycle. The last two are plotted in Figs. 8.3a and 8.3b, respectively. The same figures show the results of dilatometric tests performed on an Al-40%mass Sn C-PCM alloy. In general, the simulated strain-temperature curve shows a satisfactory agreement with the experimental one, being the thermal strains in the same order of magnitude (Fig. 8.3a). In the temperature range where both Al and Sn are solid, the slopes of the numerically derived heating and cooling curves (corresponding to the instant CTE of the C-PCM given in Fig. 8.3b) are consistent with the experimental one (Fig. 8.3a). In the transition temperature range instant CTE curves reach peak values. Differently from the experimental results, Sn melting and solidification coincide, since undercooling was not considered in

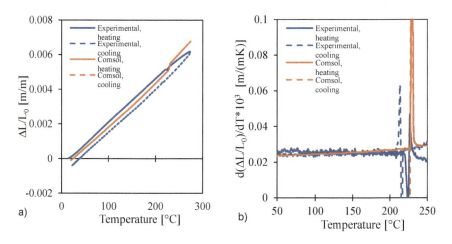

Fig. 8.3: Comparisons between numerical (Comsol) and experimental results of dilatometric thermal cycle of the investigated C-PCM: thermal strains in vertical direction (a) and their temperature derivatives (b).

the cooling stage. In the range of molten Sn, the C-PCM experimental curve (Fig. 8.3a) displays a decreasing slope which could be attributed to creep phenomena, not taken into account in the present simple version of the model. FE results were then analyzed in terms of strain and stress fields in the z direction developing at various stages of the thermal cycle in the Al and Sn phases. Representative images are given in Figs. 8.4 and 8.5, respectively. It can be clearly seen that the higher tension or compression strain of thermal origin, and the corresponding stress develop close to the interface between the two phases. Focusing the attention on these regions, it can be noticed that during heating, at 227°C, i.e. just before the onset of Sn melting (Fig. 8.4a) Sn is compressed while Al is in tension, as a result of its lower CTE with respect to solid tin. At this temperature Sn phase is plastically deformed, since at this temperature equivalent stress overcomes yield strength. Accordingly, the overall strains in z direction at 227°C (Fig. 8.5a) show a higher deformation for Sn. During heating, at 233°C (Figs. 8.4b and 8.5b), Sn compression and its relative volumetric

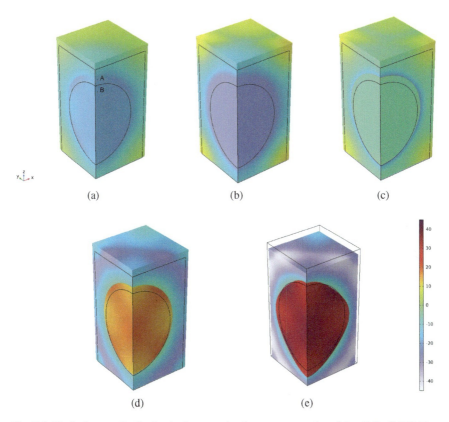

Fig. 8.4: Vertical stress distribution in the control volume representative of the Al-Sn C-PCM in various situations a) heating, 227°C, b) heating, 233°C, c) cooling, 233°C, d) cooling, 227°C, e) cooling, 20°C.

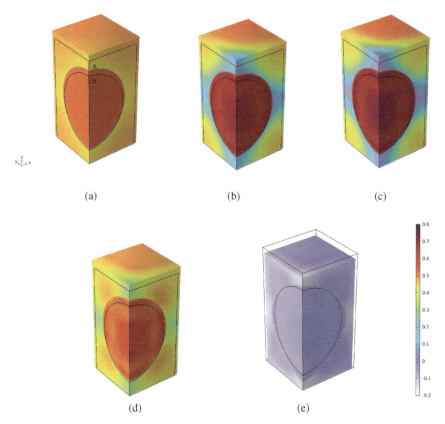

Fig. 8.5: Vertical strain distribution in the control volume representative of the Al-Sn C-PCM in various situations a) heating, 227°C, b) heating, 233°C, c) cooling, 233°C, d) cooling, 227°C, e) cooling, 20°C.

expansion soarted than in the solid range due to the steep Sn melting expansion, which corresponds to CTE peak in Fig. 8.3b. During the further heating, both the strain and stress in Al increase, but the conditions for the onset of plasticity are not reached. At first significant temperature in cooling, i.e., 233°C, just before the solidification onset, different conditions are met with respect to the same temperature in heating. In cooling, at 227°C, the completion of Sn solidification step brings about significant changes in the stress and strain distribution in the phases of C-PCMs. The high shrinkage of Sn leads to the onset of tension stress in the already solid (Sn) and compressive stresses arise in Al. Although experiencing more severe strains than Al, Sn positive strain magnitude sensibly reduces with respect to the previous stage (Fig.e 8.5d), due to the solidification shrinkage modelled as steep CTE change shown in Fig. 8.3b. In both cases, conditions for plastic deformation are reached in both phases. Further cooling to low temperature increases the tensile stress in Sn and the compressive state in Al. From the completion of its solidification, Sn equivalent stress exceeds the temperature dependent yield stress, while in the case of Al, the conditions of plastic deformation were met only in a temperature range close to the completion of solidification.

The complex strain fields developed during the thermal cycle in the Al and Sn phases are presented in Fig. 8.6 by strain in z direction for Al (A) and Sn (B) points laying at the interface along z direction o the symmetry axis. While strains in Sn are very close in heating and cooling, those developed in Al are lower during the compression cycle. While this strain appears to be relatively low, it could potentially affect the thermomechanical behaviour of the C-PCM during following cycles. Even if after a thermal cycle the overall deformation of the C-PCMs is of the order of -0.01% at point A/B Al' with %, at point A Al. Al can display far higher shrink which can affect the local thermomechanical behavior during further thermal cycles. The local strain and thus the microscopic.

8.4 Final Remarks

Some considerations can be derived on the basis of the previously discussed results. A reliable numerical model, describing the thermomechanical responses of a representative microstructural-based control volume, was developed. Finite Element Analysis results are in accordance with experimental dilatometric curves, in terms of both thermal strains and instant CTE as a function of temperature, especially below Sn melting temperature in both heating and cooling stage. In this sense, the addition of creep or undercooling to the model developed, which demonstrated to influence the thermomechanical C-PCM response, an even more reliably foresee Al-Sn dilatometric behaviour. Even in the presence of thermal loading only, thermal strain and stress, which could overcome the plastic deformation threshold, can be

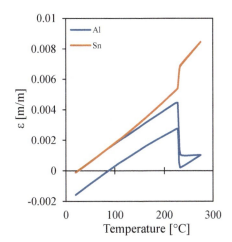

Fig. 8.6 Development of vertical strain in Al and Sn at interface point on the midplane of the control volume during the thermal cycle.

met. This occurs for both Al and Sn phases after the completion of Sn solidification in the cooling stage. An overall shrinkage of -0.01% was computed at the end of the reference geometry at the end of the set thermal cycle. However, locally, in the Al regions close to Al-Sn interface, higher strains can be experienced. The situations at which the phases of the C-PCM enter in the plastic field are strongly correlated to their temperature-dependent mechanical and thermal properties. The phases properties and the representative microstructure are thus critical features for the development of a reliable model, which can be used to describe the material behaviour also during repeated thermal cycles or under more complex situations.

References

[1] Pielichowska K, Pielichowski K (2014) Phase change materials for thermal energy storage. Progress in Materials Science **65**:67–123
[2] Wei G, Wang G, Xu C, Ju X, Xing L, Du X, Yang Y (2018) Selection principles and thermophysical properties of high temperature phase change materials for thermal energy storage: A review. Renewable and Sustainable Energy Reviews **81**:1771–1786
[3] Rawson A, Villada C, Kolbe M, Stahl V, Kargl F (2022) Suitability of aluminium copper silicon eutectic as a phase change material for thermal storage applications: Thermophysical properties and compatibility. Energy Storage **4**(2):e299
[4] Velasco-Carrasco M, Chen Z, Aguilar-Santana JL, Riffat S (2020) Experimental evaluation of phase change material blister panels for building application. Future Cities and Environment **6**(1)
[5] Fukahori R, Nomura T, Zhu C, Sheng N, Okinaka N, Akiyama T (2016) Macro-encapsulation of metallic phase change material using cylindrical-type ceramic containers for high-temperature thermal energy storage. Applied Energy **170**:324–328
[6] Zhao B, Guo Y, Wang C, Zeng L, Gao K, Sheng N, Gariboldi E, Zhu C (2023) Development of over 3000 cycles durable macro-encapsulated aluminum with cavity by a sacrificial layer method for high temperature latent heat storage. Chemical Engineering Journal **457**:141,352
[7] Chen ZH, Yu F, Zeng XR, Zhang ZG (2012) Preparation, characterization and thermal properties of nanocapsules containing phase change material n-dodecanol by miniemulsion polymerization with polymerizable emulsifier. Applied Energy **91**(1):7–12
[8] Wang S, Lei K, Wang Z, Wang H, Zou D (2022) Metal-based phase change material (PCM) microcapsules/nanocapsules: Fabrication, thermophysical characterization and application. Chemical Engineering Journal p 135559
[9] Zhang Y, Jia Z, Hai AM, Zhang S, Tang B (2022) Shape-stabilization micromechanisms of form-stable phase change materials-A review. Composites Part A: Applied Science and Manufacturing p 107047

10. Soo XY, Muiruri JK, Yeo JC, Png ZM, Sng A, Xie H, Ji R, Wang S, Liu H, Xu J, et al (2023) Polyethylene glycol/polylactic acid block co-polymers as solid–solid phase change materials. SmartMat p e1188
11. Cui H, Wang P, Yang H, Shi Y (2022) Large-scale fabrication of expanded graphite aerogel-based phase change material composite for efficient solar harvesting. Journal of Energy Storage **56**:105,890
12. Liu X, Su H, Huang Z, Lin P, Yin T, Sheng X, Chen Y (2022) Biomass-based phase change material gels demonstrating solar-thermal conversion and thermal energy storage for thermoelectric power generation and personal thermal management. Solar Energy **239**:307–318
13. Liu K, Yuan Z, Zhao H, Shi C, Zhao F (2023) Properties and applications of shape-stabilized phase change energy storage materials based on porous material support–A review. Materials Today Sustainability p 100336
14. Sugo H, Kisi E, Cuskelly D (2013) Miscibility gap alloys with inverse microstructures and high thermal conductivity for high energy density thermal storage applications. Applied Thermal Engineering **51**(1-2):1345–1350
15. Confalonieri C, Perrin M, Gariboldi E (2020) Combined powder metallurgy routes to improve thermal and mechanical response of Al- Sn composite phase change materials. Transactions of Nonferrous Metals Society of China **30**(12):3226–3239
16. Confalonieri C, Gariboldi E (2021) Al-Sn miscibility gap alloy produced by power bed laser melting for application as phase change material. Journal of Alloys and Compounds **881**:160,596
17. Bassani P, Molteni M, Gariboldi E (2023) Microstructural features and thermal response of granulated Al and A356 alloy with relevant Sn additions. Materials & Design **229**:111,879
18. Deshpande V, Sirdeshmukh D (1961) Thermal expansion of tetragonal tin. Acta Crystallographica **14**(4):355–356
19. Nix F, MacNair D (1941) The thermal expansion of pure metals: copper, gold, aluminum, nickel, and iron. Physical Review **60**(8):597
20. Kurochkin A, Popel PS, Yagodin DA, Borisenko A, Okhapkin A (2013) Density of copper-aluminum alloys at temperatures up to 1400C determined by the gamma-ray technique. High Temperature **51**(2):197–205
21. Meydaneri Tezel F, Saatçi B, Arı M, Durmuş Acer S, Altuner E (2016) Structural and thermo-electrical properties of Sn-Al alloys. Applied Physics A **122**:1–12
22. Buyco EH, Davis FE (1970) Specific heat of aluminum from zero to its melting temperature and beyond. Equation for representation of the specific heat of solids. Journal of Chemical and Engineering Data **15**(4):518–523
23. Stankus SV, Khairulin RA (2006) The density of alloys of tin-lead system in the solid and liquid states. High Temperature **44**(3):389–395
24. Sharafat S, Ghoniem N (2000) Summary of Thermo-Physical Properties of Sn, and Compounds of Sn–H. Sn–O, Sn–C, Sn–Li, and Sn–Si and Comparison of Properties of Sn, Sn–Li, and Pb–Li, Report: UCLA-UCMEP-00-31, University of California, Los Angeles

[25] Khvan A, Babkina T, Dinsdale A, Uspenskaya I, Fartushna I, Druzhinina A, Syzdykova A, Belov M, Abrikosov I (2019) Thermodynamic properties of tin: Part I Experimental investigation, ab-initio modelling of α-, β-phase and a thermodynamic description for pure metal in solid and liquid state from 0 K. Calphad **65**:50–72
[26] Alexiades V, Hannoun N, Mai TZ (2003) Tin melting: Effect of grid size and scheme on the numerical solution. Electronic Journal of Differential Equations (EJDE)[electronic only] **2003**:55–69
[27] Hopkins DC, Baltis T, Pitaress JM, Hazelmyer DR (2012) Extreme thermal transient stress analysis with pre-stress in a metal matrix composite power package. Additional Papers and Presentations **2012**(HITEC):000,361–000,372
[28] Rezaei Ashtiani H, Bisadi H, Parsa M (2011) Inhomogeneity of temperature distribution through thickness of the aluminium strip during hot rolling. Proceedings of the Institution of Mechanical Engineers, Part C: Journal of Mechanical Engineering Science **225**(12):2938–2952
[29] Davoudi K (2017) Temperature dependence of the yield strength of aluminum thin films: Multiscale modeling approach. Scripta Materialia **131**:63–66
[30] Hayashi M, Yamada H, Nabeshima N, Nagata K (2007) Temperature dependence of the velocity of sound in liquid metals of group XIV. International Journal of Thermophysics **28**:83–96
[31] Nadal MH, Le Poac P (2003) Continuous model for the shear modulus as a function of pressure and temperature up to the melting point: analysis and ultrasonic validation. Journal of Applied Physics **93**(5):2472–2480
[32] Ludwik P (1909) Elemente der technologischen Mechanik. Springer, Berlin
[33] Chinh NQ, Illy J, Horita Z, Langdon TG (2005) Using the stress–strain relationships to propose regions of low and high temperature plastic deformation in aluminum. Materials Science and Engineering: A **410**:234–238
[34] Zhu F, Zhang H, Guan R, Liu S (2006) Investigation of microstructures and tensile properties of a Sn-Cu lead-free solder alloy. Journal of Materials Science: Materials in Electronics **17**:379–384

Chapter 9
The Effect of Dynamic Loads on the Creep of Geomaterials

Andrei M. Golosov, Evgenii P. Riabokon, Mikhail S. Turbakov, Evgenii V. Kozhevnikov, Vladimir V. Poplygin, Mikhail A. Guzev, and Hongwen Jing

Abstract The samples of a geomaterial are tested under constantly applied loading. Under the combined action of a static preload and an additional dynamic load a linear increase in the axial strain of the sample is observed even at the initial stage of the experiment. Dynamic loading activates an intense creep of the geomaterial and leads to a decrease in the Young's modulus.

9.1 Introduction

Under constant loads geomaterials deform over time, this phenomenon is commonly called creep. Creep is observed in both soft [1] and hard [2] rocks. With the manifestation of creep due to the accumulation of internal structural deformations by the geomaterial, its mechanical characteristics, and the Young's modulus in particular, change. Knowledge of how the change in Young's modulus manifests itself during the creep process is especially important for engineers in the design of underground structures (such as production wells), since the continuity of the technological process depends on the reliability of the structure. There are experimental works on vari-

---

Andrei M. Golosov · Evgenii P. Riabokon · Mikhail S. Turbakov · Evgenii V. Kozhevnikov · Vladimir V. Poplygin
Perm National Research Polytechnic University, Perm, Russian Federation
e-mail: a-dune@mail.ru, riabokon.evgenii@gmail.com, msturbakov@gmail.com, kozhevnikov_evg@mail.ru, poplygin@bk.ru

Mikhail A. Guzev
Perm National Research Polytechnic University, Perm & Institute for Applied Mathematics of the Far Eastern Branch of the Russian Academy of Sciences, Vladivostok Russian Federation
e-mail: guzev@iam.dvo.ru

Hongwen Jing
State Key Laboratory for Geomechanics and Deep Underground Engineering,
China University of Mining and Technology, Xuzhou, P.R. China
e-mail: hwjing@cumt.edu.cn

able loading of sandstone [3] and limestone [4] samples, which show the dispersion (growth) of the Young's modulus of the geomaterial with increasing frequency and amplitude of the variable load. The researchers also developed classical [5] and non-classical [6] mathematical models that describe the change in the Young's modulus under the action of a variable load depending on its frequency and amplitude.

Conventional creep tests last from a few hours to months, depending on the structure and strength of the geomaterial. At the same time, in rock creep tests in the presence of cyclic (dynamic) loads (cyclic creep tests), the time and deformation of the sample before failure are reduced in comparison with the conventional creep test under constant load [7].

As for today, a large number of theoretical and experimental works have been devoted to the study of the issues of creep of geomaterials under quasi-static loading. It is generally accepted that the creep deformation of a geomaterial includes three stages such as primary creep (creep rate decelerating), secondary creep (constant creep rate) and tertiary creep (creep rate accelerating) [8]. The existing mathematical models of creep of geomaterials can be found in [9]. At the same time, the study of the change in mechanical characteristics and in particular the change in the Young's modulus in the process of dynamic creep is not widely covered.

In this regard, the purpose of this work is to study the effect of dynamic loading of a geomaterial under a constant static preload on the Young's modulus of the samples. Section 9.2 provides a description of the rocks and the research methodology. Section 9.3 presents the results and their discussion, followed by a conclusion.

9.2 Materials and Methods

9.2.1 Materials

A consolidated sandstone (sedimentary rock) was used in the study having a granular structure, confined to the lower Bobrikovsky horizon C_1bb of the Carboniferous period and occurring in the area of one of the deposits of liquid minerals in the north of the Perm region. The source material for the preparation of samples was a core with a diameter of 100 mm, extracted from a depth of 2228.4-2355.2 m to the surface while drilling one of the oil wells in the field. During manufacture, the samples were drilled (Fig. 9.1a), cut (Fig, 9.1b), ground (Fig. 9.1c), hydrocarbon extracted (Fig. 9.1d); as a result, 25 samples were prepared (Fig. 9.1e). The study performed on dry samples.

Fig. 9.1: Sample preparation including: (a) drilling the sample; (b) cutting the sample; (c) grinding the ends of the sample; (d) extracting the hydrocarbon form the samples on a Soxhlet apparatus; (e) samples prepared for examination.

9.2.2 Methods

The tests were performed on the servo-hydraulic testing system MTS 816 Rock Mechanics Test Systems (Fig. 9.2) at the Research Center for Geomechanics and Geodynamics of Highly Compressed Rocks and Masses of the Far Eastern Federal University (Vladivostok) from June 1 to November 30 in 2022. The tests were carried out in the zone of linear elasticity of the geomaterial, in which the stress σ and the axial strain ε are connected linearly through the Young's modulus E in accordance with the ratio $\sigma = E\varepsilon$. Prior to the dynamic experiments, to determine the zone of linear elasticity, the samples were subjected to quasi-static loading during which the uniaxial compressive strength was determined (Fig. 9.3). One of the limitations in studying the creep of a geomaterial under dynamic loads is the hydraulic principle of the test system, which does not allow testing with a high frequency and amplitude of dynamic load, as well as high preload. To satisfy all the constraints, the stress during testing was chosen at the beginning of the zone of linear elasticity of the geomaterial at the level of 80 MPa.

The loading program included two stages. At the first stage (time t interval from 1 to 107 s) the test system pump fills the actuator with oil to create a predetermined dynamic load. After the target load is reached and the specimen is preloaded with

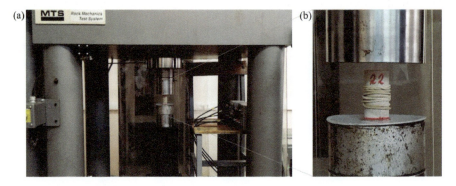

Fig. 9.2: Sample Creep Test: (a) General view of the MTS-816 testing system with the sample installed; (b) sample No. 22 between loading plates.

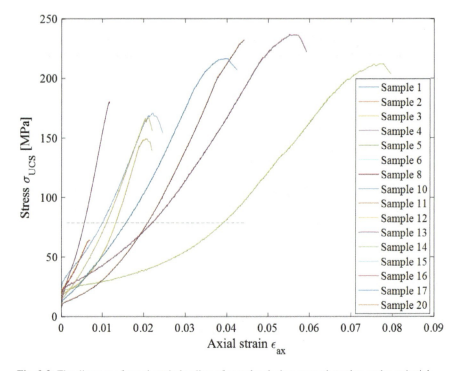

Fig. 9.3: The diagram of quasi-static loading of samples during strength testing under uniaxial compression and determination of the boundaries of the zone of linear elasticity of the geomaterial.

$F = 45$ kN, in the second stage (from 108 s) the actuator servo-valve starts and the dynamic action is activated with a variable sinusoidal load at the frequency $\omega = 2$ Hz (Fig. 9.4).

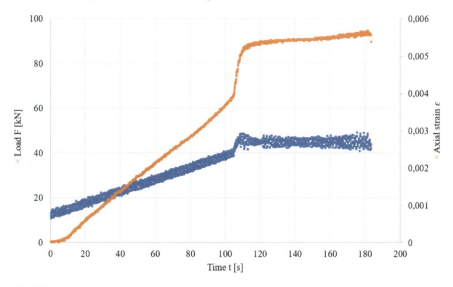

Fig. 9.4: The diagram of the load and the axial strain during the test: Stage I corresponds to the set of the required load on the sample; Stage II corresponds to dynamic loading of a statically preloaded specimen.

9.3 Results and Discussion

The results of the experiment showed that at a constant average load on the sample (Fig. 9.5a), equal to the value of the preload $F = 45$ kN together with the dynamic load with the amplitude equal to 2 kN, the stress in the sample remains constant $\sigma = 80$ MPa (Fig. 9.5b). At the same time, the longitudinal (axial) strain ε increases from 5.25×10^{-3} to 5.5×10^{-3} (Fig. 9.5c) and the Young's modulus E decreases from 15 GPa to 14 GPa over the 70 s (Fig. 9.5d). The rate of decrease in the Young's modulus was approximately 14.3 MPa/s or 7.15 MPa for one load-unload cycle of the sample under dynamic load with a frequency of 2 Hz. Experimental data were recorded at a frequency of 10 Hz, resulting in a large number of peak values on the diagram. Figure 9.5 shows the experimental data of the stage II recorded from 108 s of the test.

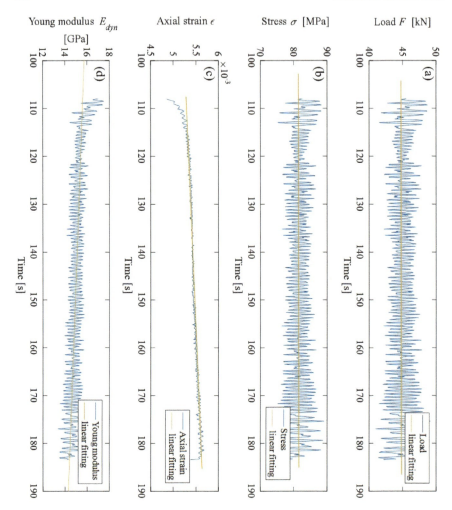

Fig. 9.5: The graphs of experimental parameters under loading of geomaterial with a frequency of 2 Hz: (a) load; (b) stress; (c) axial strain; (d) Young's modulus.

It can be seen from the Fig. 9.5 that the axial strain beyond the long-term strain increases monotonically at the same load. Creep has a monotonous (steady rate) character and could be attributed to the second stage (steady creep) [7, 10], however, practice shows that with a longer test duration, a flattening of the axial strain curve will be observed, as a result of which the resulting creep follows classify as primary creep. Note, that the experiment of a longer period of time with will be a part of future studies.

The decrease in the Young's modulus is carried out in accordance with the linear dependence on the dimensionless time ($t_0 = 1$ s), in which the free term corresponds to the value of the initial Young's modulus with the dimension of [GPa]:

$$E_{\text{dyn}} = -0.012\frac{t}{t_0} + 16.7. \tag{9.1}$$

The decrease in the Young's modulus is apparently associated with the occurrence of mesocracks and dislocations in the sample, which accumulate with increasing loading cycles and lead to a decrease in the resistance of the geomaterial and a deterioration in mechanical characteristics such as the Young's modulus.

9.4 Conclusion

The study of the geomaterial creep under the action of joint static and dynamic loading was carried out. During the study, rock samples were statically preloaded with a value of 45 kN and additionally subjected to dynamic sinusoidal action of the dynamic load with an amplitude of 2 kN and a frequency of 2 Hz. As a result of the study, the following conclusions can be drawn:

(1) dynamic loading activates intense creep of the geomaterial;
(2) with a constant value of the total load on the sample (static plus dynamic), with increasing time (number of cycles) of loading, the Young's modulus decreases at a rate of 14.3 MPa/s due to a constant increase in the axial strain.

Acknowledgements The research was supported by a grant from the Russian Science Foundation No. 22-19-00447, https://rscf.ru/project/22-19-00447/.

References

[1] Hunsche U, Cristescu ND (1997) Time Effects in Rock Mechanics. Wiley, New York
[2] Qian L, Zhang J, Wang X, Li Y, Zhang R, Xu N, Li Z (2022) Creep strain analysis and an improved creep model of granite based on the ratio of deviatoric stress-peak strength under different confining pressures. Environmental Earth Sciences **81**(4):109
[3] Guzev M, Kozhevnikov E, Turbakov M, Riabokon E, Poplygin V (2020) Experimental studies of the influence of dynamic loading on the elastic properties of sandstone. Energies **13**(23):6195
[4] Riabokon E, Turbakov M, Popov N, Kozhevnikov E, Poplygin V, Guzev M (2021) Study of the influence of nonlinear dynamic loads on elastic modulus of carbonate reservoir rocks. Energies **14**(24):8559
[5] Guzev M, Riabokon E, Turbakov M, Kozhevnikov E, Poplygin V (2020) Modelling of the dynamic young's modulus of a sedimentary rock subjected to nonstationary loading. Energies **13**(23):6461

[6] Guzev MA, Riabokon EP, Turbakov MS, Poplygin VV (2021) Non-classical model for description of the dynamic elasticity modulus of the material. Materials Physics and Mechanics **47**(5):720–726
[7] Chang KT, Lee ZZ Kevin, Yeh PT, Chang CM, Yu JY (2021) Creep behavior of cemented sand investigated under cyclic loading. Environmental Earth Sciences **80**(23):766
[8] Goodman RE (1989) Introduction to Rock Mechanics. Wiley, New York
[9] Frenelus W, Peng H, Zhang J (2022) Creep behavior of rocks and its application to the long-term stability of deep rock tunnels. Applied Sciences **12**(17):8451
[10] Maqsood Z, Koseki J, Miyashita Y, Xie J, Kyokawa H (2020) Experimental study on the mechanical behaviour of bounded geomaterials under creep and cyclic loading considering effects of instantaneous strain rates. Engineering Geology **276**:105,774

Chapter 10
A Novel Simulation Method for Phase Transition of Single Crystal Ni based Superalloys in Elevated Temperature Creep Regions via Discrete Cosine Transform and Maximum Entropy Method

Hideo Hiraguchi

Abstract Single crystal Ni based superalloys are composed of cubic γ' phases and γ phase channels. It is known that the γ' phases gradually connect with one another perpendicular to the tensile direction and change into rafting structures during the transient creep region and the steady state creep region. Moreover, the rafting structures grow or connect with one another to the tensile direction during the accelerated creep regions. There are recent reports by the author that the phenomenon of connecting and rafting of γ' phases is able to be simulated via the Discrete Cosine Transform and the Maximum Entropy Method. Therefore, in this research, it is demonstrated that the phenomenon of connecting and rafting of γ' phases can be simulated by using this novel simulation method.

10.1 Introduction

Single crystal Ni based superalloys for gas turbine blades consist of γ' phase cubes (L12 Type FCC) and γ phase channels (FCC). In the single crystal Ni based superalloys, the γ' phase cubes of the same size are arranged regularly at equal intervals in the γ phase solid solution during the transient creep region. However, when it reaches the steady state creep region, the γ' phase cubes begin to connect with one another and become rectangles with low height called rafting structures. Moreover, when it reaches the accelerated creep region, the narrow rafting structures with low height begin to connect with one another or grow their height and become rafting structures with large height. However, the dynamic metallographic changes from γ' cubes to the γ' rafting structures with low height are not well-known in detail. Moreover, the dynamic metallographic changes from the rafting structures with low height to the rafting structures with large height are not well-known in detail either. Therefore, it is

Hideo Hiraguchi
The Institution of Professional Engineers, Japan (IPEJ), 3-5-8, Shibakoen, Minato-ku, Tokyo, Japan,
e-mail: hideoh@abox2.so-net.ne.jp

convenient for researchers to be able to estimate the dynamic metallographic changes between one SEM chart before metallographic changes and another SEM chart after metallographic changes by selecting two charts before and after the metallographic changes from the published papers. To resolve the above problem, in this research, it is demonstrated that the dynamic metallographic changes such as the phenomenon of connecting and rafting of γ' phases can be estimated via the combination of the two dimensional Discrete Cosine Transform (2D-DCT) which is also used for creep equations [1] and the Maximum Entropy Method by using only two SEM charts before and after the metallographic changes selected from the published paper. By using this novel simulation method, the researchers in this field can easily grasp the approximate trends of dynamic metallographic changes at low cost. The obtained results are reported in this paper in detail.

10.2 Materials and Experiments

10.2.1 *A Single Crystal Ni Based Superalloy, CMSX-4*

A single crystal Ni based superalloy, CMSX-4, is selected as a specimen of creep test. Specifically, the SEM charts of CMSX-4 published in [2] are selected. Moreover, by using these charts, simple geometrical models of the γ' phase cubes and γ phase channels are made for calculation.

10.2.2 *Creep Tests*

Creep interruption tests at 1,273 K, 160 MPa are adopted as creep tests. The creep interruption tests mean that the metallographic structures of specimens are measured when stopping the creep tests at the target measurement points during the creep tests. The test pieces are 13 mm single crystal round bars made by the precision casting, and the longitudinal direction is [001]. In addition, the smooth specimens are cut from the above round bars. The diameter of the parallel portion of each specimen is 8 mm and the gauge length is 40 mm. The CMSX-4 SEM charts of three creep interruption specimens at (a) 1.08×10^5 s, (b) 1.08×10^6 s and (c) 2.52×10^6 s [2] are used for making the simple geometrical models.

Figure 10.1 shows the relationship between the creep strain rate and the ratio of the time to the rupture life. (a), (b) and (c) in Fig. 10.1 mean the above (a), (b) and (c). Reading values of the creep strain rate data in [2] are used in Fig. 10.1. In Fig. 10.1, (a) is a point of the transient creep region, (b) is a point of the steady state creep region and the minimum creep strain rate point, and (c) is a point of the accelerated creep region. Moreover, (d) is a point of the rupture time.

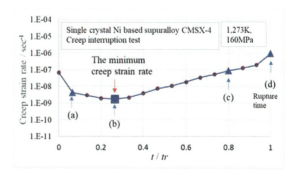

Fig. 10.1 Relationship between the creep strain rate and the ratio of the time to the rupture life [2]

Figure 10.2 (a), (b) and (c) shows the simple geometrical models [3] of CMSX-4 at (a), (b) and (c) points meshed into 21×21, which are created by measuring the length of one side of γ' phase cube and the width of γ phase channel of the metallographic SEM charts in Fig. 10.3 (Fig. 8 of [2]). In Fig. 10.2(a), the length of one side of the cube of γ' phase is 0.4 μm and the width of γ phase channel is 0.1 μm. Moreover, in Fig. 10.2(b), the height of rectangle of γ' rafting phase is 0.4 μm and the width of γ phase channel is 0.1 μm. In addition, in Fig. 10.2(c), the height of rectangle of γ' rafting phase is 0.7 μm and the width of γ phase channel is 0.2 μm. The tensile direction is the j axis direction. The lattice constant of γ phase Cγ at 1,273 K is 3.586 Å, and that of γ' phase Cγ' at 1,273 K is 3.580 Å. Therefore, the lattice misfit is "-0.17%" [4]. The equation of lattice misfit is Cγ' – Cγ/Cγ × 100%. In Fig. 10.2, for convenience, the lattice misfit is set as "+0.17%" to make γ' phase popping out.

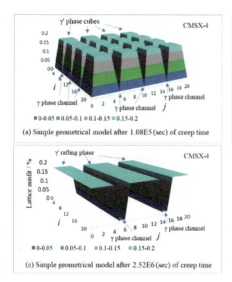

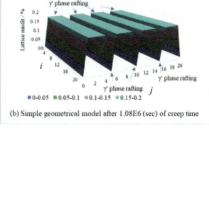

Fig. 10.2: Simple geometrical model (a), (b) and (c) for single crystal Ni based superalloy CMSX-4.

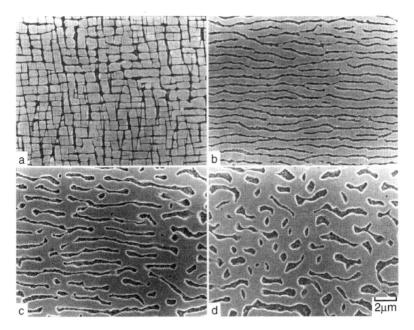

Fig. 10.3: SEM charts of the specimens at 160 MPa, 1,273 K for (a) 1.08E5 s, (b)1.08E6 s, (c)2.52E6 s and (d)creep ruptured (Fig. 8 of [2]). The tensile direction of the test is vertical.

10.2.3 Two Dimensional Discrete Cosine Transform

Two dimensional discrete cosine transform (2D-DCT) [5–8] is expressed as follows:

$$F[k,l] = \sum_{j=0}^{N-1} \sum_{i=0}^{N-1} f[i,j] \varphi k[i] \varphi l[j] \tag{10.1}$$

$$f[i,j] = \sum_{l=1}^{N-1} \sum_{k=0}^{N-1} F[k,l] \varphi k[i] \varphi l[j] \tag{10.2}$$

$$\begin{aligned} \varphi k[i] &= \frac{1}{\sqrt{N}}, \quad k = 0 \\ &= \sqrt{\frac{2}{N}} \cdot \cos\frac{(2i+1)k\pi}{2N}, \quad k = 1, 2, \ldots N-1 \end{aligned} \tag{10.3}$$

where i and j are the coordinate numbers of Fig. 10.2, k and l are the coordinate numbers of the coefficient of the 2D-DCT, $f[i,j]$ is a discrete signal (lattice misfit), which expresses the value entered in the pixel of the coordinate (i, j) in Fig. 10.2. $F[k, l]$ is the coefficient of the 2D-DCT.

The above $F[k,l]$ of Eq. (10.1) is used for the Maximum Entropy calculation incorporating the 2D-DCT [3, 9].

10.2.4 Maximum Entropy Method

Maximum Entropy Method is utilized to obtain the precise electron densities of crystals in the field of crystallography. The crystal structure factor $F(h)$ is expressed by using the Fourier transform equation as follows:

$$F(h) = V \sum_r \rho(r) \exp(-2\pi i r \cdot h) \qquad (10.4)$$

where ρ is the electron density, r is the position vector, h is the reciprocal space vector, and V is the volume of one voxel of a unit cell.

In the Maximum Entropy method, the constraint function is used to obtain the precise electron density distribution. The constraint function C is expressed as follows [10–12]:

$$C = \frac{1}{N} \sum_h \frac{|F\text{cal}(h) - F\text{obs}(h)|^2}{\sigma^2(h)} \qquad (10.5)$$

where $F\text{cal}(h)$ is the hth structure factor calculated from $\rho(r)$, $F\text{obs}(h)$ is the hth structure factor observed in the experiment, and N and $\sigma(h)$ are the total number and variance of $|F\text{obs}(h)|$, respectively.

When calculating the precise electron densities, the Shannon-Kullback relative information I [13–18] expressed in Eq. (10.6) should be minimized subject to the constraint C expressed in Eq. (10.5)

$$I = \sum_r p(r) \ln\left[\frac{p(r)}{m(r)}\right] \qquad (10.6)$$

where $m(r) = \tau(r)/\sum \tau(r)$ and $p(r) = \rho(r)/\sum \rho(r)$ are normalized prior and posterior electron densities in pixel r. The Shannon-Jaynes entropy S [19] is related to I through $S = -I$. Maximization of S under the constraint C leads to an exponential expression for $\rho_{ME}(r)$ in Eq. (10.7) [20–22]

$$\rho_{ME}(r) = \exp\left[\ln \tau(r) + \frac{\lambda F(0)}{N} \times \sum \frac{1}{\sigma^2(h)} \times \{F\text{obs}(h) - F\text{calc}(h)\} \exp\left(-2\pi i \frac{h}{r}\right)\right] \qquad (10.7)$$

where λ is Lagrange multiplier. $\rho_{ME}(r)$ in Eq. (10.7) is the electron density obtained by the Maximum Entropy Method.

In this research, the 2D-DCT coefficient $F(k, l)$ in Eq. (10.1) is utilized instead of the crystal structure factor $F(h)$ in Eq. (10.4) and Eq. (10.7) to estimate the dynamic metallographic changes of the single crystal CMSX-4 under the creep test. In this case, the obtained $\rho_{ME}(r)$ is a maximized $f[i, j]$ as a lattice misfit. Because the DCT is composed of only cosine function that does not contain the imaginary part unlike the ordinary Fourier transform, the Maximum Entropy method incorporating the 2D-DCT is comparatively simple. Dynamic metallographic changes between the two SEM charts can be estimated by extracting information from the two SEM charts by keeping Shannon-Jaynes Entropy S maximum.

10.3 Estimation of Phase Transition and Results

The phase transition or metallographic changes between the metallographic chart in Figs. 10.2(a) and 10.2(c) is calculated via the Maximum Entropy Method incorporating the 2D-DCT. In this calculation, $f(i,j)$ and $F(k,l)$ of Figs. 10.2(a) and 10.2(c) are used to derive information from the two images. Fig. 10.4 shows the estimated dynamic metallographic changes between (a) and (c). The iteration of calculations of the Maximum Entropy method incorporating the 2D-DCT is performed until C is less than or equal to 1.

In Fig. 10.4, first, the γ' phase cubes begin to connect with one another from (a) to (a-2). As shown in Fig. 10.4(a-2), the transition chart is very similar to the simple geometrical model of Fig. 10.2(b) at (b) point of Fig. 10.1. It means that the Maximum Entropy method incorporating 2D-DCT can estimate the metallographic structure of (b) point in Fig. 10.2. By using the only two images of Fig. 10.2(a) and Fig. 10.2(c), the similar metallographic structure of Fig. 10.2(b) can be obtained as the structure of Fig. 10.4(a-2) via the combination of the Maximum Entropy method and the 2D-DCT.

Second, as shown in Fig. 10.4 from (a-2) to (a-7), it shows that the number of γ phase channels of the (a-2) transition model similar to Fig. 10.2(b) is gradually changed from three to two. Finally, the estimation converges to the simple geometrical model of Fig. 10.4(c).

10.4 Discussion

As shown in Fig. 10.4 from (a) to (a-2), the γ' phase cubes begin to connect with one another perpendicular to the tensile direction. It is consistent with the experimental results. Moreover, the (a-2) transition chart is very similar to the simple geometrical model of Fig. 10.2(b) at (b) point of Fig. 10.1. It can be said that this is a strong proof of an ability of the estimation and the interpolation of the dynamic metallographic changes of the combination of the Maximum Entropy method and 2D-DCT. Consideration and resolution about the difference of the values of the lattice misfits between Fig. 10.4 (a-2) and Fig. 10.2(b) is a next challenge. Moreover, as shown in Fig. 10.4 from (a-2) to (a-7), the processes of decreasing the number of the γ phase channels and the growth of the γ' phase rafting structures can be observed in detail. The change in the number of channels from three to two is consistent with the experimental result, too. It is needed to be confirmed how small the difference between this estimation result and the real experimental data is.

10.5 Conclusion

From the above, the results of this research are as follows:

10 A Novel Simulation Method for Phase Transition of Superalloys

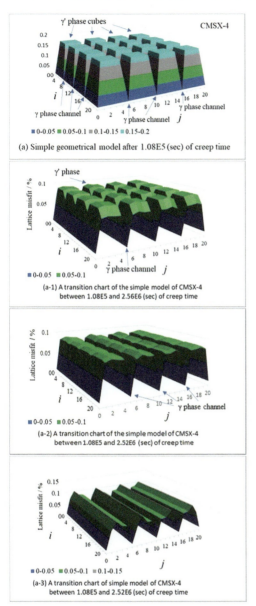

Fig. 10.4: Estimated state of dynamic metallographic changes of CMSX-4 between (a) and (c) Lattice misfit / % = $|(C\gamma' - C\gamma)/C\gamma \times 100\%|$.

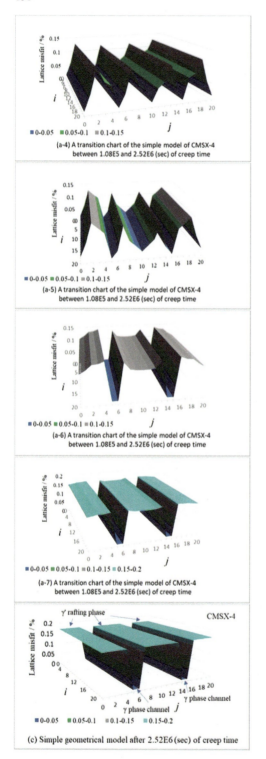

Fig. 10.4: Estimated state of dynamic metallographic changes of CMSX-4 between (a) and (c) Lattice misfit / % = $|(C\gamma' - C\gamma)/C\gamma \times 100\%|$ (Cont'd).

1. The estimated result that the γ' phase cubes begin to connect with one another perpendicular to the tensile direction in the beginning is consistent with the experimental data.
2. The (a-2) transition chart in Fig. 10.4 is very similar to the simple geometrical model of Fig. 10.2(b) at (b) point of Fig. 10.1.
3. The change in the number of γ phase channels from three to two shown in Fig. 10.4(a-2) to (a-7) is consistent with the experimental result, too.
4. It can be said that the above result is a strong proof of an ability of the estimation and the interpolation of the dynamic metallographic changes of the Maximum Entropy method and 2D-DCT.
5. Consideration and resolution about the difference of the values of the lattice misfits between Fig. 10.4 (a-2) and Fig. 10.2(b) is a next challenge.

Acknowledgements The author gratefully acknowledges the support provided by Emeritus Professor Satoshi Sasaki from the Tokyo Institute of Technology with regard to the Fourier transform. The author is also grateful to Specially Appointed Professor Osami Sakata of the Tokyo Institute of Technology (Director, Japan Synchrotron Radiation Research Institute) for valuable advice related to this work.

References

[1] Hiraguchi H (2019) Study on new creep equation using discrete cosine transform for high temperature materials. Heliyon **5**(10), e02619
[2] Miura N, Kondo Y, Matsuo T (2003) Relation between creep rate during accelerating creep stage and γ channel thickness in single crystal nickel-based superalloy, CMSX-4. Tetsu-to-hagané **89**(12):1240–1247
[3] Hiraguchi H (2021) A theoretical numerical analysis method on the relationship between creep strain rates and metallographic changes of single crystal Ni based superalloys via the discrete cosine transform and maximum entropy method. In: Proceedings of the 59th symposium on Strength of Materials at High Temperatures, The Society of Materials Science, Japan
[4] Harada H, Okazaki M (2002) Materials science toward "Super-" and "Ultra-" high temperature technologies III: High temperature strength of Ni-based superalloys and coatings for advanced gas turbines (in Japanese). Journal of the Society of Materials Science **51**(7):836–842
[5] Kekre HB, Solanki JK (1978) Comparative performance of various trigonometric unitary transforms for transform image coding. International Journal of Electronics **44**(3):305–315
[6] Rao KR, Yip P (1990) Discrete Cosine Transform: Algorithms, Advantages, Applications. Academic Press Professional, Inc., Boston, pp. 90-91
[7] Hiraguchi H (2021) Novel study on the electron density distribution projection maps calculated via the discrete cosine transform. Journal of Applied Crystallography **54**(4):1198–1206

[8] Ahmed N, Natarajan T, Rao KR (1974) Discrete cosine transform. IEEE Transactions on Computers **C-23**(1):90–93

[9] Hiraguchi H (2022) New simulation method for metallographic changes of single crystal Ni based superalloys for gas turbine blades in high temperature creep regions via Discrete Cosine Transform incorporated into Maximum Entropy Method. In: Proceedings of the 2022 Annual Meeting of the Japan Society of Mechanical Engineers, Japan Society of Mechanical Engineers, pp J192–04

[10] Gull SF, Daniell GJ (1978) Image reconstruction from incomplete and noisy data. Nature **272**(5655):686–690

[11] Collins DM (1982) Electron density images from imperfect data by iterative entropy maximization. Nature **298**(5869):49–51

[12] Livesey AK, Skilling J (1985) Maximum entropy theory. Acta Crystallographica Section A **41**(2):113–122

[13] Shannon CE (1948) A mathematical theory of communication. The Bell System Technical Journal **27**(4):623–656

[14] Kullback S, Leibler RA (1951) On information and sufficiency. The Annals of Mathematical Statistics **22**(1):79–86

[15] Hobson A, Cheng BK (1973) A comparison of the shannon and kullback information measures. Journal of Statistical Physics **7**:301–310

[16] Levine RD (1980) An information theoretical approach to inversion problems. Journal of Physics A: Mathematical and General **13**(1):91–108

[17] Steenstrup S, Wilkins SW (1984) On information and complementarity in crystal structure determination. Acta Crystallographica, Section A **40**(2):163–164

[18] Wilson AJC (ed) (1985) Structure & Statistics in Crystallography. Academic Press, New York

[19] Jaynes ET (1968) Prior probabilities. IEEE Transactions on systems science and cybernetics **4**(3):227–241

[20] Sakata M, Mori R, Kumazawza S, Takata M, Toraya H (1990) Electron-density distribution from X-ray powder data by use of profile fits and the maximum-entropy method. Journal of Applied Crystallography **23**(6):526–534

[21] Sakata M, Takata M (1990) Electron density distribution by the maximum entropy method. Journal of the Crystallographic Society of Japan **32**(3):175–183

[22] Sakata M, Sato M (1990) Accurate structure analysis by the maximum-entropy method. Acta Crystallographica, Section A **46**(4):263–270

Chapter 11
Anisotropic Creep Analysis of Fiber Reinforced Load Point Support Structures for Thermoplastic Sandwich Panels

Jörg Hohe and Sascha Fliegener

Abstract The present contribution is concerned with a numerical analysis of creep in load point support structures for sandwich panels made from different fiber reinforced thermoplastic materials. Whereas the face sheets consist of laminates of unidirectional carbon fiber reinforced plies, the support structures for the load points consist of discontinuously long fiber reinforced thermoplastics manufactured in a compression molding process. The sandwich core is a thermoplastic foam. For the numerical creep analysis of such structures under long-term loading, an anisotropic viscoelastic material model is formulated. In different versions, the model is applicable either to unreinforced thermoplastics, or to thermoplastics with discontinuous or continuous fiber reinforcement. The material model is implemented into a finite element system. The model is validated against an experimental data base on both, coupon and structural level.

11.1 Introduction

Structural sandwich panels are found in a variety of technological fields where extreme lightweight solutions are required. In addition to the classical fields of aerospace industry or the wind energy sector, sandwich structures become increasingly popular in transport applications for both the rail and road sector. In contrast to aerospace components with limited numbers of components to be manufactured, especially the automotive sector is characterized by industrial scale mass production with large numbers of components to be manufactured with an extreme demand for short cycle times. For this purpose, polymeric composite and sandwich components consisting of thermoplastic base materials are popular materials for future composite automotive designs (Bijsterbosch and Gaymans [1], Henning et al. [2]). On the other hand, one of

---

Jörg Hohe · Sascha Fliegener
Fraunhofer-Institut für Werkstoffmechanik IWM, Wöhlerstr. 11, 79108 Freiburg, Germany,
e-mail: joerg.hohe@iwm.fraunhofer.de, sascha.fliegener@iwm.fraunhofer.de

the major shortcomings of thermoplastic materials is their inherent tendency towards creep deformation even for discontinuously and in some cases also continuously fiber reinforced microstructures (Chevali et al. [3], Greco et al. [4], Brinson and Knauss [5]).

Objective of the present contribution is the evaluation of load point support structures for thermoplastic sandwich panels, involving hybrid designs of discontinuously and continuously fiber reinforced polymer matrix composites. For this purpose, creep material models for anisotropic fiber reinforced materials are developed and implemented. Based on a classical three term Kelvin-Voigt approach, a preliminary isotropic viscoelastic material model is formulated. Using a Schapery [6] type extension, a stress dependence is implemented. The isotropic base model is extended to anisotropic fiber reinforced materials in a twofold manner. For discontinuously long fiber reinforced materials, a simple generalization based on anisotropy factors is employed. To account for creep effects in continuously unidirectionally fiber reinforced materials, the isotropic base model is superimposed with an isotropic Hooke's law. In this context, the isotropic viscoelastic part represents the matrix response whereas the rate independent anisotropic Hooke's law with (almost) vanishing stiffness perpendicular to the fibers represents the response of the unidirectionally oriented continuous fibers. The different models are implemented as user-defined material models into a commercial finite element program.

In a first application, the numerical approach is validated against experimental data obtained in unidirectional coupon experiments considering unreinforced thermoplastic materials as well as discontinuously and continuously fiber reinforced materials, both tested in different spatial directions. In a second step, the model is applied to load point support structures for sandwich panels made from compression molded long fiber reinforced thermoplastics designed in a previous contribution (Fliegener et al. [7]). The sandwich face sheets are made from multidirectional laminates consisting of unidirectional carbon fiber reinforced thermoplastic plies bonded to a thermoplastic foam core. The results of the simulations are validated against experimental data obtained in creep experiments. The numerical predictions are found in good agreement with the experimental observations.

11.2 Material Model

The analyses in the present study employ the material models presented by the authors in an earlier contribution (Fliegener and Hohe [8]). Therefore, only a brief outline is given here.

11.2.1 Basic One-Dimensional Formulation

The model is based on a three element Kelvin-Voigt model using the rheological model sketched in Fig. 11.1. For this type of model the one-dimensional time dependent strains are given by the convolution integral

$$\varepsilon(t) = \int_0^t \left(\frac{1}{E^{(0)}} + \sum_{q=1}^{3} \frac{1}{E^{(q)}} \left(1 - e^{-\frac{t-t^*}{\tau^{(q)}}}\right) \right) \left. \frac{\partial \sigma}{\partial t} \right|_{t=t^*} dt^* \qquad (11.1)$$

where where the elastic moduli $E^{(q)}$ and the relaxation times

$$\tau^{(q)} = \frac{\eta^{(q)}}{E^{(q)}} \qquad (11.2)$$

together with the viscosities $\eta^{(q)}$ are material parameters.

Transforming Eq.(11.1) into a strain dependent incremental form results in

$$d\sigma(t) = \tilde{E}(t) \left(d\varepsilon(t) - \sum_{q=1}^{3} \left(1 - e^{-\frac{dt}{\tau^{(q)}}}\right) \varepsilon^{i(q)}(t - dt) \right) \qquad (11.3)$$

with the time dependent tangential stiffness

$$\tilde{E}(t) = \left(\frac{1}{E^{(0)}} + \sum_{q=1}^{3} \frac{1}{E^{(q)}} \left(1 - \frac{\tau^{(q)}}{dt}\left(1 - e^{-\frac{dt}{\tau^{(q)}}}\right)\right) \right)^{-1} \qquad (11.4)$$

and the inherited strains $\varepsilon^{i(q)}$ defined in a recursive manner by

$$d\varepsilon^{i(q)}(t) = \left(\frac{\tau^{(q)}}{E^{(q)}} \frac{\partial \sigma}{\partial t} - \varepsilon^{i(q)}(t - dt) \right) \left(1 - e^{-\frac{dt}{\tau^{(q)}}}\right) \qquad (11.5)$$

forming history variables.

Linear viscoelstic models are in many cases inadequate to model the time dependent stress-strain response observed in experiments (e.g. Fliegener et al. [9], Haj-Ali and Muliana [10], Schapery [6]). For this reason, the parameters $E^{(q)}$ in Eqns. (11.4) and (11.5) are considered as stress dependent functions $E^{(q)}(\sigma)$ rather than constant moduli. Full details on the employed material model and its definition can be found in the original contribution (Fliegener and Hohe [8]).

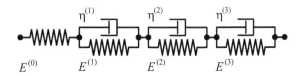

Fig. 11.1 Three element Kelvin-Voigt model.

11.2.2 Generalization to Three Dimensions

The one-dimensional viscoelastic material model (11.3) is generalized to the three-dimensional space in the same manner as Hooke's law since it must reduce to Hooke's law, if all viscous parts are deactivated by an appropriate choice of the material parameters. Therefore, the material model (11.3) is decomposed into a volumetric and a deviatoric part, resulting in the three-dimensional form

$$d\sigma_{ij}(t) = 2\tilde{G}(t)\left(d\varepsilon'_{ij}(t) - \sum_{q=1}^{3}\left(1-e^{-\frac{dt}{\tau^{(q)}}}\right)\varepsilon_{ij}^{is(q)}(t-dt)\right)$$
$$+3\tilde{\kappa}(t)\left(d\varepsilon^{v}(t) - \sum_{q=1}^{3}\left(1-e^{-\frac{dt}{\tau^{(q)}}}\right)\varepsilon^{iv(q)}(t-dt)\right)\delta_{ij} \quad (11.6)$$

with the time-dependent shear and bulk stiffnesses

$$\tilde{G}(t) = \left(\frac{1}{G^{(0)}} + \sum_{q=1}^{3}\frac{1}{G^{(q)}}\left(1 - \frac{\tau^{(q)}}{dt}\left(1-e^{-\frac{dt}{\tau^{(q)}}}\right)\right)\right)^{-1} \quad (11.7)$$

$$\tilde{\kappa}(t) = \left(\frac{1}{\kappa^{(0)}} + \sum_{q=1}^{3}\frac{1}{\kappa^{(q)}}\left(1 - \frac{\tau^{(q)}}{dt}\left(1-e^{-\frac{dt}{\tau^{(q)}}}\right)\right)\right)^{-1} \quad (11.8)$$

respectively. The generalized bulk and shear moduli are defined by

$$G^{(q)} = \frac{E^{(q)}}{2(1+v)} \quad (11.9)$$

$$\kappa^{(q)} = \frac{E^{(q)}}{3(1-2v)} \quad (11.10)$$

in the same manner as for Hooke's law. The Poisson's ratio v is assumed to be a material constant, which is not affected by viscoelasticity. The evolution equation (11.5) is generalized to the three-dimensional case by

$$d\varepsilon_{ij}^{is(q)}(t) = \frac{1}{\tilde{G}(t)}\frac{\tau^{(q)}}{dt}\left(1-e^{-\frac{dt}{\tau^{(q)}}}\right)d\sigma'_{ij} - \varepsilon_{ij}^{is(q)}(t-dt)\left(1-e^{-\frac{dt}{\tau^{(q)}}}\right) \quad (11.11)$$

$$d\varepsilon^{iv(q)}(t) = \frac{1}{\tilde{\kappa}(t)}\frac{\tau^{(q)}}{dt}\left(1-e^{-\frac{dt}{\tau^{(q)}}}\right)\frac{d\sigma_{kk}}{3} - \varepsilon^{iv(q)}(t-dt)\left(1-e^{-\frac{dt}{\tau^{(q)}}}\right) \quad (11.12)$$

where the history variables $\varepsilon_{ij}^{is(q)}$ and $\varepsilon^{iv(q)}$ are the deviatoriv and volumetric inherited strains, respectively.

11.2.3 Unidirectionally Fiber Reinforced Thermoplastics

The three-dimensional constitutive equation (11.6) constitutes an isotropic viscoelastic material law. For fiber reinforced materials, this model needs to be generalized to cover anisotropic material response as well.

For the case of a continuous unidirectional fiber reinforcement, it is assumed that stresses deriving from the deformation of the isotropic viscoelastic matrix and stresses deriving form the deformation of the linear elastic anisotropic fiber reinforcement can be superimposed by

$$d\sigma_{ij} = (1-\rho^f)d\sigma_{ij}^m + \rho^f d\sigma_{ij}^f \qquad (11.13)$$

where $d\sigma_{ij}^m$ are the stress increments in the matrix determined by Eq. (11.6) whereas the fiber stress increments

$$d\sigma_{ij}^f = C_{ijkl}\varepsilon_{kl} \qquad (11.14)$$

are obtained by means of the anisotropic Hooke's law considering the anisotropic stiffness induced to the composite by the fiber reinforcement. The parameter ρ^f denotes the fiber volume fraction.

11.2.4 Discontinuously Fiber Reinforced Thermoplastics

For the case of a discontinuous multidirectional fiber reinforcement, the effect of the anisotropy of the composite is modelled in a pragmatic manner by introduction of weight factors into the equations for the normal stresses in the different spatial directions. For this purpose, Eq. (11.6) is substituted with

$$d\sigma_{ij}(t) = 2\tilde{G}(t)\left(\bar{s}^d d\varepsilon'_{ij}(t) - \sum_{q=1}^{3}\left(1-e^{-\frac{dt}{\tau(q)}}\right)\varepsilon_{ij}^{is(q)}(t-dt)\right)$$

$$+3\tilde{\kappa}(t)\left(\bar{s}^v d\varepsilon^v(t) - \sum_{q=1}^{3}\left(1-e^{-\frac{dt}{\tau(q)}}\right)\varepsilon^{iv(q)}(t-dt)\right)\delta_{ij} \qquad (11.15)$$

where the weight factors

$$\bar{s}^d = \begin{cases} s^d & \text{if } :i=j=1 \\ 1 & \text{else} \end{cases} \qquad (11.16)$$

$$\bar{s}^v = \begin{cases} s^v & \text{if } :i=j=1 \\ 1 & \text{else} \end{cases} \qquad (11.17)$$

are factors used to introduce a proces and fiber preference orientation related anisotropy of the material response. Both factors may be dependent on the flow characteristics of the material during molding or other process parameters and thus will – in general – depend on the spatial position in the final component. Full details and a broader discussion on the formulation of the constitutive model are given in the original contribution (Fliegener and Hohe [8]).

The proposed material model for all three material types – neat isotropic polymeric material as well as anisotropic continuously and discontinuously fiber reinforced materials – are implemented as user defined materials into a commercial finite element code.

11.3 Experimental Investigation

The theoretical developments of the material model in Sect. 11.2 are complemented by an experimental investigation to provide a data base for validation and demonstration of its capabilities. Creep tests are performed on both coupon and structural level.

11.3.1 Coupon Experiments

For determination of the material parameters for the different levels of the creep model proposed in Sect. 11.2 and for a first validation, creep experiments on coupon level are performed. Three different grades of material are investigated, each supplied with a polyamide (PA) 6 matrix. The first grade is the neat matrix material. The second grade is a long glass fiber reinforced (LFT) material manufactured in a compression molding process. The material is supplied with fiber weight and volume fractions of 40wt.% and 22.5vol%, respectively. The third material grade is a unidirectionally (UD) carbon fiber reinforced material with a fiber volume fraction of 46vol%. The material was processed by Fraunhofer ICT and BASF and supplied in form of plane plates.

From the available plates, coupon specimens are manufactured using waterjet cutting. For the neat matrix material as well as for the LFT material and the UD material to be tested perpendicular to the fiber direction, dogbone specimens according to ISO 3167 [11], type A are used, supplied with an increased shoulder radius of 50 mm and – due to limited material availability – for the neat matrix material with a reduced gauge section length of 36 mm. For the UD material to be tested within the fiber direction, straight specimens according to ISO 527-5 [12], type A with an increased width of 20 mm are used.

The specimens are tested for their creep response under tensile loads according to ISO 899-1 [13] using a dead weight creep test rig. All experiments are performed in a climate chamber at ambient temperature and 62.5% relative humidity over a period of 160 h. The neat matrix material is tested in one direction only whereas the LFT material is tested within and perpendicular to the flow direction. The UD material is tested within and perpendicular to its fiber direction as well as under 45° to the fiber direction to gain information about creep effects in shear dominated loading situations. For parameter identification and validation purposes, all experiments are simulated numerically by means of the finite element method using the material models proposed in Sect. 11.2.

11.3.2 Structural Experiments

For validation on the structural level, the proposed creep model is applied to the simulation of a load point support structure for thermoplastic sandwich structures designed in a previous study (Fliegener et al. [7]). The sandwich structure consists of two CFRP fase sheets with a symmetric $[0°, 90°, \pm 45°]_s$ stacking sequence. The CFRP laminates are overmolded with a glass fiber reinfoced LFT ply on each side. In all instances, the same polyamide 6 based fiber reinforced materials as described in Subsect. 11.3.1 are employed. The face sheets are separated by a 20 mm thick thermoplastic PUR foam core. The sandwich plate is supplied with a loading point formed by a brass insert with a screw thread. To distribute the load smoothly from the brass insert into both face sheets, different types of load point support structues were investigated by Fliegener et al. [7]. The most promising support structure prooved to be the structure presented in Fig. 11.2. The support structure consists of the same LFT material as the LFT plies on top and bottom of the face sheets and is co-molded with the face sheets in a compression molding process.

For the numerical analysis, the breadboard specimen is meshed with standard displacement based finite elements considering a circular section of the square plate (Fliegener et al. [7], [8]). For the load point support structure and other LFT ranges as well as for the core, 8-node volume elements with tri-linear shape functions are used. For the CFRP laminate, 4-node first order shell elements are employed. The connection between volume and shell elements is made via appropriate constraint conditions. For a proper connection between all ranges, some triangular and tetrahedarl elements need to be used in addition. The finite element model of the breadboard specimen is modelled as clamped all around its external edges. It is loaded by a constant force

Fig. 11.2 Breadboard sandwich specimen with central load point and LFT load point support structure for validation on structural level.

acting in the downward normal direction. Due to the high stiffness of the panel, the analysis is performed within the geometrically linear framework.

The analysis is complemented by an experimental investigation, where breadboard specimens as presented in Fig. 11.2 are tested under different creep loads. The experiments are performed using a clamping system where the square specimen is placed into a tightly fitting box with a circular cutout with the same radius as the outer radius of the finite element model, thus providing similar external boundary conditions. Subsequently, a constant long term load is applied to the load point using an electromechanical Hegewald & Peschke inspekt table 250 testing machine. During the long term experiment, the crosshead displacement is continuously recorded. Since the load is constant, no strain variations in the loading rig and the machine's frame occur during the test so that crosshead displacement variations coincide with the displacement variations of the specimens load point due to creep.

11.4 Multiscale Simulation

As an additional approach for validation, a multiscale analysis is performed for the LFT material to predict the material response on the effectve level and to validate it against the experimental data available on coupon experiments. For this purpose, a representative volume element for the LFT microstructure is generated, employing the procedure presented earlier by Fliegener et al. [14]. The representative volume element is meshed using standard displacement based tetrahedral finite elements using the isotropic creep material model from Subsect. 11.2.2 for the matrix and assuming linear isotropic elasticity for the glass fibers with data sheet values for the Young's modulus and the Poisson's ratio. The finite element model is then subjected to a constant effective stress. The effective creep strain is computed as a function of time. By this means, numerical experiments for the LFT creep response on the macroscopic material level can be performed. The results are validated against the results of the physical laboratory experiments.

11.5 Results

11.5.1 Parameter Identification on Coupon Experiments

In a first validation step, the creep response of the neat matrix material is investigated. Uniaxial tensile loads of 5, 10, 15 and 20 MPa respectively are applied. The resulting creep curves are presented in Fig. 11.3 together with their counterparts obtained by a numerical simulaion based on the isotropic version (11.6) of the material model.

For all applied stress levels, the standard characteristics of the creep curves are observed. In the initial phase, a primary creep range with initially large creep rates

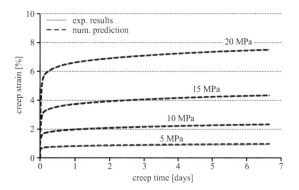

Fig. 11.3 Neat matrix material – experimental creep curves and numerical prediction by proposed viscoelastic material law.

is obtained. Subsequently, the creep rate decreases and the creep curves approach the secondary creep range featuring a constant creep rate. For all load levels, the material parameters in Eqs. (11.6) to (11.12) can be chosen such that the numerically predicted creep curves are in perfect agreement with their experimental counterparts, giving evidence for the rationale of the mathematical formulation of the constitutive model.

The isotropic viscoelastic material model with the material parameters determined before is then employed in a multiscale and homogenization analysis of the LFT material. The creep curves predicted by the multiscale analysis and the experimental counterparts for the LFT material are presented in Fig. 11.4, together with numerical predictions of the creep response using the anisotropic version (11.15) for the discontinuosly fiber reinforced LFT material with a parameter set fitted on this level of structural hierarchy. Both, the flow direction and the test direction perpendicular to the flow direction are considered.

Due to the discontinuous reinforcement by the linear elastic, non-viscous glass fibers, much lower creep strains are observed for the LFT material than for the neat matrix material, although the basic characteristics of the creep curves are maintained (see Figs. 11.3 and 11.4). For the tests within the flow direction, a rather good agreement of both numerical simulations – multiscale analysis and analysis based on the proposed LFT material model (11.15), respectively – with the experimental results is observed. Similar findings are obtained for the creep experiments perpendicular to the flow direction at the lower load levels. For the highest creep load level of 24 MPa, the proposed LFT creep model (11.15) and the experimental creep results are still found in a good agreement whereas the multiscale simulation slightly overestimates the creep rate in the initial phase, resulting in an overestimation of the creep strain in the entire creep period. This effect is probably caused by a suboptimal assumption on the fiber length and orientation distribution in the underlying representative volume element.

The results of the parameter identification for the continuously unidirectionally (UD) fiber reinforced material are presented in Fig. 11.5. The creep response is investigated within the fiber direction (0°), perpendicular to the fiber direction (90°) as well as under 45° to the fiber direction. Since the linear elastic carbon fibers carry

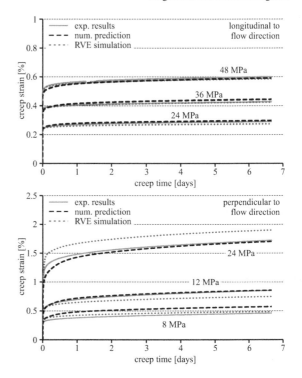

Fig. 11.4 Long fiber reinforced thermoplastics (LFT)–experimental creep curves, numerical prediction by proposed viscoelastic material law and results of multiscale analysis longitudinal and perpendicular to the flow direction.

almost the entire load within the fiber direction, almost no creep deformation develops in the experiment in the 0°-direction even under a higher load level. Perpendicular to the fiber direction and under 45° to the fiber direction, non-negligible creep effects are observed. Even for the UD CFRP material, creep effects are present when loaded in interfiber normal (90°) and shear modes (45°). Again, the experimental findings are found in a good agreement with the numerical simulation based on the proposed material model. Thus, also the UD version (11.13) of the creep model proves to provide a meaningful description of the macroscopic creep response if appropriate material parameters are selected.

Further experimental and numerical results are provided in the original publication (Fliegener and Hohe [8]).

11.5.2 Validation on Structural Level

For a validation of the proposed constitutive creep models for both LFT and UD composites together with the material parameters determined in Subsect. 11.5.1 under more complex multiaxial stress situations, the proposed material models are applied to a numerical simulation of the breadboard experiments on the load point support structure presented in Fig. 11.2.

Fig. 11.5 Unidirectionally (UD) fiber reinforced thermoplastics – experimental creep curves and numerical prediction by proposed viscoelastic material law in 0°-, 90°- and 45°-direction to the fiber orientation.

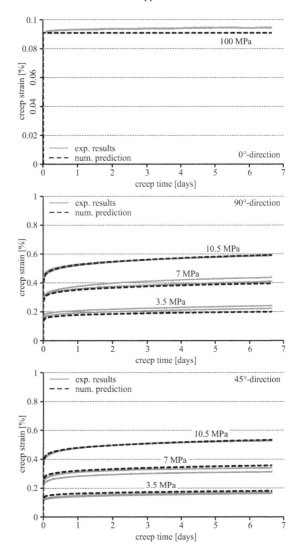

The experimental and numerical results are presented in Fig. 11.6. In the first subfigure, the stress distribution in the LFT ranges of the overmolded LFT layers on top and bottom of the CFRP laminates forming the face sheets as well as in the co-molded load point support structure are shown. The second subfigure is concerned with the creep curves in terms of the crosshead (or load point) displacement as a function of time. Two different load levels with applied forces of 2.5 kN and 5 kN are investigated.

For both load levels, the experimentally observed and the numerically predicted creep curves are found in a good agreement, especially when considering the complexity of the underlying geometry and the complex creep problem involving both,

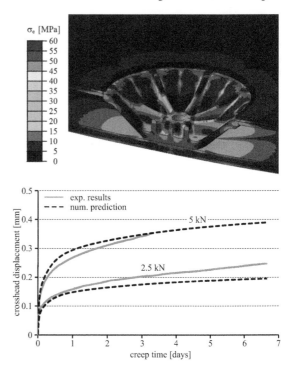

Fig. 11.6 Sandwich load point support structure – stress distribution in LFT material ranges as well as experimental creep curves and numerical prediction using proposed viscoelastic material law for the different material ranges.

UD CFRP laminates and glass fiber reinfoced LFT materials. Although the support structure design considered here proved to be the strongest among all optinons considered in the design study by Fliegener et al. [7] with a static failure load of 16.9 MPa, a significant amount of overall creep deformation is observed even at load levels of approximately 30% of the static failure load. Creep strains are found to develop especially in the co-molded LFT load point support structure and in the adjacent LFT plies on the inner side of the face sheets. Nevertheless, creep deformation is also observed in the UD plies of the face sheets, although limited to smaller amounts. The numerical simulation provides a good approximation of the experimental results, demonstrating the capabilities of the different versions of the proposed anisotropic constitutive model.

Further validations of the proposed material model on other types of load point support structures are presented in an earlier contribution by the present authors (Fliegener and Hohe [8]).

11.6 Summary and Conclusion

The present study is concerned with creep material models for fiber reinforced thermoplastics. Both, discontinuously and continuously fiber reinforced materials are considered. Based on an isotropic three element Kelvin-Voigt model, two different anisotropic versions are derived. Whereas the anisotropy for the discontinuously fiber reinforced material is introduced by introduction of anisotropy factors into the isotropic base formulation, fiber effects in the continuously finber reinforced material are introduced by superposition of the isotropic viscoelastic model representing the matrix with the anisotropic Hooke's law representing the effect of the fibers.

The model is validated against an experimental data base on coupon specimens concerning neat PA6, discontinuously long glass fiber reinforced thermoplastics and unidirectionally carbon fiber reinforced PA6 tape material. The experiments on laboratory specimens are complemented by experiments on more complex breadboard type sandwich specimens with central loading points. In all experiments on coupon and structural level, a good agrement of experimental and numerical data is observed.

The validated anisotropic material model proposed in the present study proves to provide a reliable tool for the numerical simulation of creep in glass or carbon fiber reinforced thermoplastics and structures made thereof. Both, continuously and discontinuously fiber reinforced microstructures as well as hybrid composites can be addressed. Although the development of substantial creep deformation is more likely for discontinuously fiber reinforced types, laminates consisiting of unidirectionally fiber reinforced plies might also experience non-negligible amounts of creep.

Acknowledgements The present contribution has been funded by the German Federal Department of Education and Research (BMBF) under grant no. 03X3041P as part of the collaborative "SMiLE" project. The authors are fully responsible for the contents of this contribution. The financial support is gratefully acknowledged. The authors wish to express their thanks to BASF and Fraunhofer ICT for provision of the material for the experimental studies.

References

[1] Bijsterbosch H, Gaymans RJ (1995) Polyamide 6—long glass fiber injection moldings. Polymer Composites **16**(5):363–369
[2] Henning F, Ernst H, Brüssel R (2005) LFTs for automotive applications. Reinforced Plastics **49**(2):24–33
[3] Chevali VS, Dean DR, Janowski GM (2009) Flexural creep behavior of discontinuous thermoplastic composites: Non-linear viscoelastic modeling and time–temperature–stress superposition. Composites Part A: Applied Science and Manufacturing **40**(6):870–877
[4] Greco A, Musardo C, Maffezzoli A (2007) Flexural creep behaviour of PP matrix woven composite. Composites Science and Technology **67**(6):1148–1158
[5] Brinson L, Knauss W (1991) Thermorheologically complex behavior of multiphase viscoelastic materials. Journal of the Mechanics and Physics of Solids **39**(7):859–880
[6] Schapery RA (1969) On the characterization of nonlinear viscoelastic materials. Polymer Engineering & Science **9**(4):295–310

[7] Fliegener S, Rausch J, Hohe J (2019) Loading points for industrial scale sandwich structures – A numerical and experimental design study. Composite Structures **226**:111,278
[8] Fliegener S, Hohe J (2020) An anisotropic creep model for continuously and discontinuously fiber reinforced thermoplastics. Composites Science and Technology **194**:108,168
[9] Fliegener S, Hohe J, Gumbsch P (2016) The creep behavior of long fiber reinforced thermoplastics examined by microstructural simulations. Composites Science and Technology **131**:1–11
[10] Haj-Ali RM, Muliana AH (2004) Numerical finite element formulation of the Schapery non-linear viscoelastic material model. International Journal for Numerical Methods in Engineering **59**(1):25–45
[11] N N (2003) ISO 3167: Plastics – multipurpose test specimens. Beuth, Berlin
[12] N N (2010) ISO 527-5: Plastics – determination of tensile properties – Part V: Test conditions for unidirectional fibre-reinforced plastic composites. Beuth, Berlin
[13] N N (2003) ISO 899-1: Plastics – determination of the creep behaviour – Part I: Tensile creep. Beuth, Berlin
[14] Fliegener S, Luke M, Gumbsch P (2014) 3D microstructure modeling of long fiber reinforced thermoplastics. Composites Science and Technology **104**:136–145

Chapter 12
Time-Swelling Superposition Principle for the Linear Viscoelastic Properties of Polyacrylamide Hydrogels

Seishiro Matsubara, Akira Takashima, So Nagashima, Shohei Ida, Hiro Tanaka, Makoto Uchida, and Dai Okumura

Abstract We study the linear viscoelastic properties of polyacrylamide hydrogels over a wide range of swelling and deswelling states. The experimental data of dynamic moduli demonstrate that deswollen hydrogels exhibit the linear viscoelastic behaviors while swollen hydrogels behave as a purely elastic material. To capture the linear viscoelastic nature of deswollen hydrogels, we advocate the time-swelling superposition principle inspired by the swelling-dependent linear viscoelasticity model. In this principle, the Williams-Landel-Ferry type equation and the scaling law prescribe the horizontal and vertical shift factors as a function of the volume swelling ratio, making it possible to organize the dynamic moduli in the deswelling state. The resulting master curves elucidate that the dynamic moduli have a positive power-law correlation with the angular frequency. In addition, two shift factors reveal that deswelling enhances both frequency sensitivity and elastic property in dynamic moduli. The scaling exponents of the elastic moduli in the deswelling state are, in

---

Seishiro Matsubara · Akira Takashima · So Nagashima and Dai Okumura
Department of Mechanical Systems Engineering, Nagoya University
Furo-cho, Chikusa-ku, Nagoya 464-8603, Japan,
e-mail: seishiro.matsubara@mae.nagoya-u.ac.jp, ,
takashima.akira@j.mbox.nagoya-u.ac.jp, so.nagashima@mae.nagoya-u.ac.jp,
dai.okumura@mae.nagoya-u.ac.jp

Shohei Ida
Faculty of Engineering, The University of Shiga Prefecture, 2500, Hassaka-cho, Hikone-City, Shiga, 522-8533, Japan,
e-mail: ida.s@mat.usp.ac.jp

Hiro Tanaka
Department of Mechanical Engineering, Osaka University, 2-1 Yamadaoka, Suita, Osaka, 565-0871, Japan,
e-mail: htanaka@mech.eng.osaka-u.ac.jp

Makoto Uchida
Department of Mechanical Engineering, Osaka Metropolitan University 3-3-138, Sugimoto, Sumiyoshi-ku, Osaka, 558-8585, Japan,
e-mail: uchida@osaka-cu.ac.jp

particular, larger than those in the swelling state, regardless of the contents of monomer and cross-linker. The generalized Maxwell model equipped with identified material parameters quantifies the linear viscoelastic behaviors that depend on frequency and water content.

12.1 Introduction

Hydrogels have attracted attention in biomedical and tissue engineering, such as, in particular, the reproduction of biological organs, thanks to their swellability, flexibility, and biocompatibility [1, 2]. Their swelling and deswelling-induced instabilities provide some clues to elucidate the system of morphology formation for living organisms, which plays a crucial role in reproducing high-quality artificial organs[3]. The rheology of hydrogels is also of interest to researchers who try to mimic organs [4–6] because the biological tissues constituting living organisms exhibit viscoelastic behavior [7–9]. Therefore, quantifying the relationship between water content and viscoelastic properties in hydrogels promotes a better understanding as to the morphogenesis of organs, leading to the creation of realistic biomedical products.

The common perception concerning the mechanics of swollen and deswollen hydrogels is that they exhibit swelling-dependent elastic behavior [10–12]. This is often characterized by the scaling theory [13] that relates the volume swelling ratio to the elastic modulus using the scaling exponent. Indeed, the swelling-dependent elastic behavior for various hydrogels has been extensively studied by means of evaluating their scaling exponents [14–17]. In this context, the effects of monomer and cross-linker on the scaling exponents of swollen hydrogels were comprehensively examined by Kawai et al. [18], who measured the complex shear moduli of polyacrylamide (PAAm) hydrogels in the as-prepared (AP) and equilibrium swelling (ES) states. On the other hand, the elastic modulus in the deswelling state has a higher sensitivity to the volume swelling ratio than that in the swelling state. Hence, researchers established another scaling law to determine the elastic modulus in the deswelling state, accomplishing good predictions for PAAm hydrogel [19] and Tetra-PEG gel [20, 21].

However, to our best knowledge, less experimental data have been provided enough to clarify the mechanical behaviors of deswollen hydrogels, so the possibility that viscoelastic characteristics appear in deswollen hydrogels has to be taken into account. In this regard, some studies, which focused on the gelation kinetics, discussed the viscous property of hydrogels in the AP state within linear viscoelasticity [22–27]. In summary, dynamic moduli have the power-law correlation with the angular frequency, but the contribution of viscosity diminishes with increasing the contents of monomer and cross-linker. Du and Hill [27] introduced the horizontal and vertical shift factors as functions of temperature and cross-linker, constructing the master curves of dynamic moduli for weakly cross-linked PAAm hydrogels. They also successfully quantified the linear viscoelastic behavior using the generalized Maxwell model.

Their contribution suggests that the linear viscoelastic nature of hydrogels emerges under some environmental conditions.

We study the linear viscoelastic properties of PAAm hydrogels over a wide range of swelling and deswelling states. The outline of this paper is as follows. Section 12.2 presents a series of experimental procedures, including the preparation of specimens, transient equilibrium swelling using ethanol, and dynamic mechanical analysis (DMA). Section 12.3 shows the experimental data for complex shear modulus and loss tangent as a function of the angular frequency after some discussion concerning transient equilibrium swelling. The experimental observation reveals that deswollen hydrogels exhibit linear viscoelastic behavior while swollen hydrogels behave as a purely elastic material. Section 12.4 is devoted to establishing the framework for analyzing the linear viscoelastic behavior of deswollen hydrogels. We first formulate the swelling-dependent linear viscoelasticity model, bringing out the swelling dependences of elastic modulus and relaxation time. After the advocation of the time-swelling superposition principle, their swelling dependencies are characterized using the horizontal and vertical shift factors, whose function forms are prescribed by the Williams-Landel-Ferry (WLF) type equation and the scaling law, respectively. Section 12.5 shows that the dynamic moduli of deswollen hydrogels are successfully organized using the present framework. The master curves illustrate that deswollen hydrogels exhibit the linear viscoelastic behavior following the power-law correlation between dynamic moduli and angular frequency. Owing to the power-law correlation, increasing frequency boosts the contribution of viscosity in the linear viscoelastic behavior. In contrast, increasing the contents of monomer and cross-linker weakens the frequency dependence of the dynamic moduli. In addition, the two shift factors demonstrate that deswelling enhances both frequency sensitivity and elastic property of the dynamic moduli. The scaling exponents of the elastic moduli in the deswelling state are, in particular, larger than those in the swelling state, regardless of the contents of monomer and cross-linker. Further, the quantification of linear viscoelastic nature is carried out using the generalized Maxwell model, whose material parameters are determined through the curve-fitting to the master curves. The computed complex shear moduli provide the information concerning the swelling-dependent linear viscoelasticity of hydrogels. Section 12.6 summarizes our findings and contributions.

12.2 Experiment

12.2.1 Materials

Chemically cross-linked PAAm hydrogels were synthesized through free radical polymerization using acrylamide (AAm) as monomer, N,N'-methylenebisacrylamide (BIS) as cross-linker, ammonium persulfate (APS) as polymerization initiator, and N,N,N',N'-tetramethylethylenediamine (TMEDA) as polymerization accelerator. Table 12.1 shows the constituents of hydrogels. Six types of compounds were prepared

Table 12.1: Constituents of PAAm hydrogels.

Constituents	Molar concentration [mM]
AAm	1000, 3000, 6000
BIS	2, 10
APS	5
TMEDA	10

to examine the effects of monomer and cross-linker on the material properties of hydrogels. Hereinafter, the molar concentrations of AAm and BIS in hydrogels will be denoted by X (mM) and Y (mM), respectively.

12.2.2 Mixed Solvents for Transient Equilibrium Swelling

Transient equilibrium swelling is attained using PAAm hydrogels that have different water contents in the equilibrium state, as shown in Fig. 12.1, and enables us to examine the swelling dependence of material properties for hydrogels in the final steady state of solvent diffusion. Consequently, we do not consider any mechanical and diffusion behaviors in non-equilibrium states. To accomplish transient equilibrium swelling, mixed solvents comprising water and ethanol were prepared. Since ethanol is miscible with water but a poor solvent for PAAm, it promotes dehydration and thus controls the water content of a hydrogel in the equilibrium state [28]. In this study, ten types of mixed solvents with different ethanol concentrations were prepared, as shown in Table 12.2.

Hereinafter, the ethanol concentration in a mixed solvent will be denoted by Z (%). As an exception, the hydrogel swollen solely by water ($Z = 0\%$) will be denoted by

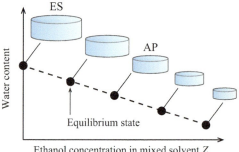

Fig. 12.1 Schematic diagram of transient equilibrium swelling. Ethanol in a mixed solvent controls the water content of a hydrogel in the equilibrium state [28]. The hydrogel swollen solely by water is denoted by ES.

Table 12.2: Ethanol concentration in mixed solvents.

Concentration [%]	0, 10, 20, 30, 40, 45, 50, 55, 60, 65

"ES". In Appendix A, we verify the validity of the present method to realize transient equilibrium swelling.

12.2.3 Measurement of Swelling

To prepare the specimens in the AP state (AP specimen), AAm and BIS were polymerized using APS and TMEDA in deionized water after the bubbling treatment using nitrogen gas, which eliminates oxygen dissolved in the aqueous solution. The polymerization liquid was poured into the silicone rubber ring (diameter, 50 mm; thickness, 1 mm) on an acrylic plate and covered with another acrylic plate along with weight. The molded precursor was left for half a day until the cure reaction was completed. Subsequently, the prepared AP specimens were placed in a flat-bottomed glass container, and the mixed solvent of 100 mL was poured over them to observe their swelling behavior. The specimens reached the specific equilibrium state depending on ethanol concentration after storage for at least one week at room temperature ($\approx 25\,°C$). The experimental data were obtained as the average of measured values for specimens whose number was three for $Z \geq 40\%$ and five for $Z \leq 30\%$ for each compound.

When the incompressibility of a hydrogel is assumed, the volume swelling ratio, J, should be obtained from the following relationship between ideally non-swollen and swollen hydrogels

$$J = \frac{V}{V_{\text{ID}}} = 1 + \left(\frac{w}{w_{\text{ID}}} - 1\right)\frac{\rho_{\text{AAm}}}{\rho_{\text{water}}}, \tag{12.1}$$

where V and w are the volume and mass of a hydrogel, respectively, and subscript ID stands for the state of ideal non-swelling. The mass densities of AAm and water are $\rho_{\text{AAm}} = 1.13\,\text{g}\cdot\text{cm}^{-3}$ and $\rho_{\text{water}} = 1.00\,\text{g}\cdot\text{cm}^{-3}$, respectively. It should be noted that the infiltration of ethanol into a hydrogel was ignored in Eq. (12.1). To verify the validity of this assumption, we confirmed that there is only a slight difference in viscoelastic properties between the AP specimen and the imitated AP specimen, whose water content was adjusted to that of the AP specimen; see Appendix A for details. Also, although w_{ID} cannot be directly measured from the specimen, the mass ratio of the AP specimen to AAm is a known value as $w_{\text{AP}}/w_{\text{ID}}$. When the ratio is expressed as

$$c = \frac{w_{\text{AP}}}{w_{\text{ID}}}, \tag{12.2}$$

the substitution of Eq. (12.2) into Eq. (12.1) yields

$$J = 1 + \left(c\frac{w}{w_{\text{AP}}} - 1\right)\frac{\rho_{\text{AAm}}}{\rho_{\text{water}}}. \tag{12.3}$$

Equation (12.3) enables us to evaluate J using the weight of a hydrogel in the equilibrium state.

12.2.4 Measurement of Dynamic Moduli

DMA was conducted on freely swollen specimens. The storage shear modulus, G', and the loss shear modulus, G'', were measured in torsional modal testing using Discovery HR-1 rotational shear rheometer (TA Instruments, Inc.). To prevent the slippage of the specimen, the water droplets left on the specimen were wiped out using the waste, and the compressive load was applied to the upper surface of the specimen. Also, to avoid the vaporization of water in the specimen, we covered the specimen with the solvent trap. Further, since the specimens comprising $(X, Y) = (1000, 2)$mM and $Z \leq 30\%$ were too soft in DMA, they were fixed using a jig made of aluminum (diameter, 40 mm). On the other hand, a jig made of steel (diameter, 20 mm) was employed for the remaining specimens. Torsional strain oscillations with angular frequencies ranging from 0.1 to 100 rad/s were applied to the specimens at room temperature. The strain amplitude was set to 0.5% such that G' and G'' fall within the proportional region. This was confirmed beforehand by conducting DMA, in which the angular frequency of torsional strain oscillation was set to 1.0 Hz after the fashion of Kawai et al. [18].

12.3 Experimenta Results

12.3.1 Transient Equilibrium Swelling

Transient equilibrium swelling of PAAm hydrogels can be seen in Fig. 12.2, where the relationships between the volume swelling ratio, J, and ethanol concentration, Z, are plotted for each compound, (X, Y). The hydrogel with lower contents of monomer and cross-linker has a larger J in the ES state. As ethanol concentration increases, J for all compounds proportionally decrease up to around $Z = 45\%$ and eventually reach almost the same value; see the approximated lines in Fig. 12.2. Accordingly, the slope, J/Z, is gentler with increasing the contents of monomer and cross-linker. The measured results until $Z = 45\%$ imply that tuning ethanol concentration and contents of monomer and cross-linker enables us to prepare a hydrogel in the desired swelling state. In contrast, once ethanol concentration exceeds 45%, J for all compounds almost sustain constant; indeed, $J/Z = -0.07$ in $50 \leq Z \leq 65\%$, whereas $J/Z = -0.18$ in $0 \leq Z \leq 45\%$ for $(X, Y) = (6000, 10)$mM. This is because hydrogels reached a semi-dry state after sufficient dehydration.

The results for the imitated AP specimens are plotted as filled yellow markers in Fig. 12.2. Note here that they were regarded as the results for the AP specimens whose ethanol concentration is unknown. Indeed, the difference in J between AP specimen and imitated AP specimen is trivial for all compounds; see Appendix A. The ethanol concentration of the imitated AP specimen decreases with increasing the content of cross-linker or decreasing the content of monomer. Also, the corresponding J is independent of the content of cross-linker but takes a smaller value with increasing

12 Swelling-Depemcent Linear Viscoelasticity of PAAm Hydrogels

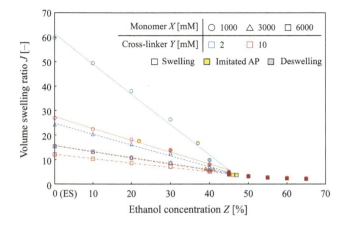

Fig. 12.2: Relationship between volume swelling ratio, J, and ethanol concentration, Z, for each compound (X, Y). J was yielded by substituting the weight swelling ratio, w/w_{AP}, shown in Fig. 12.3, into Eq. (12.3). The linear approximation lines are depicted in the range of $0 \leq Z \leq 45\%$. The hollow and filled markers indicate the results in the swelling and deswelling states, respectively. The plots for imitated AP specimens (filled yellow markers) are regarded as those in the AP state.

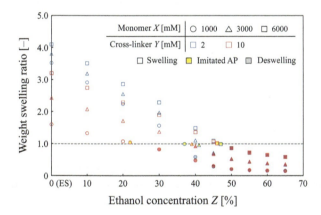

Fig. 12.3: Weight swelling ratio, w/w_{AP}, as a function of ethanol concentration, Z, for each compound, (X, Y). The hollow and filled markers denote the results in the swelling and deswelling states, respectively. Also, the plots for imitated AP specimens (filled yellow markers) are regarded as those at the AP state because Z is unknown for the AP specimen. w/w_{AP} proportionally decreases up to $Z = 45\%$. In contrast, once Z exceeds 45%, w/w_{AP} almost saturates with the specific value depending on X but independent of Y.

the content of monomer, so that the reduction of J by deswelling from the AP state increases with decreasing the content of monomer; for example, the differences in J between the AP state and $Z = 65\%$ are 15.4 for $(X,Y) = (1000, 10)$mM and 1.61 for $(X,Y) = (6000, 10)$mM. Hereinafter, the equilibrium states to which a hydrogel reaches via swelling and deswelling from the AP state will be referred to as swelling and deswelling states, respectively.

12.3.2 Linear Viscoelastic Behavior

The absolute value of complex shear modulus, $\bar{G}$, and the loss tangent, $\tan\delta$, are respectively defined as

$$\bar{G} = \sqrt{(G')^2 + (G'')^2}, \quad \tan\delta = \frac{G''}{G'} \tag{12.4}$$

using the measured dynamic moduli, G' and G'', shown in Figs. 12.4 and 12.5, to examine the linear viscoelastic behavior of a PAAm hydrogel.

The resulting $\bar{G}$ and $\tan\delta$ as a function of the angular frequency, ω, for each compound, (X,Y), and ethanol concentration, Z, are plotted in Figs. 12.6 and 12.7, respectively.

As can be seen from Fig. 12.6, $\bar{G}$ in the swelling state is almost frequency independent regardless of compound and increases with increasing the contents of monomer and cross-linker. Our experimental results are consistent with the results provided in many previous studies [10–12, 18] that claim PAAm hydrogels exhibit purely elastic behavior in the swelling state.

Although Fig. 12.7 also illustrates that hydrogels with $Y = 2$ mM can be regarded as purely elastic material in the swelling state due to their negligible $\tan\delta$, particular attention has to be paid to hydrogels with $Y = 10$ mM. In the swelling state, most $\tan\delta$ for hydrogels with $Y = 10$ mM have local maxima at $\omega \approx 2.15$ rad/s, independent of the swelling state and the content of monomer. Calvet et al. [23] also reported such a behavior for the high cross-linked hydrogel. Except for some results for $X = 3000$ mM, the local maximum increases when the content of monomer decreases, or the equilibrium state approaches to the ES state. These results might imply the existence of the relaxation process, where the thermal activities of monomeric units and molecular chains are diversified. Nonetheless, we presume that hydrogels with $Y = 10$mM exhibit purely elastic behavior in the swelling state because the corresponding $\bar{G}$ remains almost constant.

On the other hand, $\bar{G}$ and $\tan\delta$ shown in Figs. 12.6 and 12.7 imply that deswollen hydrogels need to be treated as a viscoelastic material. $\bar{G}$ for all compounds increases in response to deswelling. The sensitivity is lower with increasing the contents of monomer and cross-linker but built up with increasing the angular frequency. The appearance of viscoelastic nature in a hydrogel is more evident in $\tan\delta$, which takes a larger value as either angular frequency or degree of deswelling increases. For example, $\tan\delta$ for the hydrogel with $(X,Y) = (3000, 10)$mM and $Z = 65\%$ drastically increases

12 Swelling-Depemcent Linear Viscoelasticity of PAAm Hydrogels

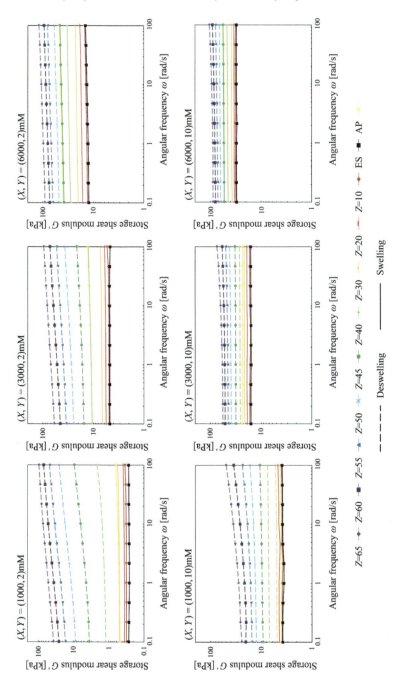

Fig. 12.4: Storage shear modulus, G', as a function of the angular frequency, ω, for each compound, (X, Y), and ethanol concentration, Z. The solid and dashed lines are the linear interpolation approximation of the measured results in the swelling and AP states and in the deswelling state, respectively. G' is almost constant in the swelling state, so that PAAm hydrogels exhibit purely elastic behavior. In contrast, deswelling develops the rate dependency of G'.

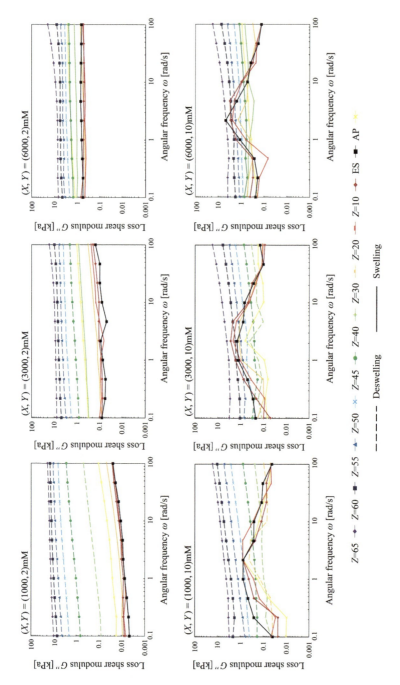

Fig. 12.5: Loss shear modulus, G'', as a function of the angular frequency, ω, for each compound, (X, Y), and ethanol concentration, Z. The solid and dashed lines are the linear interpolation approximation of the measured results in the swelling and AP states and in the deswelling state, respectively. The local maxima value of G'' are observed for $Y = 10$ mM in the swelling state, while G'' for $Y = 2$ mM is almost frequency independent regardless of X. Similar to G' in Fig. 12.4, deswelling also develops the frequency dependence of G''.

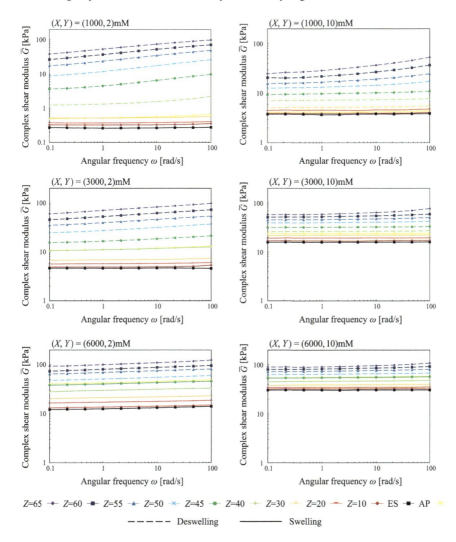

Fig. 12.6: Absolute value of complex shear modulus, $\bar{G}$, as a function of angular frequency, ω, for each compound, (X,Y), and ethanol concentration, Z. $\bar{G}$ is yielded by substituting the dynamic moduli, G' and G'', shown in Figs. 12.4, 12.5, into Eq. (12.4)$_1$. The solid and dashed lines represent the linear interpolation approximation of the measured results.

in the range of $\omega > 20$ rad/s. Also, $\tan\delta$ for the hydrogel with $(X,Y) = (1000, 2)$ mM in the range of $Z \geq 45\%$ have local maxima of $0.28 - 0.3$, and the corresponding angular frequency decreases in response to deswelling, i.e., the local maximum point shifts to the left.

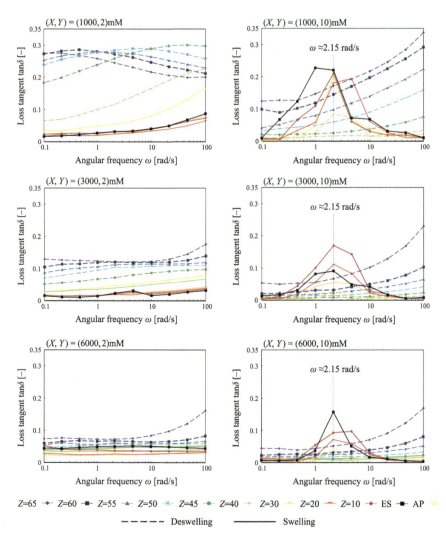

Fig. 12.7: Loss tangent, $\tan\delta$, as a function of angular frequency, ω, for each compound, (X, Y), and ethanol concentration, Z. $\tan\delta$ is yielded by substituting the dynamic moduli, G' and G'', shown in Figs. 12.4, 12.5, into Eq. (12.4)$_2$. The solid and dashed lines represent the linear interpolation approximation of the measured results.

12.4 Swelling–Dependent Linear Viscoelasticity

As mentioned earlier section, PAAm hydrogels exhibit either purely elastic or linear viscoelastic behavior in response to the equilibrium state. This section is devoted to the establishment of the framework to analyze the swelling-dependent linear viscoelastic behavior of a hydrogel, including model formulation and advocation of time-swelling superposition principle.

12.4.1 Model Formulation

The swelling-dependent linear viscoelasticity model is formulated after the fashion of Reese and Govindjee [29]. The generalized Maxwell model, assembling $N + 1$ Maxwell elements in parallel (Fig. 12.8), is introduced to quantify the linear viscoelastic characteristics in a wide range of angular frequency.

As the torsional modal testing is conducted on freely swollen hydrogels as described in Subsect. 12.2.4, we begin with introducing the volumetric-isochoric decomposition of the deformation gradient, F, as follows:

$$F = J^{\frac{1}{3}}\bar{F}, \qquad (12.5)$$

where $\bar{F}$ is the isochoric part of F, corresponding to the torsional strain applied to a freely swollen hydrogel in DMA. On the other hand, for α-th Maxwell element ($\alpha = 0, ..., N$), the decomposition of F into elastic and viscous parts is defined as

$$F = F_\alpha^e F_\alpha^v, \qquad (12.6)$$

where F_α^e and F_α^v are the elastic and viscous deformation gradients of α-th Maxwell element, respectively. Here, $\alpha = 0$ corresponds to the purely elastic element such that $F_0^v = \mathbf{1}$ and $F_0^e = F$, where $\mathbf{1}$ is the 2nd-order identity tensor. Also, we assume F_α^v to be incompressible, namely, $\det(F_\alpha^v) = 1$, because the viscous deformation is supposed to be considerably small during free swelling. Thus, F_α^e and F_α^v can be obtained as

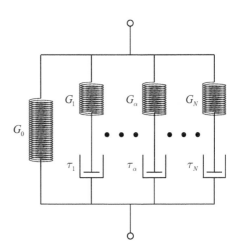

Fig. 12.8 Generalized Maxwell model to quantify the swelling-dependent linear viscoelastic behavior of a PAAm hydrogel.

$$F^e_\alpha = J^{\frac{1}{3}} \bar{F}^e_\alpha, \quad F^v_\alpha = \bar{F}^v_\alpha, \tag{12.7}$$

where $\bar{F}^e_\alpha$ and $\bar{F}^v_\alpha$ are the isochoric parts of the elastic and viscous deformation gradients of α-th Maxwell element, rerspectively, satisfying

$$\bar{F} = \bar{F}^e_\alpha \bar{F}^v_\alpha.$$

Under the assumption of viscoelastic isotropy, the strain energy of a hydrogel is defined as

$$W = \sum_{\alpha=0}^{N} W^e_\alpha (I^e_\alpha, J), \tag{12.8}$$

where W^e_α and I^e_α are the elastic strain energy of α-th Maxwell element and the first invariant of the elastic Finger tensor

$$b^e_\alpha = F^e_\alpha (F^e_\alpha)^T,$$

respectively. Here, the present study does not discuss the diffusion properties of hydrogels. From the hyperelastic constitutive law in the current configuration, the Cauchy stress is obtained as follows:

$$\sigma = \sum_{\alpha=0}^{N} \frac{\partial W^e_\alpha}{\partial J} \mathbf{1} + \frac{2}{J} \sum_{\alpha=0}^{N} \frac{\partial W^e_\alpha}{\partial I^e_\alpha} b^e_\alpha \tag{12.9}$$

On the other hand, adhering to the thermodynamic requirement, we define the law of viscosity associated with α-th Maxwell element as

$$-(b^e_\alpha)^{-1} \mathcal{L}(b^e_\alpha) = \frac{1}{\eta_\alpha} s_\alpha, \tag{12.10}$$

where

$$\mathcal{L}(b^e_\alpha) = F \overline{\left(F^{-1} b^e_\alpha F^{-T}\right)} F^T, \; \eta_\alpha,$$

and s_α are the Lie derivative of b^e_α, the viscous coefficient of α-th Maxwell element, and the deviatoric Cauchy stress of α-th Maxwell element, respectively. Here, $\dot{A}$ denotes the material time derivative of A with respect to time $t \geq t_s$, where t_s is the end time of free swelling, i.e., the start time of DMA. Note that, when

$$\eta_\alpha = \eta_{\alpha c} J^{-1}$$

with constant $\eta_{\alpha c}$, Eq. (12.10) corresponds to Newton's law of viscosity in finite strain theory. Also, $\eta_0 = +\infty$ because $\alpha = 0$ corresponds to the purely elastic element.

As the amplitude of the torsional strain in DMA is sufficiently small, $\bar{F}$ can be approximately identical to the identity tensor. Thus, Eq. (12.9) can be linearized around the isochoric elastic Finger tensor

$$\bar{b}^e_\alpha = \bar{F}^e_\alpha \left(\bar{F}^e_\alpha\right)^T = \mathbf{1}$$

as follows:

$$\sigma = \sigma^{\text{vol}}\mathbf{1} + \sum_{\alpha=0}^{N}\left(2J^{\frac{1}{3}}\left[\frac{\partial^2 W_\alpha^e}{\partial I_\alpha^e \partial I_\alpha^e}\right]_{\bar{b}_\alpha^e=1} + J^{\frac{2}{3}}\left[\frac{\partial^2 W_\alpha^e}{\partial J \partial I_\alpha^e}\right]_{\bar{b}_\alpha^e=1}\right)\left(\text{tr}\left(\bar{b}_\alpha^e\right)-3\right)\mathbf{1}$$

$$+ \sum_{\alpha=0}^{N} G_\alpha \left(\bar{b}_\alpha^e - \mathbf{1}\right) \quad (12.11)$$

in which subscript $\bar{b}_\alpha^e = \mathbf{1}$ means that $\bar{b}_\alpha^e$ involved in each term is the identity tensor. Here, G_α and σ^{vol} are the shear elastic modulus of α-th Maxwell element and the mean Cauchy stress, respectively, defined as

$$G_\alpha = 2J^{-\frac{1}{3}}\left[\frac{\partial W_\alpha^e}{\partial I_\alpha^e}\right]_{\bar{b}_\alpha^e=1}, \quad (12.12)$$

$$\sigma^{\text{vol}} = \sum_{\alpha=0}^{N}\left(G_\alpha + \left[\frac{\partial W_\alpha^e}{\partial J}\right]_{\bar{b}_\alpha^e=1}\right). \quad (12.13)$$

Similarly, Eq. (12.10) is also linearized around $\bar{b}_\alpha^e = \mathbf{1}$, as follows:

$$\dot{C}_\alpha^v = \frac{1}{\tau_\alpha}\left[\bar{C} - \frac{\text{tr}\left(\bar{b}_\alpha^e\right)}{3}C_\alpha^v\right], \quad (12.14)$$

where C_α^v and $\bar{C}$ are the right Cauchy-Green tensors with respect to F_α^v and $\bar{F}$, respectively. Also, τ_α is the relaxation time of α-th Maxwell element defined as

$$\tau_\alpha = \frac{\eta_\alpha}{G_\alpha}. \quad (12.15)$$

When linearizing $\bar{C}$, $\bar{b}_\alpha^e$, and C_α^v in Eqs. (12.11) and (12.14), we can obtain the following equations:

$$\sigma = \sigma^{\text{vol}}\mathbf{1} + \sum_{\alpha=0}^{N} 2G_\alpha e_\alpha^e, \quad \dot{e}_\alpha^v = \frac{1}{\tau_\alpha}e_\alpha^e, \quad (12.16)$$

where e_α^e and e_α^v are the elastic and viscous parts of the deviatoric small strain tensor, e, with respect to the α-th Maxwell element, respectively, i.e.,

$$e = e_\alpha^e + e_\alpha^v.$$

Note that the reference configuration of Eq. (12.16) is the freely swollen state, so that σ^{vol}, G_α, and τ_α depend on J.

To evaluate the linear viscoelastic properties of a freely swollen hydrogel, we assume that the deviatoric Cauchy stress,

$$s = s_0 \exp(i\omega t + \delta),$$

with the constant amplitude, s_0, and the angular frequency, ω, emerges against the deviatoric small strain

$$e = e_0 \exp(i\omega t)$$

with the constant amplitude, e_0. Then, Eq. (12.16) yields the following relationship between s and e.

$$s = 2(G' + iG'')e, \quad (12.17)$$

Here, the storage and loss shear moduli, G' and G'', are defined as

$$G' = G_0 + \sum_{\alpha=1}^{N} \frac{G_\alpha \omega^2 \tau_\alpha^2}{\omega^2 \tau_\alpha^2 + 1}, \quad G'' = \sum_{\alpha=1}^{N} \frac{G_\alpha \omega \tau_\alpha}{\omega^2 \tau_\alpha^2 + 1}, \quad (12.18)$$

respectively, where G_0 is the shear elastic modulus of purely elastic element ($\alpha = 0$). Note here that $\dot{J} = 0$ during DMA because hydrogels have already reached the equilibrium state at $t = t_s$. The substitution of Eq. (12.4) into Eq. (12.17) also yields

$$s = 2\bar{G} e_0 \exp\{i(\omega t + \delta)\}, \quad (12.19)$$

where δ is the phase difference between s and e related to $\tan\delta$.

12.4.2 Time-Swelling Superpostion Principle

The relaxation time, τ_α, and the elastic modulus, G_α, depend on the volume swelling ratio, J, characterizing the swelling-dependent linear viscoelastic behavior of a PAAm hydrogel. Taking advantage of the swelling dependences of τ_α and G_α, we advocate the time-swelling superposition principle to construct the master curves of the dynamic moduli, $G^* = (G', G'')$, the way of which is described in what follows. Under the assumption of rheological simplicity with respect to J, we relate the linear viscoelastic properties in the current swelling state to those in the reference swelling state, J_R, as follows:

$$\tau_\alpha(J) = \alpha_T(J, J_R) \tau_\alpha^R, \quad (12.20)$$

$$G_\alpha(J) = \beta_T(J, J_R) G_\alpha^R, \quad (12.21)$$

where τ_α^R and G_α^R are the reference relaxation time and the reference shear elastic modulus associated with α-th Maxwell element, respectively. Also, α_T and β_T are the horizontal and vertical shift factors, respectively. Then, the substitutions of Eqs. (12.20) and (12.21) into Eq. (12.18) yield the following relation with respect to G^*.

$$\beta_T G^*(\omega, J_R) = G^*(\alpha_T \omega, J) \quad (12.22)$$

As shown in Fig. 12.9, α_T and β_T shift G^* measured in each equilibrium state to the horizontal and vertical directions such that experimental data are superposed, constructing the master curves of G^* in the reference swelling state.

The WLF equation relates the thermal expansion relevant to the free volume change in the molecular network structure to the viscous property and is often introduced to a rubber material satisfying the thermo-rheological simplicity. Assuming that swelling and deswelling also induce the free volume change in a hydrogel, we define the following function form for α_T.

$$\log_{10} \alpha_T = \frac{-C_1 (J - J_R)}{C_2 + (J - J_R)}, \tag{12.23}$$

where C_1 and C_2 are the positive parameters to be identified. This equation characterizes the effect of J on the rate-dependence for the linear viscoelastic behavior of a hydrogel. Note that α_T only functions in the deswelling state because a PAAm hydrogel exhibit purely elastic behavior in the swelling state.

On the other hand, as β_T has to be directly linked to the scaling law of shear elastic modulus [13], we introduce the following elastic strain energy associated with α-th Maxwell element.

$$W_\alpha^e = \frac{E_d^\alpha}{6} J^m \left(I_\alpha^e - 3J^{\frac{2}{3}} \right), \tag{12.24}$$

where E_d^α and m are the elastic modulus of α-th Maxwell element in the ideally dry state and the exponent, respectively. Then, the substitution of Eq. (12.24) into Eq. (12.12) yields the scaling law associated with G_α.

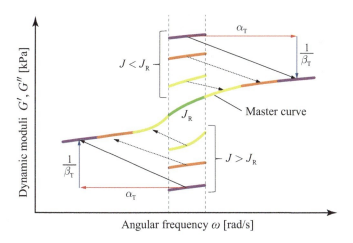

Fig. 12.9: Schematic diagram for construction of master curves. The horizontal and vertical shift factors, α_T and β_T, shift dynamic moduli to the directions of angular frequency (horizontal) and stiffness (vertical), respectively, so that the master curves of dynamic moduli in the reference swelling state, J_R, are constructed.

$$G_\alpha = \frac{E_d^\alpha}{3} J^{m-\frac{1}{3}} = \frac{E_d^\alpha}{3} J^{-\nu}, \tag{12.25}$$

where ν is the scaling exponent that has the same value for every Maxwell element. Here, Li et al. [19] demonstrated that the swelling sensitivity of the elastic modulus differs between swelling and deswelling states. Such a difference can be considered by the following function forms using positive parameters, $E_{dvs}^\alpha, E_{dvd}^\alpha, a_{vs}, a_{vd}$.

$$E_d^\alpha = \begin{cases} E_{dvs}^\alpha & \text{if } J \geq J_{AP} \\ E_{dvd}^\alpha & \text{if } J < J_{AP} \end{cases} \text{ and } \nu = \begin{cases} a_{vs} & \text{if } J \geq J_{AP} \\ a_{vd} & \text{if } J < J_{AP} \end{cases}, \tag{12.26}$$

where J_{AP} is the volume swelling ratio in the AP state. Also, our experimental results imply that the swelling sensitivity of the elastic modulus changes on the way to deswelling in the case of hydrogels with low contents of monomer and cross-linker. This can be reflected using the following function forms for E_{dvd}^α and a_{vd}.

$$E_{dvd}^\alpha = \begin{cases} E_{dvd1}^\alpha & \text{if } J_R \leq J < J_{AP} \\ E_{dvd2}^\alpha & \text{if } J \leq J_R \end{cases} \text{ and } a_{vd} = \begin{cases} b_{vd}^1 & \text{if } J_R \leq J < J_{AP} \\ b_{vd}^2 & \text{if } J \leq J_R \end{cases}, \tag{12.27}$$

in which $E_{dvd1}^\alpha = E_{dvd2}^\alpha$ and $b_{vd}^1 = b_{vd}^2$ at $J = J_R$. Further, since the reference swelling state should be set to the deswelling state where the linear viscoelastic behavior emerges, the corresponding shear elastic modulus is obtained as

$$G_\alpha^R = \left(E_{dvd}^\alpha/3\right) J_R^{-a_{vd}}$$

from Eq. (12.25). As a consequence, the following function form for β_T is obtained using Eqs. (12.21) and (12.25).

$$\log_{10}\beta_T = -\nu \log_{10}\left(\frac{J}{J_R}\right) + c, \tag{12.28}$$

where the constant

$$c = (a_{vd} - \nu)\log_{10}J_R + \log_{10}\left(E_d^\alpha/E_{dvd}^\alpha\right)$$

is assumed to be the same for every Maxwell element, being zero in the deswelling state but a parameter in the swelling state. Thus, the requisite parameters for β_T are the scaling exponents, $a_{vs}, b_{vd}^1, b_{vd}^2$, and the constant, c.

12.5 Discussion

12.5.1 Master Curves of Dynamic Moduli

For each compound (X,Y), the dynamic moduli, G' and G'', in the deswelling state were organized using two shift factors, α_T and β_T, introduced in the previous section. The reference volume swelling ratio, J_R, was set to the volume swelling ratio, J, at ethanol concentration, $Z = 50\%$, having almost the same value for any compound; see Fig. 12.2. Also, the experimental data within the ranges of $Z \leq 30\%$ for $(X,Y) = (1000, 10)$mM and $Z \leq 40\%$ for $(X,Y) = (3000, 10)$mM were omitted here because the undulation of G'' observed in Fig. 12.5 interferes with the construction of master curves. This is justified on the ground that the corresponding G' is almost frequency independent and indicates a purely elastic behavior of a PAAm hydrogel. The parameters associated with the vertical shift factor, β_T, were identified in constructing the master curves of G', G'' after those associated with the horizontal shift factor, α_T, were determined in constructing the master curve of $\tan\delta$ irrelevant to β_T.

Figure 12.10 shows the master curves of G' and G'' in the deswelling state as a function of the angular frequency, ω, for each compound, (X,Y). Although the sophisticated identification method of relative parameters needs to be established, two shift factors, α_T and β_T, successfully organize the dynamic moduli in the deswelling state, providing several findings concerning the swelling dependence on the linear viscoelastic properties of PAAm hydrogels. We emphasize that the combination of the WLF type equation in Eq. (12.23) and the scaling law in Eq. (12.28) is available for capturing the swelling-dependent linear viscoelastic behavior of a PAAm hydrogel.

We found that G' and G'' in the deswelling state have the power-law correlation with ω regardless of compounds. The power-law correlation is well-known as the characteristic linear viscoelastic nature near the sol-gel transition [27, 30–32] and indicates that the physical entanglements distributed in the imperfect network structure contribute to the mechanical behavior of deswollen hydrogel. The power of G' increases with decreasing the contents of monomer and cross-linker. This may be because the lack of monomer and cross-linker triggers the formation of an imperfect network structure in a hydrogel. Here, the power of G' for any compound is less than half that reported in previous studies [27, 30]. This can be easily understood in terms that PAAm hydrogels examined in this study are solid polymers.

The reference shear elastic modulus, G_0^R, in Fig. 12.10 is brought by shifting the storage shear modulus, G', in the swelling state (i.e., the shear elastic modulus in the swelling state) such that the master curve of G' is extended toward the low-frequency range. As we expected, G_0^R takes a larger value for the hydrogel with higher contents of monomer and cross-linker. Focusing on the master curves of G' in the vicinity of the lowest angular frequency, we can see that G' saturates to G_0^R for sufficiently low angular frequency. Hence, G' in the deswelling state consists of the purely elastic part and the contribution part of the physical entanglements, suggesting that the

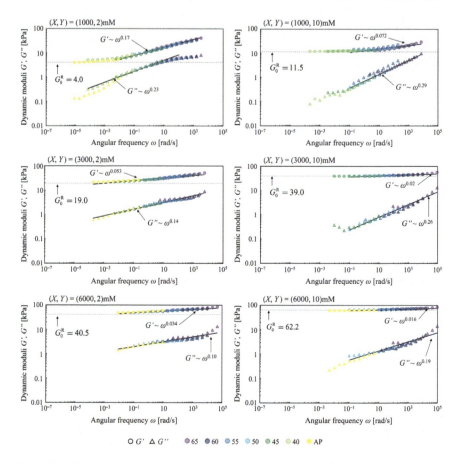

Fig. 12.10: Master curves of the storage shear modulus, G', (circle) and the loss shear modulus, G'', (triangle) for each compound, (X, Y). The reference volume swelling ratio, J_R, is set to the volume swelling ratio, J, at ethanol concentration, $Z = 50\%$, for all compounds. The reference shear elastic modulus, G_0^R, is depicted by the gray dotted line. The linear approximation lines uncover the power-law correlations between the dynamic moduli and the angular frequency, ω.

generalized Maxwell model introduced in Sect. 12.4 is available for capturing the linear viscoelastic behavior of a PAAm hydrogel.

12.5.2 Swelling Dependence of Linear Viscoelastic Properties

Figure 12.11 shows the horizontal shift factor, α_T, for each compound, (X, Y), within the range of the deswelling state. α_T monotonically increases in response to the deswelling, indicating that deswelling boosts the contribution of the physical entanglements to the dynamic moduli and thus enhances the frequency dependence of the

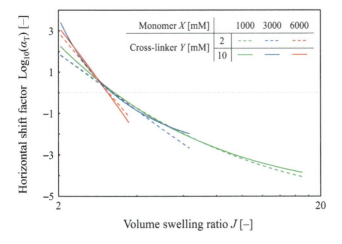

Fig. 12.11: Horizontal shift factor, α_T, as a function of the volume swelling ratio, J, for each compound, (X, Y). All the lines are drawn using Eq. (12.23).

linear viscoelastic behavior in cooperation with the positive power-law correlation of dynamic moduli. The sensitivity of α_T with respect to the volume swelling ratio, J, increases with increasing the contents of monomer and cross-linker except for the results for $(X, Y) = (3000, 10)$mM in the sufficiently deswollen state. It should be noted that since the power of the storage shear modulus, G', in Fig. 12.10 decreases with increasing the contents of monomer and cross-linker, the frequency dependence for the linear viscoelastic behavior of the hydrogel with high contents of monomer and cross-linker is not so enhanced by deswelling.

The scaling exponents, a_{vs}, b_{vd}^1, and b_{vd}^2, for each compound, (X, Y), are shown in Fig. 12.12. b_{vd}^1 has a larger value than the corresponding a_{vs} for every compound, indicating that the shear elastic modulus in the deswelling state has a higher sensitivity to the volume swelling ratio, J, than that in the swelling state. Although such a trend has already been reported in the previous studies [19, 21], we newly revealed that it is observed for a variety of compounds. Also, b_{vd}^2 has a larger value than b_{vd}^1 for compounds $(X, Y) = (1000, 2), (1000, 10), (3000, 2)$mM. This suggests that the scaling law for the hydrogel with low contents of monomer and cross-linker has some nonlinearities in the deswelling state.

12.5.3 Frequency Dependence of Complex Shear Moduli

The frequency dependence for the linear viscoelastic behavior of a PAAm hydrogel in the deswelling state is also discussed using the absolute value of complex shear modulus, $\bar{G}$, which is identified with the shear elastic modulus, G, in some previous

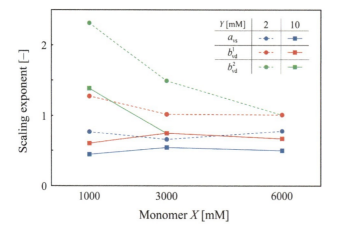

Fig. 12.12: Scaling exponents, a_{vs}, b^1_{vd}, and b^2_{vd}, as a function of the content of monomer, X, for each content of cross-linker, Y. The solid lines represent linear interpolation approximations of the results for $Y = 10$ mM, while the dotted lines represent those for $Y = 2$ mM.

studies [11, 18]. With a view to calculating $\bar{G}$ using Eqs. (12.4)$_1$ and (12.18), we identified the shear elastic modulus and the relaxation time for each Maxwell element in the reference swelling state, G^R_α and τ^R_α, by means of the curve-fitting for the master curves. To precisely capture the linear viscoelastic property, we introduced the generalized Maxwell model, in which at most 20 Maxwell elements and purely elastic element with the reference shear elastic modulus, G^R_0, are arranged in parallel. The curve-fitted results and identified parameters are shown in Fig. 12.13 and Tables 12.3, 12.4, respectively.

Figure 12.14 shows the normalized absolute value of complex shear modulus, $\bar{G}_N = \bar{G}/(\beta_T G^R_0)$, for each compound, (X, Y), in the range of the angular frequency, $10^{-4} \le \omega \le 10^2$ rad/s. $\bar{G}_N$ has 1.0 for a purely elastic hydrogel but increases with emerging linear viscoelastic nature. As $\bar{G}$ is normalized by β_T, the swelling dependence of the shear elastic modulus is eliminated, so that only frequency dependence of a PAAm hydrogel in the deswelling state is displayed in Fig. 12.14. Hydrogels with $(X, Y) = (3000, 10), (6000, 10)$mM have $\bar{G}_N$ of less than 1.5 over a whole range of deswelling states and, thus, are regarded as purely elastic material within $\omega \le 100$ rad/s. Although $\bar{G}_N$ for $(X, Y) = (1000, 10)$mM also takes nearly 1.0 in some extent of deswelling state ($J > 4.0$), the frequency dependence of $\bar{G}$ begins to be apparent in the sufficiently deswelling state. Thus, even adequately cross-linked hydrogels can exhibit frequency dependent mechanical behavior, depending on the degree of deswelling and the compound formulation. On the other hand, $\bar{G}_N$ for hydrogels with $Y = 2$mM have higher sensitivity to ω than those with $Y = 10$mM although taking small value in the early stage of deswelling. In particular, $\bar{G}_N$ for the softest hydrogel with $(X, Y) = (1000, 2)$mM drastically increases with increasing ω and the progression of deswelling, so that $\bar{G}$ at $\omega = 100$rad/s ends up having over ten times

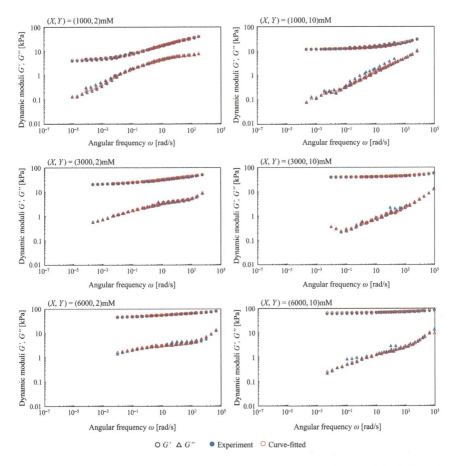

Fig. 12.13: Curve-fitted results for the master curves of the dynamic moduli, G' and G'', for each compound, (X, Y). To attain high accuracy of the curve-fitted results, we introduce the generalized Maxwell model, in which at most 20 Maxwell elements and purely elastic element with the reference shear elastic modulus, G_0^R, are arranged in parallel. G' and G'' are calculated using Eq. (12.18) with the shear elastic modulus, G_α^R, and the relaxation time, τ_α^R, enumerated in Tables 12.3, 12.4.

larger value than the corresponding G_0^R. As a result, hydrogels with low content of cross-linker have to be treated as viscoelastic materials, and the evaluation of elastic property using $\bar{G}$ is inappropriate in this case.

Table 12.3: Identified shear elastic modulus, G_α^R, and relaxation time, τ_α^R, for each content of monomer, X, at the content of cross-linker, $Y = 2$ mM.

	$X = 1000$ mM		$X = 3000$ mM		$X = 6000$ mM	
α	G_α^R [kPa]	τ_α^R [s]	G_α^R [kPa]	τ_α^R [s]	G_α^R [kPa]	τ_α^R [s]
1	1.825	1.998×10^{-5}	1.671	1.912×10^{-5}	22.922	8.286×10^{-6}
2	6.037	4.020×10^{-5}	4.701	5.880×10^{-5}	7.500	8.362×10^{-6}
3	8.675	4.013×10^{-5}	6.208	5.876×10^{-5}	5.225	2.939×10^{-5}
4	9.924	4.196×10^{-5}	6.891	5.900×10^{-5}	5.749	1.588×10^{-4}
5	6.986	4.125×10^{-4}	6.947	5.899×10^{-5}	4.434	1.067×10^{-3}
6	6.838	1.927×10^{-3}	5.096	5.916×10^{-4}	4.420	7.121×10^{-3}
7	6.162	9.540×10^{-3}	5.133	3.509×10^{-3}	4.318	5.803×10^{-2}
8	4.533	3.907×10^{-2}	4.157	1.988×10^{-2}	3.420	4.544×10^{-1}
9	3.877	1.542×10^{-1}	3.381	9.177×10^{-2}	1.182	2.413×10^{0}
10	2.882	6.021×10^{-1}	2.709	3.977×10^{-1}	1.166	2.413×10^{0}
11	1.151	2.663×10^{0}	1.324	2.398×10^{0}	1.682	8.590×10^{0}
12	1.117	2.663×10^{0}	1.236	2.398×10^{0}	1.970	3.696×10^{1}
13	1.563	1.118×10^{1}	1.673	1.469×10^{1}	0.698	2.276×10^{2}
14	1.121	4.446×10^{1}	0.967	6.481×10^{1}	1.188	2.276×10^{2}
15	0.323	1.743×10^{2}	0.416	2.398×10^{2}	1.076	1.696×10^{3}
16	0.408	1.743×10^{2}	0.601	2.398×10^{2}	0.465	1.522×10^{4}
17	0.415	1.176×10^{3}	0.732	1.553×10^{3}	0.216	1.573×10^{5}
18	0.328	9.893×10^{3}	0.763	1.189×10^{4}	0.179	2.559×10^{6}
19	0.076	1.284×10^{5}	0.345	1.118×10^{5}		
20	0.122	1.323×10^{5}	0.345	7.477×10^{5}		

12.6 Conclusion

Linear viscoelastic properties of chemically cross-linked PAAm hydrogels over a wide range of swelling and deswelling states were studied. Various hydrogels were prepared to study the effects of monomer and cross-linker on the mechanical properties. To attain transient equilibrium swelling, the water content of a hydrogel in the equilibrium state was controlled by ethanol. DMA was conducted on swollen hydrogels to evaluate the swelling-dependent linear viscoelastic properties. The obtained experiment data illustrated that PAAm hydrogels exhibit either purely elastic or linear viscoelastic behavior in the equilibrium state.

To comprehensively understand linear viscoelastic properties of hydrogels during transient equilibrium swelling, we formulated the swelling-dependent linear viscoelasticity model with the generalized Maxwell model. The distinctive feature of the present model is that the shear elastic modulus and the relaxation time of each Maxwell element depend on the volume swelling ratio. Taking advantage of this feature, we advocated the time-swelling superposition principle and originally introduced the horizontal and vertical shift factors depending on the volume swelling ratio. Two shift factors were exactly prescribed by the WLF type equation and the scaling law, respectively, and quantified the swelling dependence on the linear viscoelastic properties of a hydrogel.

Table 12.4: Identified shear elastic modulus, G_α^R, and relaxation time, τ_α^R, for each content of monomer, X, at the content of cross-linker $Y = 10$ mM.

	X = 1000 mM		X = 3000 mM		X = 6000 mM	
α	G_α^R [kPa]	τ_α^R [s]	G_α^R [kPa]	τ_α^R [s]	G_α^R [kPa]	τ_α^R [s]
1	1.755	1.941×10^{-5}	67.954	1.471×10^{-6}	2.256	1.458×10^{-5}
2	5.275	4.346×10^{-5}	3.402	4.194×10^{-6}	5.841	1.393×10^{-5}
3	7.359	4.288×10^{-5}	3.291	9.527×10^{-6}	7.819	1.400×10^{-5}
4	9.371	4.307×10^{-5}	5.844	4.427×10^{-5}	4.927	2.597×10^{-5}
5	9.965	4.371×10^{-5}	2.750	1.936×10^{-4}	2.987	1.238×10^{-4}
6	5.640	4.944×10^{-4}	0.491	7.594×10^{-4}	3.024	6.801×10^{-4}
7	3.660	2.492×10^{-3}	2.323	1.165×10^{-3}	1.703	3.994×10^{-3}
8	2.042	9.238×10^{-3}	1.795	1.182×10^{-2}	1.150	8.645×10^{-3}
9	1.669	3.622×10^{-2}	0.971	1.093×10^{-1}	1.522	4.667×10^{-2}
10	1.081	1.645×10^{-1}	0.629	9.857×10^{-1}	0.738	1.537×10^{-1}
11	0.237	6.824×10^{-1}	0.130	7.221×10^{0}	0.828	6.897×10^{-1}
12	0.285	6.824×10^{-1}	0.135	7.221×10^{0}	0.271	6.897×10^{-1}
13	0.459	2.600×10^{0}	0.243	1.165×10^{2}	0.685	5.393×10^{0}
14	0.315	1.257×10^{1}	0.440	1.165×10^{2}	0.411	3.705×10^{1}
15	0.173	1.648×10^{2}			0.147	2.119×10^{2}
16	0.163	1.648×10^{2}			0.116	2.119×10^{2}
17	0.171	2.216×10^{3}			0.513	4.047×10^{3}
18	0.007	3.364×10^{4}			0.465	1.797×10^{4}
19	0.010	3.130×10^{5}			0.230	1.587×10^{5}
20	0.115	3.599×10^{6}			0.247	3.805×10^{6}

The dynamic moduli of a hydrogel in the deswelling state were successfully organized as the master curves using two shift factors, providing several important findings concerning the linear viscoelastic properties of a hydrogel. We found that the dynamic moduli in the deswelling state have positive power-law correlation with the angular frequency regardless of compounds. Then, since the horizontal shift factor increases in response to the deswelling, deswelling and high-speed oscillation boost the contribution of the physical entanglements to the dynamic moduli and thus enhance the frequency dependence of the linear viscoelastic behavior. This was confirmed by the complex modulus calculated from the present model with identified material parameters. Also, deswelling enhanced the elastic property of a hydrogel. Indeed, the scaling exponent in the deswelling state has a larger value than that in the swelling state. In particular, it is supposed that the scaling law for the hydrogel with low contents of monomer and cross-linker has some nonlinearities in the deswelling state.

Although this paper only made an effort to evaluate the swelling-dependent linear viscoelastic properties along with the proposal of the framework, it should be mentioned here that the present constitutive model with identified linear viscoelastic properties enables us to conduct the numerical analysis for deswollen hydrogels. The study for the other representative viscoelastic behaviors such as creep, stress relaxation, and strain recovery is of significance but left for our future work.

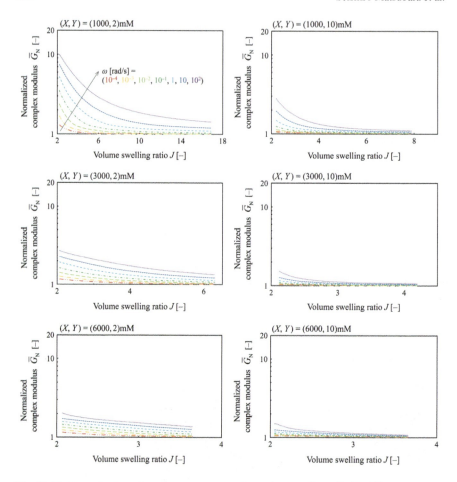

Fig. 12.14: Normalized absolute value of complex shear modulus, $\bar{G}_N = \bar{G}/(\beta_T G_0^R)$, for each compound, (X,Y), in the range of the angular frequency, $10^{-4} \leq \omega \leq 10^2$ rad/s, and in the deswelling state. $\bar{G}$ is calculated using Eqs. (12.18), (12.19) with the identified shear elastic modulus, G_α, and relaxation time, τ_α.

Appendix A: Validity for Transient Equilibrium Swelling Using Ethanol

The validity for the transient equilibrium swelling using ethanol, as described in Sect. 12.2, was verified by comparing the volume swelling ratio, J, the absolute value of complex shear modulus, $\bar{G}$, and the loss tangent, $\tan(\delta)$, between AP and imitated AP specimens. The imitated AP specimens were prepared using ethanol such that their weight swelling ratio, w/w_{AP}, falls inside the range from 0.95 to 1.05. Three specimens were prepared for each compound. Other preparation methods and settings of DMA were the same as those described in Sect. 12.2. Table 12.5 shows the volume

Table 12.5: Volume swelling ratios of the AP specimen, J_{AP}, and imitated AP specimen, J_{IAP}, for each compound, (X,Y).

	$Y = 2\text{mM}$			$Y = 10\text{mM}$		
	J_{AP} [−]	J_{IAP} [−]	$\|J_{AP} - J_{IAP}\|/J_{AP}$ [%]	J_{AP} [−]	J_{IAP} [−]	$\|J_{AP} - J_{IAP}\|/J_{AP}$ [%]
$X = 1000\text{mM}$	16.9137	16.6748	1.4124	16.9333	17.4654	3.1423
$X = 3000\text{mM}$	6.3046	6.0108	4.6601	6.3111	6.2013	1.7393
$X = 6000\text{mM}$	3.6523	3.5837	1.8768	3.6555	3.6611	0.1514

swelling ratios of the AP specimen, J_{AP}, and imitated AP specimen, J_{IAP}, for each compound, (X,Y). As the difference in J between AP and imitated AP specimens is less than 5.0%, the imitated AP specimen is in almost the same swelling state as the corresponding AP specimen regardless of compounds. Also, Figs. 12.15 and 12.16 show $\bar{G}$ and $\tan(\delta)$ as a function of the angular frequency, ω, for each compound, respectively. The results of imitated AP specimens were in good agreement with those of AP specimens.

Therefore, the validity for the transient equilibrium swelling using ethanol was verified.

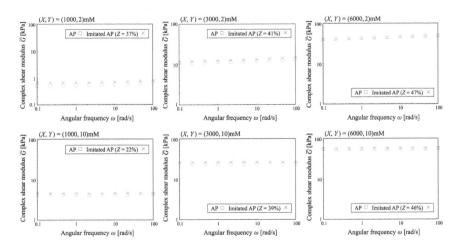

Fig. 12.15: Comparison of measured results between AP and imitated AP specimens concerning absolute values of the complex shear modulus, $\bar{G}$, as a function of angular frequency, ω, for each compound, (X,Y). Z denotes the ethanol concentration in the mixed solvent used to prepare the imitated AP specimen.

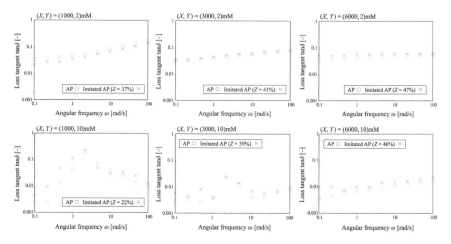

Fig. 12.16: Comparison of measured results between AP and imitated AP specimens concerning loss tangent, $\tan(\delta)$, as a function of angular frequency, ω, for each compound, (X, Y). Z denotes the ethanol concentration in the mixed solvent used to prepare the imitated AP specimen.

Appendix B: Experimental Data

The weight swelling ratio, w/w_{AP}, the storage shear modulus, G', and the loss shear modulus, G'', are shown in Figs. 12.3, 12.4 and 12.5 as the raw experimental data, respectively.

Acknowledgements This work was partially supported by The Japan Society for the Promotion of Science (JSPS) under a Grant-in-Aid for Scientific Research (A) (19H00739) and a Grant-in-Aid for Early-Career Scientists (22K14142).

References

[1] Vedadghavami A, Minooei F, Mohammadi MH, Khetani S, Rezaei Kolahchi A, Mashayekhan S, Sanati-Nezhad A (2017) Manufacturing of hydrogel biomaterials with controlled mechanical properties for tissue engineering applications. Acta Biomaterialia **62**:42–63

[2] Sharma S, Tiwari S (2020) Retracted: A review on biomacromolecular hydrogel classification and its applications. International Journal of Biological Macromolecules **162**:737–747

[3] Tallinen T, Chung JY, Rousseau F, Girard N, Lefevre J, Mahadevan L (2016) On the growth and form of cortical convolutions. Nature Physics **12**:588–593

[4] Distler T, Schaller E, Steinmann P, Boccaccini A, Budday S (2020) Alginate-based hydrogels show the same complex mechanical behavior as brain tissue.

Journal of the Mechanical Behavior of Biomedical Materials **111**:103,979
[5] Weizel A, Distler T, Schneidereit D, Friedrich O, Bräuer L, Paulsen F, Detsch R, Boccaccini A, Budday S, Seitz H (2020) Complex mechanical behavior of human articular cartilage and hydrogels for cartilage repair. Acta Biomaterialia **118**:113–128
[6] Tang S, Richardson BM, Anseth KS (2021) Dynamic covalent hydrogels as biomaterials to mimic the viscoelasticity of soft tissues. Progress in Materials Science **120**:100,738
[7] Lawless BM, Sadeghi H, Temple DK, Dhaliwal H, Espino DM, Hukins DW (2017) Viscoelasticity of articular cartilage: Analysing the effect of induced stress and the restraint of bone in a dynamic environment. Journal of the Mechanical Behavior of Biomedical Materials **75**:293–301
[8] Mahumane GD, Kumar P, du Toit LC, Choonara YE, Pillay V (2018) 3d scaffolds for brain tissue regeneration: architectural challenges. Biomaterials Science **6**:2812–2837
[9] Sattari S, Eskandari M (2020) Characterizing the viscoelasticity of extra- and intra-parenchymal lung bronchi. Journal of the Mechanical Behavior of Biomedical Materials **110**:103,824
[10] Baker BA, Murff RL, Milam VT (2010) Tailoring the mechanical properties of polyacrylamide-based hydrogels. Polymer **51**(10):2207–2214
[11] Li J, Hu Y, Vlassak JJ, Suo Z (2012) Experimental determination of equations of state for ideal elastomeric gels. Soft Matter **8**:8121–8128
[12] Subramani R, Izquierdo-Alvarez A, Bhattacharya P, Meerts M, Moldenaers P, Ramon H, Van Oosterwyck H (2020) The influence of swelling on elastic properties of polyacrylamide hydrogels. Frontiers in Materials **7**
[13] Gennes PGD (1979) Scaling Concepts in Polymer Physics. Cornell University Press, New York
[14] Obukhov SP, Rubinstein M, Colby RH (1994) Network modulus and superelasticity. Macromolecules **27**:3191–3198
[15] Kundu S, Crosby AJ (2009) Cavitation and fracture behavior of polyacrylamide hydrogels. Soft Matter **5**:3963–3968
[16] Denisin AK, Pruitt BL (2016) Tuning the range of polyacrylamide gel stiffness for mechanobiology applications. ACS Applied Materials & Interfaces **8**, 21893-21902
[17] Okumura D, Kondo A, Ohno N (2016) Using two scaling exponents to describe the mechanical properties of swollen elastomers. Journal of the Mechanics and Physics of Solids **90**:61–76
[18] Kawai R, Tanaka H, Matsubara S, Ida S, Uchida M, Okumura D (2021) Implicit rule on the elastic function of a swollen polyacrylamide hydrogel. Soft Matter **17**:4979–4988
[19] Li Z, Liu Z, Ng TY, Sharma P (2020) The effect of water content on the elastic modulus and fracture energy of hydrogel. Extreme Mechanics Letters **35**:100,617

[20] Sakai T, Kurakazu M, Akagi Y, Shibayama M, Chung Ui (2012) Effect of swelling and deswelling on the elasticity of polymer networks in the dilute to semi-dilute region. Soft Matter **8**:2730–2736

[21] Katashima T, Chung Ui, Sakai T (2015) Effect of swelling and deswelling on mechanical properties of polymer gels. Macromolecular Symposia **358**(1):128–139

[22] Grattoni CA, Al-Sharji HH, Yang C, Muggeridge AH, Zimmerman RW (2001) Rheology and permeability of crosslinked polyacrylamide gel. Journal of Colloid and Interface Science **240**(2):601–607

[23] Calvet D, Wong JY, Giasson S (2004) Rheological monitoring of polyacrylamide gelation: Importance of cross-link density and temperature. Macromolecules **37**(20):7762–7771

[24] Okay O, Oppermann W (2007) Polyacrylamide-clay nanocomposite hydrogels: Rheological and light scattering characterization. Macromolecules **40**(9):3378–3387

[25] Niţă LE, Chiriac AP, Bercea M, Neamtu I (2007) In situ monitoring the sol-gel transition for polyacrylamide gel. Rheologica Acta **46**(5):595–600

[26] Savart T, Dove C, Love BJ (2010) In situ dynamic rheological study of polyacrylamide during gelation coupled with mathematical models of viscosity advancement. Macromolecular Materials and Engineering **295**(2):146–152

[27] Du C, Hill RJ (2019) Linear viscoelasticity of weakly cross-linked hydrogels. Journal of Rheology **63**(1):109–124

[28] Sakohara S, Maekawa Y, Tateishi Y, Asaeda M (1992) Effects of gel composition on separation properties of ethanol/water mixtures by acrylamide gel membranes. Journal of Chemical Engineering Sakoharaof Japan **25**(5):598–603

[29] Reese S, Govindjee S (1998) A theory of finite viscoelasticity and numerical aspects. International Journal of Solids and Structures **35**(26):3455–3482

[30] Adibnia V, Hill RJ (2016) Universal aspects of hydrogel gelation kinetics, percolation and viscoelasticity from PA-hydrogel rheology. Journal of Rheology **60**(4):541–548

[31] Ng TSK, McKinley GH (2008) Power law gels at finite strains: The nonlinear rheology of gluten gels. Journal of Rheology **52**(2):417–449

[32] Winter HH, Chambon F (1986) Analysis of Linear Viscoelasticity of a Crosslinking Polymer at the Gel Point. Journal of Rheology **30**(2):367–382

Chapter 13
Application of Nonlinear Viscoelastic Material Models for the Shrinkage and Warpage Analysis of Blow Molded Parts

Patrick Michels, Christian Dresbach, Esther Ramakers-van Dorp, Holm Altenbach, and Olaf Bruch

Abstract The prediction of shrinkage and warpage of extrusion blow molded plastic parts is a topic of high industrial demand. Nevertheless, simulation results are still associated with uncertainties. One of the major difficulties is the description of the complex time-, temperature- and process-dependent material behavior of semi-crystalline polymers like high density polyethylene (HDPE). It is state of the art to use linear viscoelastic material models for the shrinkage and warpage analysis. However, linear viscoelastic behavior can only be assumed if the stresses are small. To increase the prediction accuracy of the current simulation models, nonlinear viscoelastic material models, such as the Abaqus Parallel Rheological Framework (PRF), are investigated. The calibration of the PRF model can be quite challenging, especially if a higher number of networks is used. Consequently, we present a calibration strategy that uses functional relations to describe the parameters along the network elements in order to reduce the dimensions of the design space for model calibration. To find the best possible solution, the global optimization algorithm Adaptive Simulated Annealing (ASA) is used. A simplified one-dimensional representation of the PRF model is implemented in Matlab to further reduce the computational effort of the

---

Patrick Michels
Bonn-Rhein-Sieg University of Applied Sciences, Grantham Allee 20, 53757 Sankt Augustin, Germany,
e-mail: `patrick.michels@h-brs.de`

Christian Dresbach · Esther Ramakers-van Dorp
Bonn-Rhein-Sieg University of Applied Sciences, Von-Liebig-Str. 20, 53359 Rheinbach, Germany,
e-mail: `christian.dresbach@h-brs.de, esther.vandorp@h-brs.de`

Holm Altenbach
Otto von Guericke University Magdeburg, Universitätsplatz 2, 39106 Magdeburg, Germany, e-mail: `holm.altenbach@ovgu.de`

Olaf Bruch
Dr. Reinold Hagen Stiftung, Kautexstr. 53, 53229 Bonn & Bonn-Rhein-Sieg University of Applied Sciences, Grantham Allee 20, 53757 Sankt Augustin, Germany,
e-mail: `o.bruch@hagen-stiftung.de, olaf.bruch@h-brs.de`

model calibration. The calibration workflow is successfully tested using a set of relaxation tests with subsequent unloading at different strain and temperature levels. A good agreement between the experimental material tests and the simulation results, using the calibrated PRF model, is observed.

13.1 Introduction

The extrusion blow molding process is one of the most economic methods for the production of hollow plastic parts like bottles, cans, fuel tanks and large containers. The process itself can be divided into three main steps. In the first step a hollow tube, which is called parison, is extruded. Once the parison has reached its final length, the mold closes and the parison is inflated against the walls of the cooled mold. The blowing pressure is then maintained until the part solidifies. During the cooling under mold constraint, thermal stresses build up which lead to shrinkage and warpage of the final part after demolding. These undesired shape deviations cause major problems for the blow molding industry. In general, higher demolding temperatures lead to higher shrinkage and warpage, whereas lower demolding temperatures lead to a higher amount of residual stresses. In practice, there are several ways to deal with these difficulties. First, the cooling time can be increased, which leads to a higher amount of residual stresses and in most cases to an uneconomical production. On the other hand, the part warpage can be reduced by specific changes to the mold design. In the latter case, the use of Computer Aided Engineering (CAE) methods at an early stage of the product development offers great potential. Nevertheless, the prediction of the process-related shrinkage and warpage is still associated with uncertainties. One of the major difficulties is the modeling of the complex time-, temperature- and process-dependent material behavior of semi-crystalline polymers like HDPE. During processing, the polymer passes various stages in which its mechanical and thermal behavior drastically changes. The extrusion and inflation takes place at temperatures above the crystallite melting temperature T_m. In this temperature range, the material can be assumed as an amorphous melt with low structural stiffness. Below T_m the material behaves like a thermo-viscoelastic solid.

In literature, only a few research groups have dealt with the shrinkage and warpage analysis of blow molded parts. A first simulation approach considering the complete blow molding process including parison formation, clamping, inflation, solidification, and warpage was introduced by Laroche et al. [1]. They assumed the material to behave like an isotropic thermorheologically simple solid during the cooling. A linear viscoelastic material model (fluid-like generalized Maxwell) with three relaxation times was used in conjunction with the reduced time concept to model the temperature dependency [1, 2]. The shift function was approximated by the WLF-equation according to Williams, Landel and Ferry [3]. A good qualitative agreement between the warpage simulation and experimental measurements of a plastic fuel tank (PFT) was observed. Debergue et al. [4] investigated the influence of a small and large displacement approach on the warpage analysis of a blow molded automotive part. A fluid-like generalized Maxwell model with 4 relaxation times was used. The investigation showed that the large displacement approach played a rather subordinate role in the warpage analysis. In contrast, the authors stated that part positioning after demolding in conjunction with gravity plays a critical role for the shrinkage and warpage prediction. A comparison of the simulation results with experimental warpage measurements under varying process conditions showed inaccuracies both qualitatively and quantitatively. Further investigations based on a fluid-like generalized Maxwell model with six relaxation times were carried out by Benrabah et al. [5]. The main focus of their investigation was the influence of deflashing on the component warpage. In [6], the implementation of a solid-like generalized Maxwell model was presented and compared to the fluid-like model. The warpage deformation of the solid-like model was less than the deformation of the fluid-like model, which the authors suggested was due to the lower structural stiffness of the fluid like model below the melting temperature. Finally, in a validation case study Benrabah et al. [7] observed a good qualitative agreement between the warpage simulation and experimental measurements of a blow molded PFT. A fluid-like generalized Maxwell model with five relaxation times was used for the analysis.

In [8], the shrinkage behavior of simple blow molded parts was investigated under varying process conditions and compared to simulation results. Anisotropic shrinkage behavior of the investigated parts was observed at which the level of anisotropy increased with higher degrees of stretching. Experimental measurements of Ramakers-van Dorp et al. [9] and Grommes et al. [10] on extrusion blow molded parts showed a rather small anisotropy of the elastic modulus, whereas experiments of Ramakers-van Dorp [11] showed a pronounced anisotropy of the coefficient of thermal expansion (CTE). Consequently, an isotropic generalized Maxwell solid model with 19 relaxation times in conjunction with the WLF-equation and orthotropic process- and temperature-dependent CTE was used in [8]. The simulation results matched the anisotropic shrinkage values reasonably well.

In summary, the use of linear viscoelastic models can be seen as state of the art for the shrinkage and warpage simulation of blow molded parts. However, in case of semi-crystalline polymers like HDPE, linear viscoelastic material behavior can only be assumed for small stresses and strains. At higher stresses and strains, HDPE reacts nonlinear viscoelastically. Creep experiments of Lai and Bakker [12] on HDPE samples indicate a strong nonlinear behavior even at very small stresses. They suggested that linearity exists only at vanishing small stresses [12]. In literature, several constitutive equations for the description of nonlinear viscoelastic material behavior are presented, among others in [13–16]. But only recently, nonlinear viscoelastic models like the Abaqus (Dassault Systèms) PRF model [16–18] have become available as standard in commercial finite element software products. However, the use of these models in the shrinkage and warpage analysis is quite challenging. To cover the extensive time and temperature range of the shrinkage and warpage analysis, the model calibration might involve a huge number of material parameters which need to be identified. Consequently, we present a calibration strategy which reduces the dimension of the design space by the use of functional relations between the material parameters. The Abaqus nonlinear viscoelastic PRF model is used, but the general procedure can also be applied to similar models like the Parallel Network (PN) model [19] which is provided by the PolyUMod library (PolymerFEM LLC) [20].

13.2 Material Models

In this study, the Abaqus Parallel Rheological Framework model will be integrated into the shrinkage and warpage anlysis presented in [8] and compared with the linear viscoelastic model. In the following, the basic equations and the most important features of the material models will be discussed. We start with the linear viscoelastic model implemented in Abaqus which has already been used in [8]. Thereafter, the nonlinear viscoelastic PRF model will be discussed.

13.2.1 Linear Viscoelastic Material Model

The stress response of a linear viscoelastic material can be described by the following integral equation:

$$\sigma(t) = \int_0^t E_R(t-s)\dot{\varepsilon}(s)\,ds, \qquad (13.1)$$

where $\sigma(t)$ is the stress at time t, $\dot{\varepsilon}$ is the strain rate and $E_R(t)$ is the time dependent relaxation modulus. The relaxation modulus is often used in a normalized form so that we obtain:

$$\sigma(t) = E\int_0^t g_R(t-s)\dot{\varepsilon}(s)\,ds, \qquad (13.2)$$

where E is the instantaneous modulus and $g_R(t)$ is the dimensionless relaxation function. In Abaqus, the normalized relaxation function $g_R(t)$ is approximated by a Prony series [18]

$$g_R(t) = \frac{E_R(t)}{E} = 1 - \sum_{i=1}^{N} g_i\left(1 - e^{-t/\tau_i}\right). \qquad (13.3)$$

In Eq. (13.3) g_i and τ_i are material parameters, the so called Prony values and relaxation times. Assuming isotropic material behavior, Eq. (13.2) can be generalized to multiaxial loading by separating the strain tensor $\boldsymbol{\varepsilon}$ into deviatoric and volumetric parts. For the time dependent Cauchy stress tensor, $\boldsymbol{\sigma}$ applies [19]:

$$\boldsymbol{\sigma}(t) = 2G\int_0^t g_R(t-s)\dot{\boldsymbol{\varepsilon}}_{\text{dev}}(s)\,ds + K\int_0^t \kappa_R(t-s)\dot{\boldsymbol{\varepsilon}}_{\text{vol}}(s)\,ds, \qquad (13.4)$$

with the instantaneous shear modulus G, the normalized shear relaxation function $g_R(t)$, the instantaneous bulk modulus K, the normalized bulk relaxation function $\kappa_R(t)$, and the time derivatives of the deviatoric ε_{dev} and volumetric ε_{vol} parts of the strain tensor ε. Thus, the three-dimensional material behavior can be defined by two independent relaxation functions, the shear relaxation function, and the bulk relaxation function.

Temperature effects can be included by the use of the time temperature superposition (TTS). Therefore, the reduced time $\xi(t)$ is used in Eq. (13.5) [18]

$$\sigma(t) = 2G \int_0^t g_R(\xi(t) - \xi(s)) \dot{\varepsilon}_{\text{dev}}(s) \, ds + K \int_0^t \kappa_R(\xi(t) - \xi(s)) \dot{\varepsilon}_{\text{vol}}(s) \, ds. \quad (13.5)$$

The reduced time is defined by [18]:

$$\xi(t) = \int_0^t \frac{ds}{\alpha(\theta(s))}, \quad (13.6)$$

where $\alpha(\theta(t))$ is the shift function which can be approximated using the WLF equation (Eq. (13.7)) or the Arrhenius equation (Eq. (13.8))

$$\log_{10}(\alpha) = -\frac{C_1(\theta - \theta_{\text{Ref}})}{C_2 + (\theta - \theta_{\text{Ref}})}, \quad (13.7)$$

$$\ln(\alpha) = \frac{E_A}{R}\left(\frac{1}{T} - \frac{1}{T_{\text{Ref}}}\right). \quad (13.8)$$

The variables C_1 and C_2 are material parameters, E_A is the activation energy, R the universal gas constant, θ the temperature, and θ_{Ref} the reference temperature. For the Arrhenius equation the temperatures T and T_{Ref} must be specified in Kelvin. In addition, the instantaneous modulus can also be defined as a function of temperature [18].

To define thermal expansion behavior of the linear viscoelastic model, Abaqus allows the use of isotropic and orthotropic thermal expansion coefficients which can be constant or a function of temperature and field variables [18].

Instead of the previously described integral equation, the linear viscoelastic material model can also be derived by differential equations, which in fact is equivalent to the integral form [19]. The differential form is often used to build rheological models, which can be constructed using simple rheological spring and dashpot elements. Therefore, the linear viscoelastic model defined by the Prony series (Eq. (13.3)) is equivalent to a generalized Maxwell model (Fig. 13.1) [19]. It consists of an arbitrary number of Maxwell elements (series of spring and dashpot) in parallel. If an additional equilibrium network (spring) is used (Fig. 13.1), the material model will represent a solid behavior [21]. In this case, the stress in a stress relaxation experiment would relax to a non zero plateau, which is defined by the stress in the equilibrium network. Without the equilibrium network, the stress would relax to zero, which

Fig. 13.1 Illustration of the generalized Maxwell solid model which consists of a series of Maxwell elements in parallel and one equilibrium network.

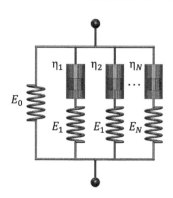

represents a fluid-like behavior [21]. Rheological models, such as the generalized Maxwell model are often used as a starting point for the development of nonlinear viscoelastic models [19]. In Subsect. 13.4.2, the setup of the generalized Maxwell model is used to develop a simplified one-dimensional model which represents the nonlinear viscoelastic PRF model.

13.2.2 Abaqus Parallel Rheological Framework Model

Similar to the generalized Maxwell model (Fig. 13.1), the Abaqus PRF model consists of an arbitrary number of viscoelastic networks in parallel and an optional equilibrium network [17]. The main difference to the generalized Maxwell model is that nonlinear hyperelastic models are used for the springs and nonlinear creep laws for the dashpots. The response of the equilibrium network can be purely elastic or elastoplastic [17]. For each viscoelastic network $i = 1, 2, 3, \ldots, N$, a multiplicative split of the deformation gradient into an elastic and an inelastic part is assumed [17]

$$F = F^{\text{el}} \cdot F^{\text{in}}. \tag{13.9}$$

The elastic response of the PRF model can be represented by any hyperelastic model implemented in Abaqus [18]. Similar to the Prony values g_i of the linear viscoelastic model, the stiffness of each network is represented by a stiffness ratio s_i where the sum of all stiffness ratios must be less or equal to one. In case the stiffness ratio is equal to one, the model is defined without equilibrium network. The same hyperelastic model is used for all networks. Using Abaqus, all hyperelastic models are described by an energy potential $U(\varepsilon)$ as a function of the strain [18].

Considering the shrinkage and warpage analysis, the strains are rather small so that a linear elastic model would do well. Thus a simple Neo-Hookean hyperelastic model is used in this study. The strain energy potential is given by [18]:

$$U = C_{10}(\bar{I}_1 - 3) + \frac{1}{D_1}(J^{\text{el}} - 1)^2. \tag{13.10}$$

Thereby C_{10} and D_1 are material parameters, $\bar{I}_1$ is the first deviatoric strain invariant and J^{el} is the elastic volume ratio. The viscous behavior for each network is defined using the creep potential G^{in} [18]. The creep potential is given by the equivalent deviatoric Cauchy stress $\bar{p}$, so that the flow rule can be described as follows [18]:

$$\boldsymbol{D}^{in} = \frac{3}{2\tilde{q}} \dot{\bar{\varepsilon}}^{in} \bar{\boldsymbol{\tau}}, \qquad (13.11)$$

with

$$\tilde{q} = J\bar{p}, \qquad (13.12)$$

where $\boldsymbol{D}^{in}$ is the symmetric part of the velocity gradient $\boldsymbol{L}^{in}$, $\bar{\boldsymbol{\tau}}$ is the deviatoric Kirchhoff stress, J is the determinant of the deformation gradient $\boldsymbol{F}$, the so called Jacobien, and $\dot{\bar{\varepsilon}}^{in}$ the equivalent creep strain rate. For the evolution of the creep strain rate, several models are available in Abaqus. These are the power law model, the strain hardening model, the hyperbolic-sine model, the Bergstrom-Boyce model, and a user-defined creep model which can be implemented by a user subroutine [18]. In this study, the strain hardening model (Eq. (13.13)) is used for the evaluation of the creep strain rate [18]

$$\dot{\bar{\varepsilon}}^{in} = \left(A\,\tilde{q}^n\,[(m+1)\,\bar{\varepsilon}^{in}]^m \right)^{\frac{1}{m+1}}, \qquad (13.13)$$

where $\bar{\varepsilon}^{in}$ is the equivalent inelastic strain and A, m and n are material parameters. If parameter n is set equal to one and m is set equal to zero, the evolution of the creep strain rate $\dot{\bar{\varepsilon}}^{in}$ is linearized [17].

Similar to the linear viscoelastic model, the reduced time concept can be used to model temperature effects [18]. In contrast to the linear viscoelastic model where the same WLF or Arrhenius parameters are used for all networks, the PRF model allows the specification for each network individually. Alternatively, all parts of the material model, the instantaneous modulus, the creep model, and the stiffness ratios s_i can be defined temperature-dependently using tabular values [18]. For temperatures between the specified temperatures, the material parameters are interpolated. In the shrinkage and warpage analysis, the temperature changes continuously from a high temperature to room temperature (RT). Therefore, the reduced time concept is used in this study to realize a continuous function for the temperature dependence of the material model.

Currently [18], only isotropic thermal expansion can be used with the Abaqus PRF model. This excludes the use of orthotropic process-dependent expansion coefficients as they were used in [8].

13.3 Shrinkage and Warpage Analysis

The simulation workflow for the shrinkage and warpage analysis of extrusion blow molded parts has been described in detail in [8], so that we will only recall the most important parts for this study. Considering a complex extrusion blow molded part, at least three simulation steps are necessary for the shrinkage and warpage prediction. The first step is the process simulation of the parison inflation to determine the process-related wall thickness distribution. After the process simulation, the cooling of the part inside the closed mold and at ambient air after demolding is analyzed. The transient temperature field is then used in a subsequent shrinkage and warpage analysis to determine the part deformation. For the process simulation, currently B-SIM of Accuform, a finite element based process simulation software for blow molding applications, is used. The resulting wall thickness distribution as well as local degrees of stretching and their orientation are mapped to the finite element mesh of the following analysis steps using the MpCCI Mapper of Fraunhofer SCAI. Since blow molded parts are thin-walled components, shell elements are used for the cooling and warpage analysis. The part cooling is analyzed in two steps using Simulia Abaqus (Dassault Systèms). In the first step, the part cools down under mold constraint. Due to the contact with the cooled mold, the outer surface cools down rapidly whereas the inner surface cools down much slower. After demolding, further cooling takes place at ambient air until RT is reached. Similarly to the cooling simulation, the shrinkage and warpage analysis is also carried out in two steps using Abaqus. In the first step, all degrees of freedom of all nodes are fixed so that thermal stresses will build up during the cooling under mold constraint. In the second step, the boundary conditions are changed so that the part can shrink and warp freely due to the accumulated thermal stresses and further temperature changes at ambient air.

In this study, we will focus on a one-element simulation of the shrinkage analysis which sufficiently represents a local area of a complex blow molded part. The cooling simulation will be carried out using the simulation model published in [8]. A wall thickness of 2 mm and a cooling time of 60 s is used. The required temperature-dependent material data for density, thermal conductivity, and heat capacity as well as the heat transfer coefficients are taken from [8, 22, 23]. For the shrinkage and warpage analysis, the thermal expansion behavior and the mechanical material behavior need to be defined. Because the thermal expansion behavior of HDPE is highly temperature-dependent, a temperature-dependent CTE is taken from literature [22, 23] (Fig. 13.2). The CTE was determined from Pressure-Volume-Temperature (P-V-T) data at a pressure of approximately 0.1 N/mm^2 [22, 23]. The peak at 130 °C marks the crystallite melting temperature where the volume of the polymer drastically changes.

For the definition of the mechanical behavior, a master curve was obtained from dynamic mechanical analysis (DMA) using frequency sweeps in the temperature range of $-20 - 120$ °C [8]. The WLF equation was used to shift the isothermals of the temperature-frequency-sweeps with RT as θ_{Ref} to obtain a continuous master curve at RT. The storage modulus E' was then converted from the frequency domain to the time domain using the following approximation formula [24, 25]:

$$t = \frac{1}{2\pi f}. \tag{13.14}$$

Considering the part cooling, a large temperature range from about 200 °C to RT needs to be considered. After demolding, a viscoelastic retardation occurs, due to

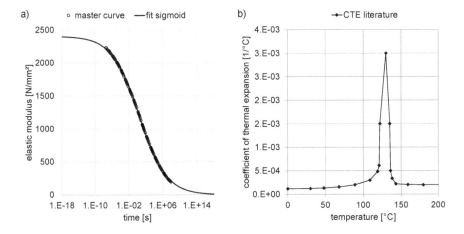

Fig. 13.2: Material data for the shrinkage and warpage analysis. a) Experimental master curve obtained from DMA experiments [8]. The master curve is fitted by a sigmoid function and extrapolated to cover the whole time range; b) Temperature-dependent CTE [22, 23].

the accumulated thermal stresses [26]. This retardation can take several hours to days. To cover the extensive time and temperature range in the material modeling, the experimental master curve is extrapolated. Due to the typical s-formed shape, a sigmoid function is used to fit and extrapolate the experimental data. The extrapolated master curve is given by Eq. (13.15) (Fig. 13.2)

$$f(t) = -1201.79 \tanh\left(\frac{\log_{10}(t) - 0.449}{6.37}\right) + 1201.79. \quad (13.15)$$

Using one relaxation time per decade, the linear viscoelastic model is calibrated using Eq. (13.3) with 37 prony terms (Fig. 13.3). The same relaxation function is used for the normalized shear and bulk relaxation function $g_R(t)$ and $\kappa_R(t)$. The Poissons ratio is set to 0.5, so that incompressibility is assumed. For the temperature dependence of the material model, the reduced time concept is applied, using the WLF equation for the approximation of the shift function. The WLF coefficients of the master curve creation are used in the simulation model.

At the crystallite melting temperature T_m, the material behavior changes from a thermo-viscoelastic solid to a thermo-viscoelastic fluid with a structural stiffness of almost zero. It is therefore assumed that the evolution of thermal stresses in the shrinkage analysis starts at temperatures below 130 °C. To ensure that no thermal stress is stored in the material model at temperatures above 130 °C, the CTE is set to zero (Fig. 13.3). Since the demolding temperatures are usually below 130 °C, free shrinkage at temperatures above 130 °C can be ruled out.

If the nonlinear viscoelastic PRF model is used with the reduced time concept to model temperature effects, the short relaxation times will cause convergence problems at high temperatures. This is because the already short relaxation times will be further

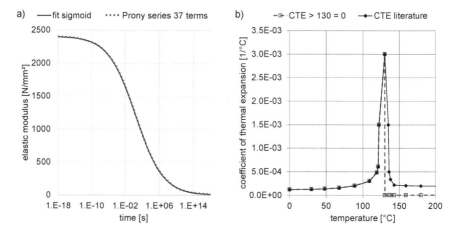

Fig. 13.3: Material data for the shrinkage and warpage analysis. a) Experimental master curve obtained from DMA experiments [8] which is extrapolated using a sigmoid function. A Prony series is used to fit a linear viscoelastic material model (generalized Maxwell solid model) to the sigmoid function; b) Temperature-dependent CTE according to [22, 23], and modified CTE which is set to zero for temperatures which are above the crystallite melting temperature.

shortened by the use of the WLF or Arrhenius equation. However, in a shrinkage and warpage analysis, high strain rates will only occur at very high temperatures at the outer surface where the cooling rate is high. Below 130 °C, the cooling rate is much lower. For the use of the PRF model, the short time behavior between 10^{-18} and 1.0 s is neglected. The prony series is then modeled at RT with an instantaneous modulus of 1450 N/mm^2 and the first relaxation time is set to 1.0 s. 19 Prony terms are used to model the material behavior in the range 1.0 s until 10^{18} s. To validate this approach, a shrinkage analysis using a linearized version of the PRF model with 19 networks is carried out and compared to a shrinkage analysis using the linear viscoelastic model with 19 and 37 networks respectively.

For small strains, the response of the Neo-Hookean model will be similar to a linear elastic model. At these small strains, the nonlinear behavior of the PRF model results mainly from the viscous flow which is modeled by the strain hardening model (Eq. (13.13)). To obtain a linear viscoelastic representation of the PRF model, the strain hardening model needs be converted to linear flow. This can be achieved by setting the n-parameter of each viscoelastic network i equal to one and the m-parameter equal to zero. The parameter A_i for each network can then be calculated using the instantaneous elastic modulus E, the Prony values g_i, and relaxation times τ_i of each network of the linear viscoelastic model as follows:

$$A_i = \frac{1}{E \, g_i \, \tau_i}. \tag{13.16}$$

For the elastic part of the PRF model, the parameter D_1 of the Neo-Hookean model (Eq. (13.10)) is set to zero to obtain incompressibility. The parameter C_{10} can then be determined using the instantaneous elastic modulus and the Poissons ratio of the linear viscoelastic model (Eq. (13.17)). The stiffness ratios will be equal to the Prony values of the linear model ($s_i = g_i$).

$$C_{10} = \frac{E}{2[2(1+\nu)]} = \frac{E}{6}. \tag{13.17}$$

The results of the two linear viscoelastic models with 37 and 19 prony terms are compared to the linearized PRF model in Fig. 13.4. Fig. 13.4a shows the thermal stress accumulated during the cooling under mold constraint. The stress at the outer surface is much higher due to the higher cooling rate. Fig. 13.4b shows the viscoelastic retardation for a period of 24 h after demolding. The response of the linear viscoelastic model is exactly the same if the prony series is reduced from 37 to 19 terms. The linearized PRF model deviates slightly from the linear viscoelastic representation (Fig. 13.4). However, since the differences are quite small, numerical reasons are suspected.

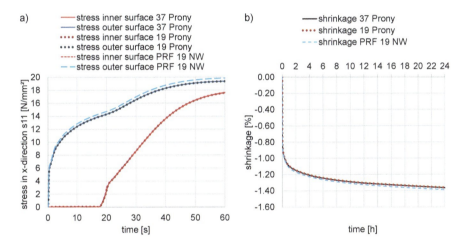

Fig. 13.4: Comparison of a one-element shrinkage analysis using a linear viscoelastic model with 37 Prony terms, a linear viscoelastic model with 19 Prony terms and a linearized PRF model with 19 networks (NW). a) Build up of thermal stresses of the inner and outer surface during cooling under mold constraint; b) Shrinkage for a period of 24 h after demolding.

13.4 Calibration Strategy

Due to the complexity of the PRF model, an elaborate calibration strategy is needed to find suitable material parameters to match a set of experimental measurement data. Using the strain hardening model, each network is described by the four material parameters s_i, A_i, n_i and m_i. In addition, the modulus of the entire network and the activation energy of the Arrhenius equation also have to be determined. Even if only three networks are used, the total amount of parameters to be identified is 14. For six networks 26 parameters and for 12 networks 50 material parameter need to be identified. It can be expected that for a nonlinear viscoelastic model, less networks are needed compared to the linear viscoelastic model. However, considering a large time and temperature range to cover, several networks might be necessary for an accurate representation of the experimental data. This involves the determination of a large set of material parameters. The associated objective function of the optimization problem to solve might have many local minima. To find the best possible solution, the use of global optimization methods is necessary.

In the following, a calibration strategy is presented, which uses functional relations to describe the material parameters of the individual networks. Thus, the dimension of the design space is reduced to a manageable number. However, the use of a global optimization strategy is still associated with high computational effort if a large number of optimization loops is involved. In each optimization loop, a numerical model of the experiment is computed and the results are compared to the measurement data. The most expensive part is the numerical simulation of the material test. If a finite element analysis using Abaqus is carried out, even if just one element is used, several seconds are needed for the verification of license and the interpretation of the input deck. To save computation time, a simplified numerical model which represents the PRF model is implemented in Matlab. In the following, the material data for the model calibration as well as the reduction of the material parameters and the computation time is explained in detail.

13.4.1 Experimental Data

The shrinkage and warpage analysis can be divided into a loading phase in which thermal stresses will build up and relax during the cooling under mold constraint, followed by an unloading (demolding) where the part is free to shrink. For the model calibration, a set of experiments is used which involves a loading and relaxation phase followed by an unloading phase. Therefore, relaxation tests with subsequent unloading at several temperature and strain levels are carried out using a Zwick Kappa Multistation. Material samples of type 1A (DIN EN ISO 527-2) of the blow molding HDPE grade Lupolen 5021DX (LyondellBasell) were taken from compression molded plates (polystat 200 T/2, Servitec, at 210 °C and 200 bar ($2\,10^7$ Pa) for 70 s) with 2 mm thickness. At RT, three different strain levels (0.5%, 1.0% and 2.0%) were investigated. The highest strain level was also tested at different temperatures (RT, 40 °C, 60 °C, 80°C and 100 °C). For each test point, the average of three measurements is taken. At the beginning of the loading phase, a constant strain rate of $0.0015\,\text{s}^{-1}$ was applied to the sample until the specified strain level was reached. The strain level was then held constant (strain controlled) for a period of 600 s to measure the stress relaxation response. At the end of the loading phase, the sample was unloaded (force controlled) with a force rate of 300 N/s. The force was then held constant at zero N for a period of 1800 s to measure the viscoelastic retardation. Fig. 13.5 shows the measured stresses and strains of the relaxation tests with subsequent unloading.

13.4.2 Implementation of a One-Dimensional Model to Reduce Computation Time

To develop an efficient numerical model which represents the PRF equations of a relaxation test with subsequent unloading, four steps are considered. The first step covers the loading, assuming a constant strain rate. Once the final strain is reached, it is held constant over a defined period of time. The third step covers the unloading, assuming a constant stress rate. Once the stress approaches zero, it is held at zero for a defined period. The Matlab implementation will be a one-dimensional representation of the material equations of the PRF model. In contrast to the Abaqus model, only the material equations are solved, so that no spatial discretization is needed. This reduces the system of partial differential equations (PDE) to a system of ordinary

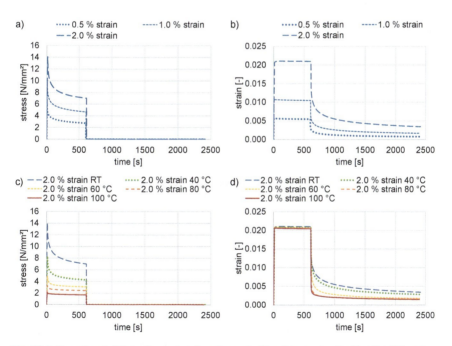

Fig. 13.5: Experimental data of a stress relaxation test with subsequent unloading for different strain and temperature levels. a) Stresses for three different strain levels at RT; b) Strains of the experiments for three different strain levels at RT; c) Stresses of a strain level of 2.0% at different temperatures; d) Strains of a strain level of 2.0% at different temperatures.

differential equations (ODE) which can be efficiently solved using a suitable ODE solver.

For the model we start with the total stress σ_{total} of the framework, which is the sum of all network stresses σ_i

$$\sigma_{\text{total}} = \sigma_0 + \sum_{i=1}^{N} \sigma_i. \tag{13.18}$$

The total strain is equivalent in each network element due to the parallel arrangement (Fig. 13.1) and is decomposed additively into an elastic part (spring) and an inelastic part (dashpot)

$$\varepsilon_{\text{total}} = \varepsilon_i^{\text{el}} + \varepsilon_i^{\text{in}}. \tag{13.19}$$

Since the strains in the performed material tests are small, it can be assumed that a linear elastic material model is sufficient for the description of the springs. The stress of each network is then calculated by Hooke's law:

$$\sigma_i = \varepsilon_i^{\text{el}} E_i, \tag{13.20}$$

where E_i is the elastic modulus of the i'th network. For the stress of the equilibrium network applies:

$$\sigma_0 = \varepsilon_{\text{total}} E_0. \tag{13.21}$$

The strain rate of the dashpots is represented by the strain hardening model.

$$\dot{\varepsilon}_i^{\text{in}} = \left(A_i \sigma_i^{n_i} [(m_i + 1) \varepsilon_i^{\text{in}}]^{m_i} \right)^{\frac{1}{m_i+1}}, \tag{13.22}$$

where $\dot{\varepsilon}_i$ is the inelastic strain rate of the network, and A_i, n_i and m_i are material parameters. By taking the time derivative of Eq. (13.19) and (13.20), we obtain:

$$\dot{\varepsilon}_{\text{total}} = \dot{\varepsilon}_i^{\text{el}} + \dot{\varepsilon}_i^{\text{in}}, \tag{13.23}$$

and

$$\dot{\sigma}_i = \dot{\varepsilon}_i^{\text{el}} E_i. \tag{13.24}$$

Substituting Eq. (13.22) and Eq. (13.24) into Eq. (13.23) and rearranging gives an ordinary differential equation of the form $f(t, \sigma_i, \dot{\sigma}_i) = 0$ for the i'th network

$$\dot{\sigma}_i = \dot{\varepsilon}_{\text{total}} E_i - E_i \left(A_i \sigma_i^{n_i} [(m_i + 1) \varepsilon_i^{\text{in}}]^{m_i} \right)^{\frac{1}{m_i+1}}. \tag{13.25}$$

Equation (13.25) still contains the unknown inelastic strain $\varepsilon_i^{\text{in}}$. By replacing $\varepsilon_i^{\text{in}}$ by the expression $\varepsilon_{\text{total}} - \frac{\sigma_i}{E_i}$, a differential equation is obtained which contains only known quantities

$$\dot{\sigma}_i = \dot{\varepsilon}_{\text{total}} E_i - E_i \left(A_i \sigma_i^{n_i} \left[(m_i + 1) \left(\varepsilon_{\text{total}} - \frac{\sigma_i}{E_i} \right) \right]^{m_i} \right)^{\frac{1}{m_i+1}}. \tag{13.26}$$

13 Shrinkage and Warpage Analysis using Nonlinear Viscoelastic Models

To model the temperature dependency of the material model, the inelastic strain rate of each network is divided by the shift factor $\alpha(T)$

$$\dot{\sigma}_i = \dot{\varepsilon}_{\text{total}} E_i - \frac{E_i \left(A_i \sigma_i^{n_i} \left[(m_i+1)\left(\varepsilon_{\text{total}} - \frac{\sigma_i}{E_i}\right)\right]^{m_i}\right)^{\frac{1}{m_i+1}}}{\alpha(T)}. \quad (13.27)$$

The shift factor is calculated using the Arrhenius equation (Eq. (13.8)). The stresses σ_i of the N individual networks are obtained by solving Eq. (13.27) with an ODE solver. The Matlab solver ode15s turned out to be very efficient for this kind of problem.

Depending on the choice of the parameters m_i and n_i, numerical difficulties can occur if the network stress σ_i is negative. This can in some cases lead to complex numbers. In order to deal with these difficulties, Eq. (13.27) is modified as follows:

$$\dot{\sigma}_i = \dot{\varepsilon}_{\text{total}} E_i - \frac{\sigma_i}{|\sigma_i|} \frac{E_i \left(A_i |\sigma_i|^{n_i} \left[(m_i+1)\left|\varepsilon_{\text{total}} - \frac{\sigma_i}{E_i}\right|\right]^{m_i}\right)^{\frac{1}{m_i+1}}}{\alpha(T)}. \quad (13.28)$$

By taking absolute values of σ_i and $\varepsilon_{\text{total}} - \frac{\sigma_i}{E_i}$, complex numbers are ruled out. The term $\frac{\sigma_i}{|\sigma_i|}$ is introduced to define the direction of the viscous flow.

For the unloading phase and the subsequent holding phase, the external stress rate respectively the external stress (which is zero) is specified but the internal network stresses and stress rates are unknown. Therefore, instead of the equation for the entire system, the differential equations of the dashpots are considered (Eq. (13.22)). The unknown network stress σ_i is replaced by the expression $E_i(\varepsilon_{\text{total}} - \varepsilon_i^{\text{in}})$. One obtains:

$$\dot{\varepsilon}_i^{\text{in}} = \frac{(E_i(\varepsilon_{\text{total}} - \varepsilon_i^{\text{in}}))}{|E_i(\varepsilon_{\text{total}} - \varepsilon_i^{\text{in}})|} \frac{\left(A_i |E_i(\varepsilon_{\text{total}} - \varepsilon_i^{\text{in}})|^{n_i} [(m_i+1)|\varepsilon_i^{\text{in}}|]^{m_i}\right)^{\frac{1}{m_i+1}}}{\alpha(T)}. \quad (13.29)$$

To solve this ODE, it is necessary to replace the total strain with known quantities. For the equilibrium network applies:

$$\sigma_0 = \sigma_{\text{total}} - \sum_{i=1}^{N} (\varepsilon_{\text{total}} E_i - \varepsilon_i^{\text{in}} E_i). \quad (13.30)$$

If we substitute this expression into Eq. (13.21) and rearrange, we get the following equation:

$$\varepsilon_{\text{total}} = \frac{\frac{\sigma_{\text{total}}}{E_0} + \sum_{i=1}^{N} \varepsilon_i^{\text{in}} \frac{E_i}{E_0}}{1 + \sum_{i=1}^{N} \frac{E_i}{E_0}}. \quad (13.31)$$

By substituting $\varepsilon_{\text{total}}$ in Eq. (13.29) by Eq. (13.31), the inelastic strain of the i'th network can be computed using the elastic moduli and inelastic strains of all N networks. The system of coupled differential equations is also solved with the Matlab solver ode15s.

So far, the moduli of the individual networks are specified directly. Alternatively, stiffness ratios as used by Abaqus can be specified to define the stiffness of each network. In this case, the total elastic modulus E_{total} of all networks and a stiffness ratio s_i for each network will be specified. The moduli E_i of each network is then calculated by the following equations:

$$E_i = s_i E_{\text{total}}. \tag{13.32}$$

For the modulus of the equilibrium network applies:

$$E_0 = E_{\text{total}} - \sum_{i=1}^{N} s_i E_{\text{total}}. \tag{13.33}$$

13.4.3 Reduction of Material Parameters

To reduce the amount of parameters in the model calibration, we start with the prony fit of the linear viscoelastic model to the extrapolated sigmoid curve. The sigmoid curve was fitted by the prony series using one relaxation time τ per decade. If we plot the Prony values over the network number, they can be approximated by a normalized Gaussian function (Eq. (13.34), Fig. 13.6)

$$s_i = \frac{1}{12.9} e^{\left(-\frac{1}{2}\left(\frac{i-19.8}{5.15}\right)^2\right)}. \tag{13.34}$$

Using the Gauss function for the description of the stiffness ratio of each network "i", a normalization is needed to ensure that the sum of all stiffness ratios is always smaller or equal to one. We start with the Gauss function without normalization. The parameters p_{s1} and p_{s2} are used to modify the distribution of the stiffness ratios s_i

$$s_i = e^{\left(-\frac{1}{2}\left(\frac{i-p_{s1}}{p_{s2}}\right)^2\right)}. \tag{13.35}$$

The parameter p_{s1} is used to shift the Gauss function on the abscissa. Parameter p_{s2} can be used to change the curvature. In the next step, the sum of the stiffness ratios of all networks is calculated

$$s_{\text{sum}} = \sum_{i=1}^{N} s_i. \tag{13.36}$$

The normalized Gauss function is obtained as follows:

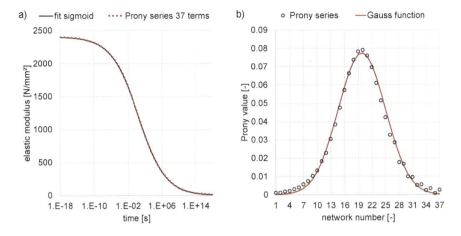

Fig. 13.6: Approximation of the Prony values by a Gauss function. a) Extrapolated master curve which is approximated using a Prony series with 37 terms; b) Distribution of the Prony values with respect to the network number. The Prony values are approximated using a Gaussian function.

$$s_{i_{\text{norm}}} = \frac{p_{s3}}{s_{\text{sum}}} e^{\left(-\frac{1}{2}\left(\frac{i-p_{s1}}{p_{s2}}\right)^2\right)}. \tag{13.37}$$

Using the parameter p_{s3}, the sum of the stiffness ratios can be modified to obtain values less than one. Thus, the stiffness ratio of an arbitrary number of networks is described by just three parameters. Also the sum of the stiffness ratios is always less or equal to one, so that no restrictions are necessary within the optimization. As presented in Sect. 13.3, the short-time behavior of the master curve will be neglected to overcome convergence problems at higher temperatures. The first relaxation time τ_i is set to 1.0 s, so that 19 networks are used in total. Therefore, the Gauss function is shifted using parameter p_{s1}.

The relaxation times τ_i of the prony series are held constant to ensure that the distribution of the stiffness ratios follows Eq. (13.37). For the PRF model, we assume that the A-values can be calculated by Eq. (13.16) using the stiffness $s_i E$ and the relaxation time τ_i. For the linearized PRF model ($n_i = 1$ and $m_i = 0$), the inelastic strain rate of the strain hardening model (Eq. (13.22)) is proportional to the stress σ_i. However, for $n > 1$ and $m < 0$ the strain hardening law becomes nonlinear. As the network stress σ_i is raised to the power of n_i, the inelastic strain rate increases nonlinearly with increasing stress. At stresses below $1\,\text{N/mm}^2$ the strain rate will be even slower compared to a linearized model and it drastically increases for higher n-values if the stress increases. If higher stresses are involved it might be necessary to adjust the A-values of the networks. However, the stresses for the investigated HDPE are rather moderate (less than $15\,\text{N/mm}^2$, Fig. 13.5). Using 19 networks, the individual network stress σ_i will be close to one for most networks. In this case, the inelastic strain rate will be a nonlinear function of the applied stress but it won't differ too much from the strain rate of the linearized model. Therefore, it seems reasonable to use constant relaxation times τ of the linear model to calculate the A-values of the PRF-model via Eq. (13.16). The A-values are thus excluded from the design space of the parameter optimization.

Similarly to the stiffness ratios, n- and m-parameters of all networks will also be described using a suitable functional relation. In order to ensure stability of the model over the entire time and temperature range, the following boundaries are set:

$$1.0 \leq n_i \leq 5.0, \tag{13.38}$$

$$-0.7 \leq m_i \leq 0.0. \tag{13.39}$$

In contrast to the stiffness ratios, a suitable distribution of the n- and m-parameters with respect to the network number is unknown. However, it is assumed that the individual n- and m-parameters can be represented by a monotonically increasing or decreasing function. A sigmoid function using four parameters is used for the distribution. Depending on the parameter selection, the sigmoid function can describe an s-curve, a curve or even a constant (Eq. (13.40), (13.41)). Some possible distributions for the parameters s, n, and m are shown in Fig. 13.7.

$$n_i = p_{n1} + ((p_{n1} - 1) p_{n2}) \tanh\left(\frac{i - p_{n3}}{p_{n4}}\right), \tag{13.40}$$

$$m_i = p_{m1} + (p_{m1} p_{m2}) \tanh\left(\frac{i - p_{m3}}{p_{m4}}\right). \tag{13.41}$$

Using the described Gauss and sigmoid functions, the design space for the model calibration is reduced to 11 parameters independent of the amount of networks.

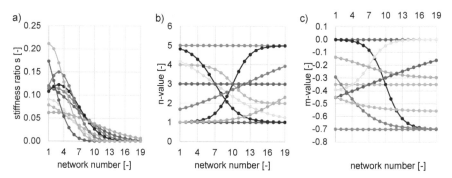

Fig. 13.7: Distribution of the material parameters of the individual networks. The curves show different possible distributions based on the three parameter Gauss function and the four parameter sigmoid functions. a) Example distributions of stiffness ratios s_i using a three parameter description of a Gaussian function; b) Example distributions of the n-values using a four parameter sigmoid function; c) Example distributions of the m-values using a four parameter sigmoid function.

13.4.4 Calibration Workflow

For the calibration of the PRF model, the process automation and design exploration software tool Simulia Isight (Dassault Systèmes) is used. The complete optimization workflow is illustrated in Fig. 13.8.

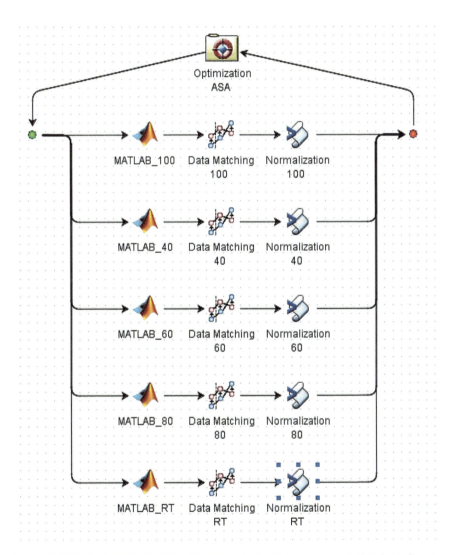

Fig. 13.8: Optimization workflow using the process automation and design exploration software Simulia Isight.

The global optimization algorithm Adaptive Simulated Annealing is used with a total amount of 10,000 design loops. This algorithm is well suited for the search of a global optimum especially for highly nonlinear problems with short calculation times [18]. In each design loop of the optimization, a Matlab component is executed which runs a numerical model of the experimental test. The Matlab model which represents the material tests at RT is used to simulate all three strain levels. All testing temperatures run in parallel, so that all 10,000 runs are calculated in less than three hours on a personal computer.

The data matching components are used for the calculation of the sum of the absolute difference between points on the simulated curves and the experimental curves. The simulation data is therefore interpolated to the measurement data. The error of the loading and unloading phase is calculated separately. After the error calculation, a script component is used to normalize the data to obtain an error value in percent. This is necessary in order to achieve equal weighting to loading and unloading. Thus the error of each virtual experiment phase is calculated as follows:

$$\mathrm{Err} = \frac{\sum_{j=1}^{N} |Y_{j_{\mathrm{exp}}} - Y_{j_{\mathrm{sim}}}|}{\sum_{j=1}^{N} Y_{j_{\mathrm{exp}}}} \cdot 100\,\%, \tag{13.42}$$

where Err is the error value in %, $Y_{j_{\mathrm{exp}}}$ is the experimental value at data point j, and $Y_{j_{\mathrm{sim}}}$ is the simulated value at data point j. To obtain the total error of all experiments, the average of all error values is calculated.

The calibration workflow using the Gauss and sigmoid functions for the parameter reduction is tested and compared to an optimization where the parameters are varied freely between boundaries. Using the free parameter variation, the stiffness of the networks is specified directly by the elastic modulus of the network to avoid violation of the condition that the sum of all stiffness ratios must be less or equal to one. The parameters as well as their boundaries of the free parameter variations are given by Table 13.1. For all model calibrations, the Arrhenius function using the same activation energy E_A for all networks is used. For the calibration model which uses the Gauss and sigmoid functions to describe the parameters s_i, n_i, and m_i, the elastic modulus of the whole network E_{total} is also used as a parameter in the calibration process. The lower boundary is set to $1,000\,\mathrm{N/mm^2}$ and the upper boundary is set to $2,000\,\mathrm{N/mm^2}$. The free optimization is tested with different numbers of networks, that is to say three, six and 12. Table 13.2 gives an overview over the different models which are used in the calibration process.

13.5 Results

Using the model variations shown in Table 13.2 the PRF model was calibrated using the experimental measurement data illustrated in Fig. 13.5. The overall errors

13 Shrinkage and Warpage Analysis using Nonlinear Viscoelastic Models

Table 13.1: Boundaries of the free parameter variations.

Parameter	lower bound	upper bound
E_0	$1\,\frac{\text{N}}{\text{mm}^2}$	$700\,\frac{\text{N}}{\text{mm}^2}$
E_i	$1\,\frac{\text{N}}{\text{mm}^2}$	$700\,\frac{\text{N}}{\text{mm}^2}$
A_i	$10^{-15}\left(\frac{\text{N}}{\text{mm}^2}\right)^{-n}\text{s}^{-m-1}$	$10^{-3}\left(\frac{\text{N}}{\text{mm}^2}\right)^{-n}\text{s}^{-m-1}$
m_i	-0.7	0.0
n_i	1.0	5.0
EA	$100{,}000\,\frac{\text{J}}{\text{mol}}$	$300{,}000\,\frac{\text{J}}{\text{mol}}$

Table 13.2: Model variations which are tested in the model calibration.

NW	Model	Parameter Variation	Parameter Amount
3	PRF	free	14
6	PRF	free	26
12	PRF	free	50
19	PRF	function	13

of the model calibrations are shown in Table 13.3. Comparing the results of the free parameter variation with the use of functional relations between the individual network parameters, it can be seen that the latter approach achieves the best results with an error of about 5.9 % (Table 13.3). The poorest result is obtained by the PRF model with just three networks, with an error of 20.3 %. Doubling the number of networks from three to six halves the error. A further duplication from six to 12 networks leads to slightly poorer results.

Figure 13.9 shows the distribution of s_i, n_i, and m_i of the best design point. It is clearly visible that the stiffness ratios become zero after the ninth network. In this case the amount of networks can be reduced to nine. The results are identical to the results using 19 networks. The n-values show a monotonically increasing trend, whereas the m-values lie on a decreasing s-curve.

Figure 13.10 shows a comparison between the results of the Matlab models and Abaqus finite element simulations for the best design point. The results are in very good agreement. Only at the highest strain level, a negligible deviation in the unloading phase can be observed.

Table 13.3: Representation of the error value of the free model calibration using three, six and 12 networks compared to the error of the model calibration using functional relations between the network parameters.

NW	Model	Parameter Variation	Parameter Amount	Error
3	PRF	free	14	20.3%
6	PRF	free	26	10.8%
12	PRF	free	50	11.5%
19	PRF	function	13	5.9%

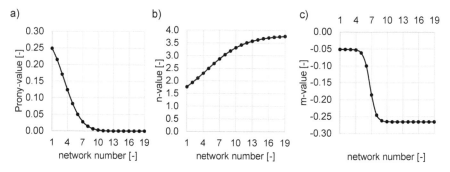

Fig. 13.9: Distribution of the material parameters of the individual networks for the best design point. a) Distributions of the stiffness ratios s_i using the three parameter description of the Gaussian function; b) Distributions of the n-values using the four parameter sigmoid function; c) Distributions of the m-values using the four parameter sigmoid function.

Comparing the Abaqus results of the best design point to the experimental data, a good agreement is observed (Fig. 13.11). The stress relaxation curve at 2% and 100°C with an error of 24%, the stress relaxation curve at 2% strain at RT with 19% error, and the unloading curve at 2% strain and 80°C with an error of about 13% show the largest deviation. All other curves are in good agreement with the experimental data.

In Fig. 13.12, the PRF model using the parameters of the best design point is integrated in the shrinkage model and the results are compared to the results of the linear viscoelastic model which was calibrated using the master curve. The stress history of the PRF model during the cooling under mold constraint is similar to the linear viscoelastic model. However, the stresses of the PRF model are a slightly lower than the stresses of the linear viscoelastic model. Moreover, the difference in stress between the inner surface and the outer surface is smaller for the PRF model. Comparing the shrinkage behavior in the first 24 h, the PRF model shows a stronger retardation.

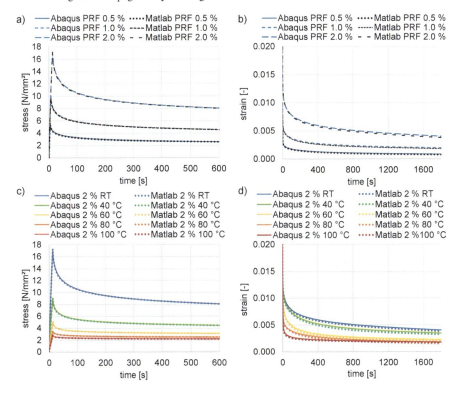

Fig. 13.10: Comparison of the results of the numerical model implemented in Matlab with an Abaqus finite element simulation at the best design point. a) Stresses of the loading phase for the three different strain levels at RT; b) Strains of the unloading phase for the three different strain levels at RT; c) Stresses of the loading phase at a strain level of 2.0% at different temperatures; d) Strains of the unloading phase at a strain level of 2.0% at different temperatures.

13.6 Discussion and Outlook

As shown in Table 13.3, the calibration workflow using functional relations between the parameters of the individual networks achieved the best results so far. This could be explained by the fact that a sufficient number of networks is used, whereas the amount of material parameters which need to be identified is still low. The poor results of the free calibration using only three networks (error of 20.3%) indicate that three networks in conjunction with the Arrhenius function seem insufficient to cover the entire time and temperature range of the experiments. Increasing the amount of networks to six improves the result significantly, but the error is still twice as large as the best solution of the calibration workflow using functional relations between the parameters of the individual networks. It can be assumed that at least up to a certain point, an increasing number of networks improves the prediction accuracy of the model. However, the use of more networks is always associated with a higher amount of material parameters which need to be identified. Using a global optimization

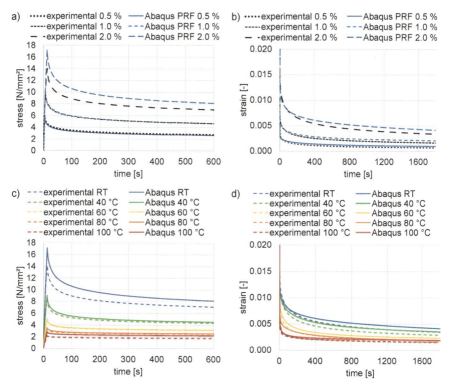

Fig. 13.11: Comparison of the Abaqus finite element simulation at the best design point with the experimental measurement data. a) Stresses of the loading phase for the three different strain levels at RT; b) Strains of the unloading phase for the three different strain levels at RT; c) Stresses of the loading phase at a strain level of 2.0 % at different temperatures; d) Strains of the unloading phase at a strain level of 2.0 % at different temperatures.

approach, the global minimum might be found with a certain probability but there is no guarantee that it is actually found by a finite number of function evaluations [27]. The fact that the use of six networks achieved a slightly better result than the use of 12 networks (Table 13.3) indicates that the number of material parameters might be too high to find a good solution using 10,000 design evaluations. In this case, a higher amount of design evaluations could be necessary. The calibration approach, which uses functional relations to describe the parameters of the networks, requires only 13 parameters regardless of the number of networks. Furthermore, a lot of parameter permutations are excluded because the parameters of the individual networks are described by continuous functions which are either increasing or decreasing. Another interesting fact is that the calibration using functional relations seems to reduce the amount of networks. In this case, the optimization was started with 19 networks but it seems that nine networks are sufficient to match the experimental data with high accuracy.

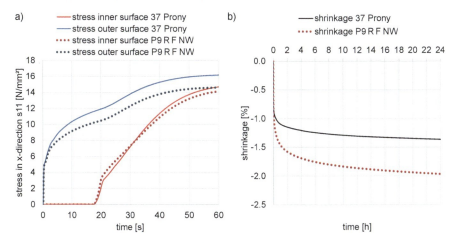

Fig. 13.12: Comparison of a one-element shrinkage analysis using a linear viscoelastic model with 37 Prony terms and the calibrated PRF model of the best design point. a) Build up of thermal stresses of the inner and outer surface during cooling under mold constraint; b) Shrinkage for a period of 24 h after demolding.

Alternatively to the use of the reduced time concept for the modeling of temperature effects, the instantaneous modulus, the stiffness ratios, and the creep model can be defined as temperature dependent by tabular values. The amount of networks necessary to cover the entire time and temperature range might be reduced in this case. However, the main advantage of the reduced time concept is that the temperature dependency is described by a continuous shift function. Thus, an extrapolation to higher temperatures seems reasonable, since the shift functions are based on experimental observations. This is essential, since experimental investigations at temperatures between 100 °C and 130 °C are difficult due to low structural stiffness at this temperature range. Nevertheless, the alternative temperature-dependent modeling using tabular values should be investigated in future work. Therefore, the use of functional relations between the parameters of the different temperature levels could also be a promising approach for a successful model calibration, especially if extrapolation to higher temperatures is necessary. In this case, the more temperature levels are tested, the better. Even if not all temperature levels are used in the model calibration, the temperature levels between the calibrated curves can be used for validation purposes. Furthermore, alternative nonlinear viscoelastic material models like the PN model provided by the PolyUMod library (PolymerFEM LLC) [20] should be investigated. One of the main disadvantages of the PRF model is that currently only isotropic thermal expansion is supported. The PN model also supports the use of constant orthotropic thermal expansion [28]. Piece-wise linear thermal expansion, which can be used to model the temperature dependency of the CTE, is currently only supported for isotropic behavior [28].

The integration of the calibrated PRF model in the shrinkage and warpage analysis leads to lower stresses during the cooling under mold constraint compared to the linear viscoelastic model (Fig. 13.12a). This is reasonable, since the linear viscoelastic model usually overestimates the stresses at higher strain levels. Moreover, the smaller stress difference between the inner and outer surface at the time of demolding (60 s) can be explained by the fact that the inelastic strain rate increases at higher stresses, so that the relaxation will be accelerated. Nevertheless, the experimental data base which was used for the model calibration in this study is relatively small. For a successful calibration covering the entire time and temperature range of the shrinkage and warpage analysis, an extensive experimental database is needed. Additionally, the various strain levels should be investigated at all temperature levels. For a comparison of the PRF model and the linear viscoelastic model in terms of prediction accuracy, an extensive experimental database considering shrinkage and warpage is needed. The database used in [8] was limited and the part shrinkage was measured six days after demolding, so that it does not provide information about the viscoelastic retardation in the first hours after demolding. Experimental data of the dynamic shrinkage behavior of simple blow molded parts for an extensive set of process conditions, as well as experimental warpage data of complex blow molded parts could provide valuable information for the model validation.

Acknowledgements This work was funded by the German Federal Ministry of Education and Research (BMBF) within the project "Resource Optimized Forming" (ROForm) in the program FH-Kooperativ under the funding code 13FH514KX9. We would also like to thank the Graduate Institute and the TREE Institute of Hochschule Bonn-Rhein-Sieg University of Applied Sciences for supporting this work by granting a scholarship.

References

[1] Laroche D, Kabanemi KK, Pecora L, Diraddo RW (1999) Integrated numerical modeling of the blow molding process. Polymer Engineering and Science **39**(7):1223–1233

[2] Kabanemi KK, Vaillancourt H, Wang H, Salloum G (1998) Residual stresses, shrinkage, and warpage of complex injection molded products: Numerical simulation and experimental validation. Polymer Engineering and Science **38**(1):21–37

[3] Williams ML, Landel RF, Ferry JD (1955) The temperature dependence of relaxation mechanisms in amorphous polymers and other glass-forming liquids. Journal of the American Chemical Society **77**(14):3701–3707

[4] Debergue P, Massé H, Thibault F, DiRaddo R (2003) Modelling of solidification deformation in automotive formed parts. SAE Transactions **112**:359–365

[5] Benrabah Z, Debergue P, DiRaddo R (2006) Deflashing of automotive formed parts: Warpage and tolerance issues. SAE International

[6] Benrabah Z, Mir H, Zhang Y (2013) Thermo-viscoelastic model for shrinkage and warpage prediction during cooling and solidification of automotive

blow molded parts. SAE International Journal of Materials and Manufacturing 6(2):349–364
[7] Benrabah Z, Bardetti A, Ilinca F, Ward G (2018) Numerical simulation of shrinkage and warpage deformation of an intermittent-extrusion blow molded part: validation case study. In: ANTECH 2018 Conference Proceedings, Society of Plastic Engineering. Blow Molding Division, ANTECH 2018 Conference, May 7-10, 2018, Orlando, FL, USA
[8] Michels P, Bruch O, Evers-Dietze B, Grotenburg D, Ramakers-van Dorp E, Altenbach H (2022) Shrinkage simulation of blow molded parts using viscoelastic material models. Materialwissenschaft und Werkstofftechnik 53(4):449–466
[9] Ramakers-van Dorp E, Blume C, Haedecke T, Pata V, Reith D, Bruch O, Möginger B, Hausnerova B (2019) Process-dependent structural and deformation properties of extrusion blow molding parts. Polymer Testing 77:105,903
[10] Grommes D, Bruch O, Geilen J (2016) Investigation of the influencing factors on the process-dependent elasticity modulus in extrusion blow molded plastic containers for material modelling in the finite element simulation. AIP Conference Proceedings 1779(1):050,013-1–050,013-5
[11] Ramakers-van Dorp E (2019) Process-induced thermal and viscoelastic behavior of extrusion blow molded parts. PhD thesis, Tomas Bata University in Zlin, Zlin
[12] Lai J, Bakker A (1995) Analysis of the non-linear creep of high-density polyethylene. Polymer 36(1):93–99
[13] Schapery RA (1969) On the characterization of nonlinear viscoelastic materials. Polymer Engineering and Science 9(4):295–310
[14] Michaeli W, Brandt M, Brinkmann M, Schmachtenberg E (2006) Simulation des nicht-linear viskoelastischen Werkstoffverhaltens von Kunststoffen mit dem 3D-Deformationsmodell. Zeitschrift Kunststofftechnik 2(5)
[15] Bergstrom JS, Bischoff JE (2010) An advanced thermomechanical constitutive model for UHMWPE. International Journal of Structural Changes in Solids – Mechanics and Applications 2(1):31–39
[16] Lapczyk I, Hurtado JA, Govindarajan SM (2012) A parallel rheological framework for modeling elastomers and polymers. In: American Chemical Society (ed) 182nd Technical Meeting of the Rubber Division of the American Chemical Society, pp 1840–1859
[17] Hurtado J, Lapczyk I, Govindarajan S (2013) Parallel rheological framework to model non-linear viscoelasticity, permanent set and mullins effect in elastomers. In: Alonso A (ed) Constitutive Models for Rubber VIII, CRC Press, pp 95–100
[18] Dassault Systèmes (2023) SIMULIA User Assistance 2023
[19] Bergstrom J (2015) Mechanics of Solid Polymers: Theory and Computational Modeling. Elsevier, Amsterdam
[20] PolymerFEM LLC (2023) PolyUMod
[21] Gutierrez-Lemini D (2014) Engineering Viscoelasticity. Springer, New York and Heidelberg and Dordrecht and London
[22] Kipping A (2003) Thermomechanische analyse der kühlphase beim extrusionsblasformen von kunststoffen. PhD thesis, Universität Siegen, Siegen

[23] Kunststoffmaschinen VDMAFG (1979) Kenndaten für die Verarbeitung thermoplastischer Kunststoffe. Hanser, München
[24] Koppelmann J (1959) Über den dynamischen Elastizitätsmodul von Polymethacrylsäuremethylester bei sehr tiefen Frequenzen. Kolloid-Zeitschrift **164**(1):31–34
[25] Sommer W (1959) Elastisches Verhalten von Polyvinylchlorid bei statischer und dynamischer Beanspruchung. Kolloid-Zeitschrift **167**(2):97–131
[26] Kulik M (1974) Ein beitrag zur analyse des kontinuierlichen extrusionsblasformens. PhD thesis, Fak. f. Maschinenwesen, RWTH Aachen, Achen
[27] Harzheim L (2014) Strukturoptimierung: Grundlagen und Anwendungen, 2nd edn. Verlag Europa-Lehrmittel, Haan
[28] PolymerFEM LLC (2022) PolyUMod: A Library of Advanced User Materials: Version: 7.0.1

Chapter 14
Modeling Solid Materials in DEM Using the Micropolar Theory

Przemysław Nosal and Artur Ganczarski

Abstract The discrete element method (DEM) is a widely accepted method used in simulations of the powder sintering process, where the real object is replaced by a set of particles representing elastic-plastic material. Recently, this method has also been used to model solid materials. In this work, we present a local constitutive model for the determination of the interaction forces, which is based on micropolar theory. The model shows an elasto-viscoplastic behavior, thus an adaptation of the Johnson–Cook description of flow stress, which allows the analysis of solid materials subjected to high strain rates. The results of the DEM simulations received after calibration of the model parameters agree well with the experimental data from literature.

14.1 Introduction

The concept of the discrete element method was introduced by Cundall [1] in 1971 and was further developed by Cundall and Strack [2]. The increase in computing power of computers has contributed to the development of DEM in terms of both the complexity of the contact models used and the number of new phenomena analyzed using this method. The main area of application of the discrete element method is broadly understood geomechanics, where it is used to simulate soil behavior or analyze rock cracking [3–7]. The discrete element method is also a widely accepted method used in simulations of the powder sintering process [8–11], and currently more and more attention is being paid to adapting this method to structural analysis, where the real object is replaced by a set of particles representing elastic plastic

Przemysław Nosal
AGH University of Science and Technology, Mickiewicza Av. 30, 30-059 Cracow, Poland e-mail: pnosal@agh.edu.pl

Artur Ganczarski
Cracow University of Technology, Jana Pawła II Av. 37, 31-864 Cracow, Poland e-mail: artur.ganczarski@pk.edu.pl

material [12–15], including phenomena such as fracture [16] and crack propagation analysis [17].

The main difference between the discrete element method and the finite element method is the discretization of the object considered. In the case of DEM, a material is represented by an agglomeration of rigid or deformable particles that interact with each other by means of contact forces and moments. In turn, in FEM, the object considered is divided into finite elements that are interconnected at individual nodes. This is a significant difference because the lack of bonds between molecules in DEM allows for the analysis of problems related, e.g. with the formation of discontinuities in the material, or problems in which significant plastic deformations occur. In addition, the need to store large stiffness matrices, which are the basis for implicit integration, disappears. In general, discrete elements can have any shape [18], however, due to the simplicity in formulating the mathematical description and efficiency of the calculations, spherical particles [19] are most often used.

In this work, we present a local constitutive model for the determination of interaction forces, which is based on micropolar theory [20, 21]. The model shows an elasto-viscoplastic behavior, thus an adaptation of the Johnson-Cook description of flow stress [22], which allows the analysis of solid materials subjected to high strain rates. The model is validated in the analysis of uniaxial tensile tests of an aluminum alloy sample. The tests considered two constant values of strain rate and two values of discrete element radius, additionally the consideration of classical Cauchy theory is presented by neglecting the coupled stress component. The results of the DEM simulations received after calibration of the model parameters agree well with the experimental data provided by [23].

14.2 Formulation of the Thermo-Elasto-Viscoplastic Contact Model

14.2.1 Short introduction to DEM basics

As mentioned above, the interaction of the particles causes the contact force vector F (Fig. 14.1). As a result of the interaction, in addition to the translational movement, there is also a rotational movement of the bodies, where the point c is treated as a temporary center of rotation. The angle of rotation is determined by the moment vector M, which is related to the resulting force F. The reliability of this method depends mainly on the proper definition of the contact model, which can be seen as a micromechanical model of the material. However, it is not trivial to define material properties on the micro scale, and they are usually not related to macro properties. Therefore, to identify these parameters, the constitutive model used must be calibrated.

The discrete element method was developed to analyze the dynamic interaction of a set of particles in contact with each other. In general, these particles are treated as

14 Modeling Solid Materials in DEM Using the Micropolar Theory

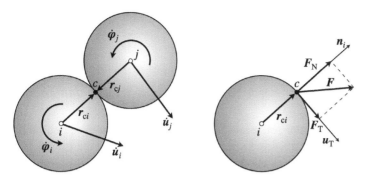

Fig. 14.1: Schematic illustration of the particle-particle interaction with force decomposition on normal F_N and tangential F_T direction.

ideally rigid bodies that have six degrees of freedom. The interaction forces between particles F are modeled using a system of fictitious springs with stiffness k_N, k_T, and for more complex models, additional elements are used in the form of dampers c_N or friction pairs μ_T are used. The resultant interaction force is the sum of the normal and tangential components

$$F = F_N + F_T = F_N n + F_T \tag{14.1}$$

where n is the unit normal vector of the plane located at the point of contact and pointing outward from the discrete element

$$n = \frac{x_j - x_i}{|x_j - x_i|}. \tag{14.2}$$

Then the resultant force is used to calculate the position and orientation of a given discrete element, using Newton–Euler dynamic equilibrium equations

$$m_i \ddot{u}_i = F_i = \sum_{n=1}^{N_i^c} F_n, \qquad I_i \ddot{\phi}_i = M_i = \sum_{n=1}^{N_i^c} r_{ci} \times F_n \tag{14.3}$$

where N_i^c is the number of interactions of i-th element, and r_{ci} is the vector connecting the center of mass of the particle with the contact point c (Fig. 14.1). While integrating the first equation of motion gives us information about the position of the element, integrating the second equation allows us to determine the rotation of the local coordinate system in relation to the reference system.

At the moment of interaction, the initial length l_0 of the fictitious element in the form of a spring, contained between the particles, is defined as

$$l_0 = \left| x_j^0 - x_i^0 \right|. \tag{14.4}$$

This length is treated as a reference value used to calculate other geometric relationships.

14.2.2 DEM Interaction Force Model Based on Micropolar Theory

At the beginning of this section, we recall the basic constitutive relations that describe Cosserat continua. The modified versions of these relations, where the strain tensor and the curvature tensor are decomposed into a symmetric () and non-symmetric <> part [21], are presented

$$\sigma_{ij} = 2\mu\varepsilon_{(ij)} + 2\upsilon\varepsilon_{<ij>} + (\lambda\varepsilon_{kk} - \alpha'\theta)\delta_{ij},$$
$$\mu_{ij} = 2\gamma\kappa_{(ij)} + 2\eta\kappa_{<ij>} + \Upsilon\kappa_{kk}\delta_{ij} \quad (14.5)$$

where σ_{ij}, μ_{ij} are the stress and coupled stress tensors respectively, μ, λ are the Lamé constants, υ, γ, η are new elastic constants, α' and $\Upsilon = 0$ are thermomechanical parameters, and δ_{ij} is a Kronecker delta. Since in (14.5) the symmetric and non-symmetric parts of these tensors were used, we recall the description of the strain tensors

$$\varepsilon_{(ij)} = \frac{1}{2}(u_{j,i} + u_{i,j}), \qquad \varepsilon_{<ij>} = \frac{1}{2}(u_{j,i} - u_{i,j}) - \epsilon_{ijk}\,\omega_k \quad (14.6)$$

as well as the curvature tensors

$$\kappa_{(ij)} = \frac{1}{2}(\omega_{j,i} + \omega_{i,j}), \qquad \kappa_{<ij>} = \frac{1}{2}(\omega_{j,i} - \omega_{i,j}). \quad (14.7)$$

In the present model, the description is limited to a simplified theory of Cosserats' brothers known as the pseudo-Cosserat continuum. This theory assumed that

$$\varepsilon_{<ij>} = \frac{1}{2}(u_{j,i} - u_{i,j}) - \epsilon_{ijk}\,\omega_k = 0,$$

which leads to the relation

$$\omega_k = \frac{1}{2}\epsilon_{klm}\,u_{m,l} \quad (14.8)$$

associating the rotation with the displacement field. Although the non-symmetric part of the strain tensor is equal to zero, the corresponding non-symmetric part of the stress tensor $\sigma_{<ij>} = 2\upsilon\varepsilon_{<ij>}$ does not vanish and can be determined from the equilibrium equation of momentum conservation

$$\epsilon_{ijk}\,\sigma_{jk} + \mu_{ji,j} = 0. \quad (14.9)$$

Taking into account the particle-particle interaction in the discrete element method, the forces and the corresponding moment are located in the plane Π determined by vectors n and e (Fig. 14.2a) in each time step, where

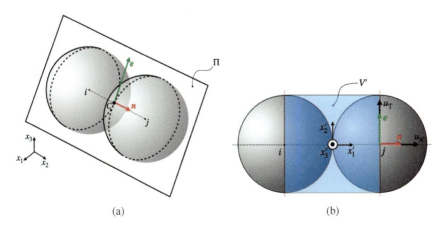

Fig. 14.2: Illustration of the displacement plane Π determined by the vectors $\boldsymbol{n}$ and $\boldsymbol{e}$ a), geometry of the virtual volume element V' b).

$$\boldsymbol{e} = \frac{\boldsymbol{u}_T}{|\boldsymbol{u}_T|}. \tag{14.10}$$

We assumed that the virtual volume element (Fig. 14.2b) is subjected to deformation by prescribed displacements $\boldsymbol{u}_N$ and $\boldsymbol{u}_T$ in this particular plane. We can rewrite those displacements as $\boldsymbol{u}_N = \boldsymbol{n} u_N$ and $\boldsymbol{u}_T = \boldsymbol{e} u_T$, and change the indices to $u_1 = u_N, u_2 = u_T$. Now we consider the deformation of the virtual element in a plane as a combination of pure shear deformation and rigid rotation, where

$$\omega_3 = \frac{1}{2} u_{2,1} = \frac{u_2}{2 l_0}. \tag{14.11}$$

This results in definitions of displacement and rotation vectors as follows

$$\boldsymbol{u} = \{u_1, u_2, 0\}^T, \quad \boldsymbol{\omega} = \{0, 0, \omega_3\}^T. \tag{14.12}$$

Based on the description widely used to determine the strains in DEM models, we can determine the corresponding strain components as

$$\varepsilon_{11} = \frac{u_1}{l_0}, \quad \varepsilon_{12} = \varepsilon_{21} = \frac{1}{2}\frac{u_2}{l_0} \tag{14.13}$$

then the curvature components

$$\kappa_{(13)} = \frac{1}{2}(\omega_{3,1} + \underbrace{\omega_{1,3}}_{=0}) = \frac{1}{2}\kappa_{13} = \frac{1}{4}u_{2,11},$$

$$\kappa_{<13>} = \frac{1}{2}(\omega_{3,1} - \underbrace{\omega_{1,3}}_{=0}) = \frac{1}{2}\kappa_{13} = \frac{1}{4}u_{2,11},$$

$$\kappa_{(23)} = \frac{1}{2}(\omega_{3,2} + \underbrace{\omega_{2,3}}_{=0}) = \frac{1}{2}\kappa_{23} = \frac{1}{4}u_{2,12},$$

$$\kappa_{<23>} = \frac{1}{2}(\omega_{3,2} - \underbrace{\omega_{2,3}}_{=0}) = \frac{1}{2}\kappa_{23} = \frac{1}{4}u_{2,12}.$$

(14.14)

The use of such a deformation, which is the same as the widely used approach in the DEM method, makes it impossible to obtain higher-order derivatives required in the current model. For this reason, a shape function was introduced in the form that describes the deformation of the virtual element, depending on the relative displacement of the centers of the particles. It was assumed that the surfaces perpendicular to the direction of the x_1' axis do not subject to warp (Fig. 14.3). The third-degree polynomial $N(x_1') = c_1 + c_2 x_1' + c_3 x_1'^2 + c_4 x_1'^3$ was used to determine the form of the function, where x_1' is the coordinate in the local coordinate system, where, for simplicity, the origin of the system is located in the center of the particle i. The assumed degree of the polynomial guarantees the differentiability of the function in the required range determined by the constitutive equations of the micropolar medium. In order to determine the constants occurring in the polynomial, boundary conditions in the form of

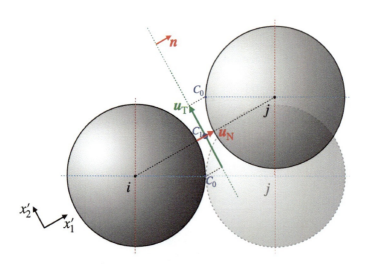

Fig. 14.3: A scheme illustrating the relative displacement of a discrete element with the normal u_N and tangent u_N components of the displacement vector in a local coordinate system x_1', x_2'.

$$\mathcal{N}(x_1')|_{x_1'=0} = 0, \qquad \mathcal{N}(x_1')|_{x_1'=l_0} = u_2 \qquad (14.15)$$

while for the first derivative $d\mathcal{N}(x_1')/dx_1' = c_2 + 2c_3 x_1' + 3c_4 x_1'^2$, due to the kinematic condition (14.8) adopted

$$\left.\frac{d\mathcal{N}(x_1')}{dx_1'}\right|_{x_1'=0} = \frac{u_2}{2l_0}, \qquad \left.\frac{d\mathcal{N}(x_1')}{dx_1'}\right|_{x_1'=l_0} = \frac{u_2}{2l_0}. \qquad (14.16)$$

Finally, the interpolation function will take the form of

$$\mathcal{N}(x_1') = \frac{1}{2l_0} x_1' + \frac{3}{2l_0^2} x_1'^2 - \frac{1}{l_0^3} x_1'^3. \qquad (14.17)$$

Figure 14.4 shows a graphical interpretation of the function (14.17) and its derivatives, where the $u_2 \in \{0,1\}$ range was assumed to illustrate the dependence on the displacement of the discrete element. The reference length was assumed as the unit $l_0 = 1$. Now, we can compute the curvature components and their derivatives, where the quantity κ_{kk} vanishes due to div$\omega = 0$. Since $\mathcal{N}$ is a function of only x_1', the second rank component of tensor κ will be zero

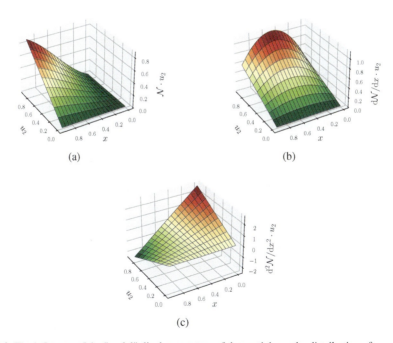

Fig. 14.4: The influence of the "nodal" displacement u_2 of the particle on the distribution of: displacements a), strains b) and curvature c) along the x_1' coordinate of the virtual element V'.

$$\kappa_{13} = \frac{1}{2} u_{2,11} = -\frac{3}{2 l_0^2} u_2, \qquad \kappa_{23} = \frac{1}{2} u_{2,12} = 0. \tag{14.18}$$

Determining the moment and force of two particles that contact each other requires an analysis of the state of stress in the plane of interaction Π. Due to the deformation, a stress state will be created in the virtual element, for which it is assumed that the surfaces $x_3' = \pm r$ are stress-free, therefore $\sigma_{13} = \sigma_{23} = \sigma_{31} = \sigma_{32} = \sigma_{33} = \mu_{31} = \mu_{32} = \mu_{33} = 0$. It can be seen that in the considered element, the stress state is reduced to a plane problem, and due to the kinematics of the system, the coupled stress components μ_{11}, μ_{12} and the stress σ_{22} are equal to zero. Since the nonsymmetric strain components are zero $\varepsilon_{<12>} = \varepsilon_{<21>} = 0$, therefore $\varepsilon_{12} = \varepsilon_{(12)}$ and $\varepsilon_{21} = \varepsilon_{(21)}$

$$\begin{aligned}
\sigma_{11} &= 2\mu\varepsilon_{11} + \frac{\nu E}{1+\nu}\varepsilon_{11} - E\alpha\theta = E\left(\frac{u_1}{l_0} - \alpha\theta\right), \\
\sigma_{12} &= 2\mu\varepsilon_{12} - \frac{1}{4}\xi\left(\kappa_{13,1} + \kappa_{23,2}\right) = \mu\frac{u_2}{l_0} + \frac{3}{2}\xi\frac{u_2}{l_0^3}, \\
\sigma_{21} &= 2\mu\varepsilon_{21} + \frac{1}{4}\xi\left(\kappa_{13,1} + \kappa_{23,2}\right) = \mu\frac{u_2}{l_0} - \frac{3}{2}\xi\frac{u_2}{l_0^3}, \\
\mu_{13} &= \xi\kappa_{13} = -\frac{3}{2}\xi\frac{u_2}{l_0^2}.
\end{aligned} \tag{14.19}$$

The constant ξ is a combination of previous constants $\gamma + \eta$, and was introduced after simple manipulations in Eq. (14.5).

14.2.3 Visco-Plasticity

During the material deformation process, which is accompanied by high deformation rates and high temperature, in addition to the plastic flow of the material, there are also viscous phenomena, that is, creep or relaxation of the material. The system of constitutive equations (14.19) can be represented as in [24] by vectors containing only non-zero components of the corresponding tensors. To maintain the appropriate dimension of a given quantity, the characteristic length ℓ was introduced. We will write the stress and the strain column vectors as

$$\widehat{\sigma} = \{\sigma_{11}, \sigma_{12}, \sigma_{21}, \mu_{13}/\ell\}^T, \qquad \widehat{\varepsilon} = \{\varepsilon_{11}, \varepsilon_{12}, \varepsilon_{21}, \kappa_{13}\ell\}^T \tag{14.20}$$

and the corresponding stiffness matrix in the form

$$\widehat{C} = \begin{bmatrix} 2\mu+\lambda & 0 & 0 & 0 \\ 0 & 2\mu & 0 & -\frac{1}{4}\xi \\ 0 & 0 & 2\mu & \frac{1}{4}\xi \\ 0 & -\frac{1}{4}\xi & \frac{1}{4}\xi & \xi \end{bmatrix}. \tag{14.21}$$

As in the classical formulation of plasticity, the strain tensor and the curvature tensor can be represented as the sum of the elastic and plastic parts [25, 26], and the strains resulting from the thermal effect. Whereas the ε_{11} strain in the discrete element method is calculated based on the penetration value of the two spheres (and not incrementally), the incremental Hooke's law will be expressed by the formula

$$\widehat{\sigma}_{r+1} = \widehat{\sigma}_{r+1}^{\text{trial}} - \Delta \widehat{\sigma}^{\text{p}} - \Delta \widehat{\sigma}^{\theta} = \widehat{C} : \left(\widehat{\varepsilon}_{r+1} - \widehat{\varepsilon}_{r+1}^{\text{p}} - \varepsilon_{r+1}^{\theta} \right) \tag{14.22}$$

The algorithm was built based on the forward Euler scheme and the use of the cutting plane method [27]. The return mapping algorithm assumes an increase in elastic strain in step $r+1$, which, using the constitutive relation, results in trial stress $\widehat{\sigma}^{\text{trial}}$. For the stress calculated in this way, the yield condition $F(\widehat{\sigma}, R) \leq 0$ is checked, if the criterion is met, then the stress vector is inside the yield surface and the test stress is the stress at the end of step $r+1$, that is, $\widehat{\sigma}_{r+1} = \widehat{\sigma}_{r+1}^{\text{trial}}$. Otherwise, when $F(\widehat{\sigma}, R) > 0$ a viscoplastic corrector is required to bring the stress vector to the current yield surface. This procedure involves calculating the increment of the plastic part of the strain tensor $\Delta \widehat{\varepsilon}^{\text{p}}$.

The yield function, assuming only isotropic hardening, can be represented as

$$F_{r+1}^{v} = f_{r+1} - \sigma_{y,r+1} \left(1 + C \ln \frac{\Delta \lambda_{r+1}}{\dot{p}_0 \Delta t_{r+1}} \right) \left(1 - \left[\frac{T_{r+1} - T_{\text{ref}}}{T_{\text{melt}} - T_{\text{ref}}} \right]^{m} \right) \tag{14.23}$$

where the flow stress is described by the Johnson–Cook model [22]. Expanding the plasticity function in a Taylor series

$$F_{r+1}^{i+1} = F_{r+1}^{i} + \frac{\partial F}{\partial \widehat{\sigma}_{r+1}} \delta \widehat{\sigma}_{r+1} + \frac{\partial F}{\partial \sigma_{y,r+1}} \delta \sigma_{y,r+1} + \frac{\partial F}{\partial v_{p,r+1}} \delta v_{p,r+1} = 0 \tag{14.24}$$

where $\delta \widehat{\sigma}$, $\delta \sigma_y$ and δv_p are increments of individual variables in the next iteration step i. Considering that the internal state variable $\dot{p} = \dot{\lambda} = \frac{\Delta \lambda}{\Delta t}$ controls the effect of viscoplasticity $v_p(\dot{p})$ will take the form

$$v_p(\Delta \lambda_{r+1}) = \left(1 + C \ln \frac{\Delta \lambda_{r+1}}{\dot{p}_0 \Delta t_{r+1}} \right) \tag{14.25}$$

while the state variable responsible for thermal effects will be written as a function

$$\Gamma_p(T_r) = \left(1 - \left[\frac{T_r - T_{\text{ref}}}{T_{\text{melt}} - T_{\text{ref}}} \right]^{m} \right). \tag{14.26}$$

Finally, after differentiation, equation (14.24) will take the form

$$\begin{aligned} F_{r+1}^{i+1} = F_{r+1}^{i} &- \frac{\partial F}{\partial \widehat{\sigma}_{r+1}} \widehat{C} \frac{\partial F}{\partial \widehat{\sigma}_{r+1}} \delta \lambda_{r+1}^{i+1} - v_{p,r+1} \Gamma_{p,r} H_{R,r+1} \delta \lambda_{r+1}^{i+1} \\ &- \sigma_{y,r+1} \Gamma_{p,r} \frac{\partial v_{p,r+1}}{\partial \dot{\lambda}_{r+1}} \delta \dot{\lambda}_{r+1}^{i+1}. \end{aligned} \tag{14.27}$$

Using the substitution previously presented for λ, the increase in the plasticity multiplier in the next iteration step will be expressed in the form

$$\delta\lambda_{r+1}^{i+1} = \frac{f_{r+1}^i}{\frac{\partial F}{\partial\widehat{\sigma}_{r+1}}\widehat{C}\frac{\partial F}{\partial\widehat{\sigma}_{r+1}} + v_{p,r+1}\Gamma_{p,r}H_{R,r+1} + \sigma_{y,r+1}\Gamma_{p,r}\frac{\partial v_{p,r+1}}{\partial\Delta\lambda_{r+1}}} \quad (14.28)$$

where

$$\frac{\partial v_p(\Delta\lambda_{r+1})}{\partial\Delta\lambda} = \frac{C}{\Delta\lambda_{r+1}},$$

$$H_{R,r+1} = \frac{dR_{r+1}}{dp_{r+1}} = Bn\lambda_{r+1}^{n-1}, \quad (14.29)$$

$$\frac{\partial F}{\partial\widehat{\sigma}}\widehat{C}\frac{\partial F}{\partial\widehat{\sigma}} = \frac{3}{2J_2}\left\{\frac{2}{9}E\sigma_{11}^2 + \mu\left(\sigma_{12}^2 + \sigma_{21}^2 + \frac{1}{8\ell^2 l_0}\xi(\sigma_{21} - \sigma_{12})\mu_{13}\right) + \frac{1}{2\ell}\xi\mu_{13}^2\right\}.$$

Within each iterative step, the thermodynamic forces are updated with respect to the current state of stress along with the calculation of a new value for the yield function

$$F_{r+1}^{v,i+1} = f_{r+1}^{i+1} - \sigma_{y,r+1}^{i+1}\left(1 + C\ln\frac{\Delta\lambda_{r+1}^{i+1}}{\dot{p}_0\Delta t_{r+1}}\right)\left(1 - \left[\frac{T_r - T_{\text{ref}}}{T_{\text{melt}} - T_{\text{ref}}}\right]^m\right) \quad (14.30)$$

and checking the convergence of the solution

$$\left|\frac{F^v(\Delta\lambda_{r+1}^{i+1})}{\sigma_{y,r+1}v_p\Gamma_p}\right| \leq \text{TOL} \quad (14.31)$$

for a fixed tolerance value. If this value is within the tolerance range, the iterative procedure is stopped, and the individual model parameters are updated.

It should be taken into account that not all the energy associated with plastic deformation is dissipated in the form of thermal energy. A certain part of it is used for structural changes that take place in the material, and hence

$$\mathfrak{D}^p = \chi\sigma\dot{\varepsilon}^p = \chi\sigma_{\text{eq}}\dot{p} \quad (14.32)$$

where the Taylor-Quinney coefficient χ is usually 0.9.

14.3 Model Verification

In this section, the verification of the model presented is discussed based on stress-strain plots. This is done by comparing the plots from the DEM simulations and the analytical model based on the parameters of the J–C model obtained from the experiment.

14.3.1 Simulation Set-up

In the current study, the introduced thermo-elato-viscoplastic model was implemented in open-source YADE software [28] and validated using an analytical solution based on the parameters presented in [23]. The numerical dynamic tensile tests were carried out on an axisymmetric sample (Fig. 14.5a) with dimensions of: $d_1 = 20$ mm, $d_2 = 10$ mm, $R = 10$ mm, $L_1 = 30$ mm, $L_2 = 15$ mm and $L_3 = 80$ mm. To analyze the influence of the discrete element size on the results, two cases of sample geometry discretization were checked. In the first case, the radius of the spherical elements was 1 mm (Fig. 14.5b) and in the second case $r = 0.75$ mm (Fig. 14.5c). The number of particles in the case of $r = 1$ mm was 1842 and for $r = 0.75$ was 4873. Performed DEM simulations do not take into account temperature softening at this moment. The parameters of the material are presented in Table 14.1. Some of the material properties were not calibrated, including Young's modulus $E = 70$ GPa, Poisson's ratio $\nu = 0.3$, material density $\rho = 2710$ kg/m^3 and reference strain rate $\dot{\varepsilon}_{ref} = 0.0001$ 1/s. Additional parameters that we introduced in the TEVP[1] model and are related to coupled stresses were the characteristic length $\ell = 0.025$ mm and the elastic constant $\xi = Er^2/3$. The iterative parameters of the return mapping algorithm were: TOL $= 1 \cdot 10^{-1}$ and MXITER $= 100$.

To illustrate the sensitivity of the model to dynamic behavior, two constant strain rates were investigated: 1 and 10000 s^{-1}.

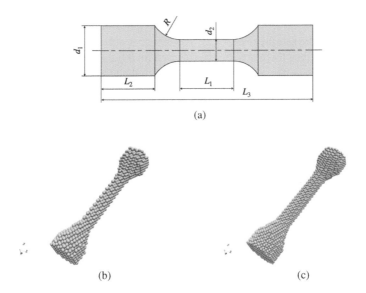

Fig. 14.5: Sample geometry used in DEM simulations a), discretized by spherical elements with radius value of 1 mm b) and 0.75 mm c).

---

[1] thermo-elasto-viscoplastic

Table 14.1: Material properties of the 6082-T6 aluminum alloy used in the simulations..

Parameter	Value from literature [23]	Calibrated value	Unit
A	277.33	178.67	MPa
B	307.93	25.73	MPa
n	0.69	0.43	-
C	0.0032	$5 \cdot 10^{-10}$	-

14.3.2 Results and Discussion

Figure 14.6a shows the comparison of the analytical solution with the stress-strain curve from DEM simulation where the initial values of the material parameters were used. There is no apparent correlation with the experimental model. The discrete element method shows a yield strength about twice as high as in reality. In addition, the observable strengthening effect is much greater than in the case of the analytical model. This is a disadvantage of the DEM formulation, where a calibration of the model must be performed in the initial step, before the right simulation [29]. In this work, we use a trial-and-error method for the calibration of the model material parameters. Due to the specific formulation of this method, the DEM model shows high sensitivity to material parameters. The initial yield strength that describes the parameter A was calibrated first, but, unlike the analytical model, the DEM model does not exhibit ideal plastic flow (Fig. 14.6b). This could be explained by the dynamic behavior of the material. During tension, we can observe the stress wave propagating along the sample (Fig. 14.7). This effect is different from the analytical model, which describes the material in terms of a material point. As a result of the material response, the effect of clear non-linearity in the initial range of plastic flow is visible. Also visible is the effect of stress gradation, which is not observed in the

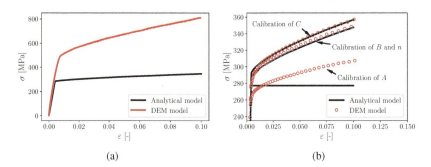

Fig. 14.6: Comparison of a non-calibrated model with an analytic solution a) and calibration of the parameters of the J-C model used in the DEM model b).

14 Modeling Solid Materials in DEM Using the Micropolar Theory

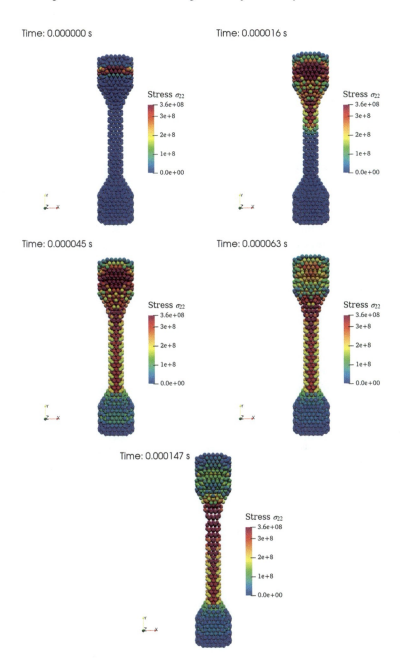

Fig. 14.7: Stress map of σ_{22} in the sample analyzed at a particular time during the loading process.

analytical model. Calibration of the parameters B and n, taking into account the value of A, results in an increase in stress in relation to strain.

Appropriate calibration of the DEM model parameters results in a proper response to the applied load in the form of the strain rate. At speeds of $1\ s^{-1}$ and $10000\ s^{-1}$, the DEM model shows good accuracy compared to the analytical model (Fig. 14.8a). A difference in the plastic behavior of the material is also observable, which in the case of the discrete element method shows a certain non-linearity in determining the yield point. It is also noticeable depending on the value of the strain rate. The graphic shows the influence of the size of the discrete element on the stress value during the sample loading process (Fig. 14.8b). Reducing the radius of the particles results in an increase in yield strength and associated stress. This effect can be related to the decrease in the stress value with an equivalent strain value due to a decrease in particle size. Changing the model by excluding coupled stresses, in the case of an analysis with a 1 mm element size, results in an increase in stresses and yield strength. This allows us to conclude that the coupled stresses result in an increase in the effective stress and therefore in a decrease in the yield strength of the material.

14.4 Concluding Remarks

In this work we have presented a new local constitutive model for the DEM obtained by using micropolar theory. The model takes into account not only the elasticity but also the viscoplastic behavior of the material, where the Johnson–Cook model was adopted. The results obtained from the uniaxial tensile tests confirm that the model correctly reproduces the mechanism of strain rate sensitivity.

These findings confirm the potential use of the TEVP model in analysis of metallic materials subjected to high strain rates. The applicability of DEM to the linear and non-linear analysis of ductile material has been proved.

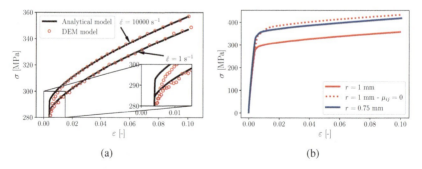

Fig. 14.8: Response of the material to strain rates of $1\ s^{-1}$ and $10000\ s^{-1}$ a), stress-strain relation depending on the size of the element and the existence of coupled stress b).

There is still a need to understand the sensitivity of the model to initial material parameters and the influence of particle size on stress response. The work on those problems is in progress and the results will be published in a succeeding paper.

Acknowledgements The financial support from subsidy granted by the Polish Ministry of Science and Higher Education within project no. 16.16.130.942 is acknowledged.

References

[1] Cundall PA (1971) A computer model for simulating progressive large scale movements in blocky rock systems. In: Proceedings of the Symposium of the International Society for Rock Mechanics, Society for Rock Mechanics (ISRM), France, pp II–8
[2] Cundall PA, Strack ODL (1979) A discrete numerical model for granular assemblies. Géotechnique **29**(1):47–65
[3] Oñate E, Zárate F, Miquel J, Santasusana M, Celigueta MA, Arrufat F, Gandikota R, Valiullin K, Ring L (2015) A local constitutive model for the discrete element method. application to geomaterials and concrete. Computational Particle Mechanics **2**(2):139–160
[4] Rojek J (2014) Discrete element thermomechanical modelling of rock cutting with valuation of tool wear. Computational Particle Mechanics **1**(1):71–84
[5] He Y, Zhang J, Andriollo T, Hattel J, Zhao W (2019) Investigation of the elastoplastic and fracture behavior of solid materials considering microstructural anisotropy: A discrete element modeling study. Computational Materials Science **170**, 109164
[6] Fathipour-Azar H, Wang J, Jalali SME, Torabi SR (2014) Numerical modeling of geomaterial fracture using a cohesive crack model in grain-based DEM. Computational Particle Mechanics **7**(4):645–654
[7] Joulin C, Xiang J, Latham JP (2020) A novel thermo-mechanical coupling approach for thermal fracturing of rocks in the three-dimensional FDEM. Computational Particle Mechanics **7**(5):935–946
[8] Martin CL, Schneider LCR, Olmos L, Bouvard D (2006) Discrete element modeling of metallic powder sintering. Scripta Materialia **55**(5):425–428
[9] Nosewicz S, Rojek J, Pietrzak K, Chmielewski M (2013) Viscoelastic discrete element model of powder sintering. Powder Technology **246**:157–168
[10] Rojek J, Nosewicz S, Pietrzak K, Chmielewski M (2013) Simulation of powder sintering using a discrete element model. Acta Mechanica et Automatica **7**(3):175–179
[11] Rojek J, Nosewicz S, Maździarz M, Kowalczyk P, Wawrzyk K, Lumelskyj D (2017) Modeling of a sintering process at various scales. Procedia Engineering **177**:263–270, xXI Polish-Slovak Scientific Conference Machine Modeling and Simulations MMS 2016.September 6-8, 2016, Hucisko, Poland

[12] André D, Iordanoff I, luc Charles J, Néauport J (2012) Discrete element method to simulate continuous material by using the cohesive beam model. Computer Methods in Applied Mechanics and Engineering **213-216**:113–125

[13] Nguyen V, Tieu A, André D, Su L, Zhu H (2020) Discrete element method using cohesive plastic beam for modeling elasto-plastic deformation of ductile materials. Computational Particle Mechanics **8**:437–457

[14] Lei Z, Bradley CR, Munjiza A, Rougier E, Euser B (2020) A novel framework for elastoplastic behaviour of anisotropic solids. Computational Particle Mechanics **7**(5):823–838

[15] Jebahi M, André D, Terreros I, Iordanoff I (2015) Discrete Element Method to Model 3D Continuous Materials, vol 1. ISTE, London

[16] Kozicki J, Tejchman J (2008) Modelling of fracture process in concrete using a novel lattice model. Granular Matter **10**(5):377–388

[17] Leclerc W, Haddad H, Guessasma M (2017) On the suitability of a discrete element method to simulate cracks initiation and propagation in heterogeneous media. International Journal of Solids and Structures **108**:98–114

[18] O'Sullivan C (2011) Particulate Discrete Element Modelling. A Geomechanics Perspective. Spon Press, Abingdon

[19] Rojek J, Zubelewicz A, Madan N, Nosewicz S (2018) The discrete element method with deformable particles. International Journal for Numerical Methods in Engineering **114**(8):828–860

[20] Cosserat E, Cosserat F (1909) Théorie des corps déformables. Herman, Paris

[21] Nowacki W (1986) Theory of Asymmetric Elasticity. Pergamon, Oxford

[22] Johnson GR, Cook WH (1983) A constitutive model and data for metals subjected to large strains, high strain rates and high temperatures. In: Proceedings of the 7th International Symposium on Ballistics, The Hague, pp 541–547

[23] Chen X, Peng Y, Peng S, Yao S, Chen C, Xu P (2017) Flow and fracture behavior of aluminum alloy 6082-T6 at different tensile strain rates and triaxialities. PLoS One **12**, e0181983

[24] de Borst R (1993) A generalisation of j_2-flow theory for polar continua. Computer Methods in Applied Mechanics and Engineering **103**(3):347–362

[25] Forest S, Sievert R (2003) Elastoviscoplastic constitutive frameworks for generalized continua. Acta Mechanica **160**(1-2):71–111

[26] Dietsche A, Steinmann P, Willam K (1993) Micropolar elastoplasticity and its role in localization. International Journal of Plasticity **9**(7):813–831

[27] Ortiz M, Simo JC (1986) An analysis of a new class of integration algorithms for elastoplastic constitutive relations. International Journal for Numerical Methods in Engineering **23**(3):353–366

[28] Yade (2015) Yade documentation. www.yade-dem.org/publi/documentation_2nd_ed/YadeBook.pdf. Cited 22 Mar 2023

[29] Yan Z, Wilkinson SK, Stitt EH, Marigo M (2015) Discrete element modelling (DEM) input parameters: understanding their impact on model predictions using statistical analysis. Computational Particle Mechanics **2**(3):283–299

Chapter 15
The Development of a Cavitation-Based Model for Creep Lifetime Prediction Using Cu-40Zn-2Pb Material

Mbombo Amejima Okpa, Qiang Xu, and Zhongyu Lu

Abstract The occurrence of creep induced cavitation can considerably shorten the lifespan of numerous high-temperature applications. A contemporary problem in structural mechanics and materials science is the inadequate mathematical description of creep deformation and rupture time. This situation stems not only form a lack of accurate quantification and incorporation of cavitation damage in current theoretical models, but it is compounded by the strong stress level dependency of the creep lifetime.

Cavitation is the rate-controlling mechanism during creep. To this end, this study has developed a cavitation-based method for creep rupture lifetime prediction. For accuracy and a representative data, cavitation data measured using x-ray synchrotron tomography, are chosen for the study. Cavitation damage modelling precisely cavity nucleation, growth and size distribution are presented. Functional relationships between creep exposure time and cavitation damage are developed to aid creep lifetime prediction. This approach has the advantage of traceability as it is developed based on quantifiable physical changes in the material (cavity nucleation and growth).

This study reports the latest progress in the development of a cavitation model for a specific material under testing condition. It is planned to incorporate it, to develop a creep lifetime prediction and extrapolation model. This paper offers a theoretical foundation for a time-based extrapolation method to predict creep lifetime. Furthermore, the cavitation modelling approach used in this study may be applied in other failure modes like fatigue.

Mbombo Amejima Okpa · Qiang Xu · Zhongyu Lu
Department of Technology and Engineering, School of Computing and Engineering, University of Huddersfield, Huddersfield, HD1 3DH, UK,
e-mail: Mbombo.okpa@hud.ac.uk, Q.Xu2@hud.ac.uk, j.lu@hud.ac.uk

15.1 Introduction

Creep damage is a crucial factor in high-temperature applications such as power plant components (e.g., boilers and steam piping), gas turbine applications, and military aircraft [1]. These applications demand a thorough understanding of creep and its governing principles as well as the conditions under which it affects the structural integrity of the components. The lifetime of components in these applications is limited by creep-induced cavitation, and premature breakdown of such critical components can result in severe financial loss and even loss of life. Predicting the creep rupture strength and lifetime of these components is a significant contemporary concern in materials science and structural integrity research, as existing predictions have often proven to be an overestimation. This issue is well-documented in the literature, as material scientists and researchers are experimenting with various paradigms to improve the accuracy of their predictions.

Several empirical equations have been developed to forecast creep lifetime by parameterising the creep curve, using strain as a function of stress and temperature. It is one of the early creep life assessment models as demonstrated by Norton [2], and subsequently by Arrhenius [3]. Another popular approach is the continuum damage mechanics (CDM), which was first proposed by Kachanov [4]. The creep rupture life prediction is based on the analysis of creep behaviour during the tertiary stage. Kachanov's original model has undergone various modifications, with Dyson's work [5] being the most significant, where internal variables were introduced to describe macroscopic behaviour and categorize damage. However, it is widely acknowledged that the existing creep lifetime prediction models are still inadequate.

15.2 Stress Breakdown and Creep Lifetime

To prevent excessive deformation and a premature rupture of critical components in high-temperature industries, safe operating limits in terms of acceptable strain, temperature limits, and expected lifespan must be defined. To achieve this, historical creep data for these components are necessary. However, obtaining long-term creep data that exceeded 100,000 hours is time-consuming and expensive, resulting in a scarcity of such data. In recognising this issue, the current approach for creep rupture time prediction, involves various techniques in which an accelerated or short-term creep test (usually with high applied stress) is conducted, and the data are analysed and extrapolated to predict long term creep behaviour and rupture time. Currently, creep life prediction via this extrapolation approach is still below expectations and extrapolation of short-term data has been reported as one of the major challenges in material science in the UK [6].

The stress breakdown phenomenon refers to the variation in the values of the stress exponent n and activation energy Q_c, when extrapolating short term creep data to predict long-term creep behaviour, resulting in an overestimated long term creep rupture life [7]. The power law equation proposed by Arrhenius is the traditional and one of the widely used methods for creep life prediction [3]. The proposed equation relates time to fracture and stress:

$$t_f = t_0 \sigma^{-n} \left(\frac{Q_c}{RT} \right) \qquad (15.1)$$

where: Q_c, is the activation energy for creep; n, is stress exponent; R is the gas constant; t_f, is the time to fracture, t_0, is a constant; and T, is the absolute temperature.

Originally, the values of Q_c and n, were assumed to be constant. However, alternative views came to light which very backed by experimental evidence that Q_c and n values are stress level dependent [7, 8]. Ennis et al. [8], adopted Norton's equation and observed that for a high Cr martensitic steel, the stress exponent n reduces from 16 at stresses above 150 MPa to 6 for stresses lower than 110 MPa. It was implied that a possible change in deformation mechanism is the reason for the variation in the value of n as stress level changes and should be given great consideration during long term extrapolation attempts. Subsequently, Lee et al. [7] adopted Arrhenius equation (15.1) and observe that for an ASTM grade 92 steel, Q_c and n changes from 667 kJ/mol and 17 respectively for a short-term creep and high stress condition, to 624 kJ/mol and 8.4 in a long-term creep and lower stress condition.

In addition to the empirical models of creep life prediction, the Continuum Damage Mechanics (CDM) is also a popular method of predicting creep lifetime. Dyson [5] through experimental observations was able to further group creep damage into broad categories including strain-induced, environmentally induced, and thermal induced damage. Thus, the relationship between creep cavity damage and creep strain was described as follows:

Dyson's Concept

$$\dot{D}_n = \frac{k_N}{\varepsilon_{f_u}} \dot{\varepsilon} \qquad (15.2)$$

where $\dot{D}_n$ is the rate of creep cavity damage, k_N, is the cavitation damage coefficient, ε_{f_u} is the strain at failure, and $\dot{\varepsilon}$ is the creep strain rate.

There have been various modifications of Dyson's framework, including studies by Yin et al. [9], Basirat et al. [10], and Chen et al. [11] in which damage coefficient A was introduced. A was originally proposed to be a constant [9]. However, Basirat et al. [10], observed that A was strongly dependent on stress and temperature. Yang et al. [12] observed that Yin's concept could not address stress breakdown phenomenon discussed earlier. With aim to address this issue, Basirat et al. [10] adopted Yin's model but relaxed the definition of A by making it stress level dependent. However, Xu et al. [13] using of 9Cr-1Mo steel data, pointed out that there was no clear trend for A when tested over a wide range of stress and temperature.

Hence, it is evident that the approach to correlate creep damage/creep lifetime to creep strain is not satisfactory. It is our view that it is better and scientifically sounder to model the cavitation directly, though the creep cavitation and/or creep lifetime is very strongly co-related to creep strain [13].

15.3 Creep Cavitation and Cavitation Data Concerns

Creep fracture at low stress and high temperature is commonly said to be intergranular [14–16]. This is due to the formation, growth, and coalescence of cavities at the grain boundary during the tertiary stage of a creep process. Eventually, the cavitation process becomes the primary source of the rapid increase in creep rate at the tertiary stage of a creep process. Therefore, for a reliable lifetime prediction, this dominate damage mechanism(cavitation) should be treated with great importance.

Formation of cavities may start early in the creep life, likely around the primary stage and their effect at this stage is negligible but as these cavities grow in terms of numbers and size, they progressively weaken the material and eventually lead to ultimate failure [16]. Cavity nucleation and growth stages have long been identified as the rate controlling stages and occupies about 80% of the creep process [17]. Therefore, the nucleation and growth stages has been a subject of intense interest.

The idea of quantifying the damage in a material for rupture time prediction has been conveyed for over six decades now [18]. Quantifiable physical characteristics of cavitation like Cavity nucleation, growth and size distribution can be analysed to enable creep lifetime predictions. The approach offers simplicity and a statistical method for evaluating microstructural evolution and rupture time prediction. One primary concern with this approach is the lack of relevant cavitation data and the ambiguity associated with the available ones [19]. This situation primarily results from a lack of an effective tool to measure and characterise creep cavities. The popular traditional methods for studying creep cavitation in materials are mostly destructive in nature; the most widely used techniques in that category are scanning electron microscopy (SEM) and transmission electron microscopy (TEM). These methods have been used extensively to study microstructural features in materials and for quantitative analysis of cavities [20–23]. The newer techniques are said to be non-destructive in nature and have gained popularity in recent years. Most notable are; the Replica metallography, commonly used for quantitative analysis of cavities on the surface of a material [24]; and the Small Angle Neutron Scattering (SANS), typically used to yield quantitative information on the size distribution of cavities as well as the morphology of voids and precipitates [24].

It is therefore obvious that cavitation damage characterisation has mostly been done using destructive two-dimensional methods, surface replica or a scattering method. Several literatures have shown that these methods are not efficient in characterising cavities. The SEM and TEM for example, produce 2D images of the microstructures and are not able to reveal the true size and shape of cavities. The surface replica is only able to detect the surface cavities shortly before fracture and the information obtained at the surface differs from those observed in the bulk of the material [25–27]. The SANS is not truly non-destructive as samples must be physically detached from their components and taken for testing [24].

The past two decades have witnessed a rapid rise in the application of X-ray synchrotron micro-tomography for studying fracture and fatigue behaviour in engineering materials [28]. It is a non-destructive technique capable of offering a 3-dimensional visualisation of a material's microstructure [28, 29]. This technique

15 The Development of a Cavitation-Based Model for Creep Lifetime Prediction   253

has a significant advantage over the traditional one as it can reveal the true size and shape of cavities [30]. Yadav et al. [31] investigated the difference in damage quantification between 2D analysis and 3D serial sectioning. It was demonstrated that the 2D method was not able to quantify complex shaped cavities and hence, overestimated the number density and volume fraction of creep cavities by almost a factor of 4. The tomography technique has been recommended in several literatures are the most efficient technique to measure and characterise cavity profiles [32, 33]. The idea to make use of such cavitation data for the development of a cavitation model was advocated by Xu [13]. To this end, this study will seek for cavitation data examined using X-ray synchrotron micro-tomography to aid cavitation damage modelling.

15.3.1 How to Use Cavitation Data

Nucleation data, precisely the evolution in number of cavities with creep time, offers information on how fast cavities proliferate in a material. Such data can be calibrated and a model that describes the nucleation rate can be ascertained. The cavity growth data on the other hand is more complex to analyse in an isolated manner, due to the significance of continuous cavity nucleation. Therefore, the cavity growth data should be analysed alongside the cavity size distribution data. This is to tackle the complexities associated with continuous cavity nucleation [16]. A meaningful cavity growth data will be one that reveals the evolution of total overall cavity size with creep time. It is important to note here that a reliable conclusion cannot be drawn on cavity growth and growth rates by analysing the evolution of the average size of cavities with creep time. This is because continuous cavity nucleation significantly influences the average size of cavities [34]. In addition, the size distribution data are essential for developing damage criteria such as the cavitated area and volume fractions. The damage criteria are generally used as indicative parameters for rupture time prediction.

15.3.2 Current Approach to Cavitation Modelling and Creep Life Prediction

Several literatures in the later part of 2010's and early 2020's highlighted the need to deviate from the empirical models to a physically based model [13, 33, 35, 36]. Generally, the models are either strain or stress driven.

Sandström and He [37] developed a cavitation model that is strain based. It was put forward that both cavity nucleation and growth are directly proportional to strain. They merged the equations for both cavity cavity nucleation and growth to create a damage criterion using the cavitated area fraction concept. It was observed that the final fracture occurred when the cavitated area fraction surpassed 0.25. Davanas

[36] proposed a connection between nucleation rate and minimum strain rate, by linking the minimum strain rate to the rupture time through the Monkman-Grant ductility. The proposed model is based on strain, it is assumed that the cavity growth stage is less significant and, therefore, not considered. Yu et al. [38] studied the potential of predicting creep rupture using a constrained diffusional cavity growth model on G115 Martensitic heat-resistant steel. Huang et al. [39] combined Nicking and Riedel's continuous nucleation models to forecast the creep rupture time on Alloy 800. They used Hart's criterion to determine the start of necking and assumed that cavity formation is directly related to the creep strain rate.

Xu et al. [13] pointed out several limitations in the current cavitation damage models and proposed a new damage criterion for a P91 material based on Riedel's cavitation theories. Using the cavitated area fraction denoted as w, a relationship between cavitation damage and time to fracture t_f was established. It was concluded that rupture occurred when $\bar{w}$ reached a critical value denoted as w_f. Therefore, time to fracture is described as follows:

$$w_f = U' t_f^2 \tag{15.3}$$

where: w_f is the critical value for the cavitated area fraction; U' is a stress dependent variable; and t_f is the time to fracture.

Xu's model demonstrates a good potential for success as the kinetics of cavity nucleation and growth have been incorporated. In addition, results demonstrated a clear trend between stress and stress-dependent variables U' and A_2. This is significant as the trend can be analysed and extrapolated into lower stress regimes. Zheng et al. [40] extended the application of this model to other materials (P92 and E911) and obtained a similar trend between stress and U' and A_2 for those materials. More recently, Zheng [41] and Fu and Xu [42] demonstrated that the model could be used to calibrate cavitation damage at various stages of a creep process, not just at the time of fracture, using type 316H steel creep data. This demonstrates and proves the feasibility of extrapolating early-stage creep cavitation data to predict rupture time, resulting in the formal proposal of early creep lifetime prediction approach [43].

15.4 Aims

Creep lifetime modelling and prediction has been established recently, and its applications include the extrapolation from higher stress to lower stress and preliminary early creep lifetime prediction approach. The primary aim of this study is to develop and apply the latter.

15.5 Experimental Data and Method

15.5.1 Experimental Data

An in-situ examination of creep cavitation process leading to creep fracture on Cu-40Zn-2Pb, subjected to 25MPa and at 375°C using X-ray micro-tomography [44]

1. Creep data one: Evolution of number of cavities (number of cavities vs creep time)
2. Creep data two: Cavity size histogram at different creep time
3. Creep data three: Evolution of the total cavity volume (total cavity volume vs creep time)

15.5.2 Method

A link between experimental data the theories of cavity nucleation and growth is the cavity size distribution function proposed by Riedel (1987), and it is denoted by $N(R,t)$:

$$N(R,t) = \frac{A_2}{A_1} R^\beta t^{\alpha+\gamma} \left(1 - \frac{1-\alpha}{1+\beta} \frac{R^{\beta+1}}{A_1 t^{1-\alpha}}\right)^{(\alpha+\gamma)/(1-\alpha)} \tag{15.4}$$

A general solution to these cavity nucleation and growth theories can be summarised in a power law form:

$$\dot{R} = A_1 R^{-\beta} t^{-\alpha} \tag{15.5}$$

$$J^* = A_2 t^\gamma \tag{15.6}$$

(4.2) where: $\dot{R}$ is the non-stationary growth rate of cavity radius; J^* is the nucleation rate of cavity; A_1, A_2, γ, α and β are all unknown material constants that may be dependent on stress.

To determine the extent of material damage, one can use the absolute cavitated area w as a measure. This quantity is computed by summing up the areas of individual cavities πR^2 and then multiplying by their number density NdR. This procedure is repeated for all cavity sizes

$$w = \sum_{R_{min}}^{R_{max}} \pi R^2 N(R,t) dR \tag{15.7}$$

where: R_{max} and R_{min} are maximum and minimum cavity radius, R is the cavity radius, and $N(R,t)$ is the cavity size distribution function.

The primary task in the cavitation modelling is to find a solution for a set of the five unknown material parameters A_1, A_2, γ, α and β over a series of creep times. Functional relationships between the absolute cavitated area and creep time are then developed to aid creep lifetime prediction.

15.5.3 Determination of Cavitation Constants

Experimental data (set 1) can provide insights into the evolution of cavity number over the creep process. Equation (15.6) can be integrated to obtain the number of cavities, J

$$J = A_2 \frac{t^{\gamma+1}}{\gamma+1} + C_2 \quad \text{for} \quad \delta \neq -1. \tag{15.8}$$

where: J is the number of cavities; t, time in minutes; A_2, material parameter associated with nucleation rate; and C_2, is an integration constant.

To determine the values of A_2, C_2 and γ, two reference points are selected from both ends of the experimental data: J_{366} and J_{52}, which correspond to the cavity time at $t = 366$ and $t = 52$ min, respectively. The values of A_2 and C_2 can be obtained from these reference points, and γ can be determined through optimization or trial-and-error methods. It's important to note that the correct values for these constants should accurately describe both the cavity size distribution and nucleation data. Therefore, the process of obtaining γ involves a relaxation technique and was not restricted to the cavity nucleation data.

$$A_2 = \frac{(J_{366} - J_{52})\gamma + 1}{t_{366}^{\gamma+1} - t_{52}^{\gamma+1}}, \tag{15.9}$$

$$C = J_{366} - \left(\frac{A_2 t_{366}^{\gamma+1}}{\gamma+1}\right). \tag{15.10}$$

By solving Eqs. (15.9) and (15.10), the following material parameters were obtained: $A_2 = 6.218$ cavities/min$^{1.5}$, integration constant $C = 49.62$, and material parameter $\gamma = 0.5$. These values provide a good fit when compared with the experimental data, as shown in Fig. 15.1.

After 400 min of creep time, the effect of cavity coalescence on the modelling results becomes apparent as the total number of cavities gradually decreases over time. This observation suggests that some of the cavities may have merged to form larger cavities, which highlights a concern with using the number of cavities per unit area as a parameter for assessing the level of creep damage in a material.

15.5.4 Cavity Size Distribution Modelling

This section presents the results of modelling the cavity size distribution and compares them with corresponding experimental data. Equation (15.5) is relevant to this analysis. It is necessary to mention that the experimental data are a lot denser, and a few points have been picked for clarity.

Given the known values of A_2 and γ, time t is constant since only one deformation stage is investigated [45], $\alpha = 1$ can be assumed for continuum cavity nucleation,

15 The Development of a Cavitation-Based Model for Creep Lifetime Prediction

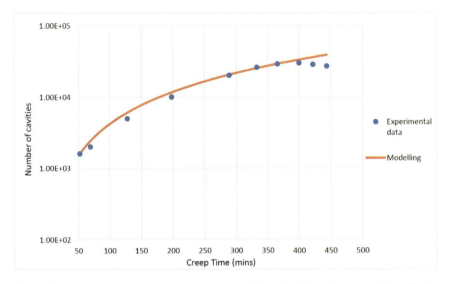

Fig. 15.1: Modelling result compared with experimental data. Evolution of creep cavities with time. Experimental data retrieved from [44].

as per Riedel's theory. This assumption is critical since cavity nucleation is said to be continuous. With this information, the remaining constants A_1 and β can be determined from the cavity size distribution function $N(R,t)$. The results of this modeling are presented in Table 15.1, where C_1 is an integration constant related to the cavity growth rate.

The cavitation constants, which consist of the parameter β, are believed to be solely influenced by stress, and this stress-related dependency should show a clear pattern in accordance with the second law of thermodynamics [43]. The findings indicate that β decreases as the creep time increases. The present modelling technique used is limited by the manual calibration procedures used to determine the cavitation constants and this could be a possible explanation for the variability of β. Nevertheless, the consequence of a variation in β is not significant as it only influences the predicted

Table 15.1: Modelling results for the values of A_1, C_1 and β.

Time (min)	A_1	C_1	β
52	1.080	1.781	0.50
110	1.103	1.822	0.50
137	0.983	1.527	0.30
196	1.099	1.795	0.30
307	1.007	1.330	0.00
440	1.083	0.548	-0.4

pattern in which cavities grow to make up the total cavitated area or total cavitated volume. The volumetric damage will be reported in a future study.

Figures 15.2-15.4 shows the results of modelling the cavity size distribution and compares them with corresponding experimental data.

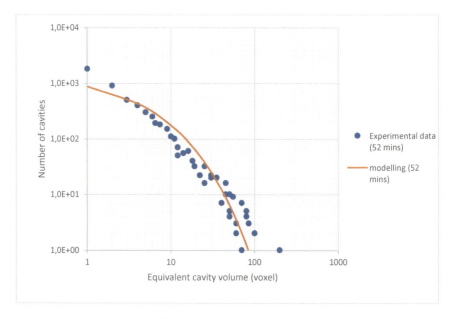

Fig. 15.2: Cavity size distribution data at $t = 52$ min. Modelling result compared with experimental data from [44].

Based on the obtained values of the cavitation constants, the predicted relationship between cavity growth and creep time is illustrated Fig. 15.5, using Eq. (15.11).

$$\beta = -0.4, \qquad R^{\frac{5}{3}} = \frac{5}{3}(A_1 \ln t + C_1). \qquad (15.11)$$

The evolution of cavity growth and nucleation rates, A_1 and A_2 respectively, are extracted from the cavity size distribution modelling results and are presented graphically Figs. 15.6 and 15.7. It is observable that the coefficients remain relatively stable over varying creep times. The outcomes indicate a distinct pattern for the cavitation coefficients and establish a reliable foundation for predicting creep lifetime through extrapolation techniques. This provides assurance and groundwork for potential applications of the model in predicting creep lifetime.

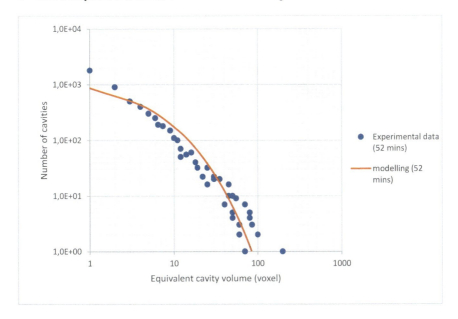

Fig. 15.3: Cavity size distribution data at $t = 196$ min. Modelling result compared with experimental data from [32].

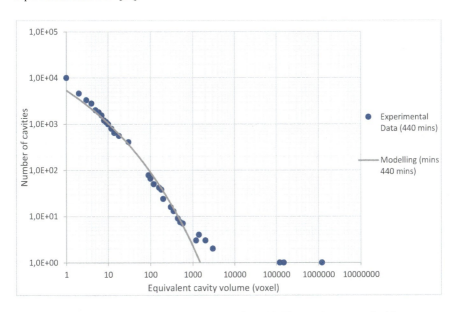

Fig. 15.4: Cavity size distribution data at $t = 440$ min. Modelling result compared with experimental data from [44].

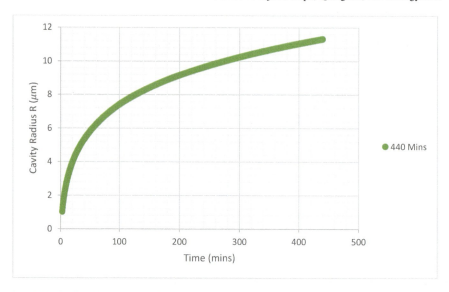

Fig. 15.5: Predicted cavity growth pattern.

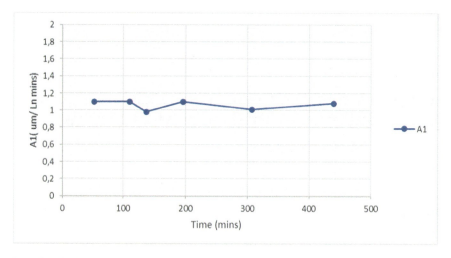

Fig. 15.6: Evolution of growth rate coefficient with time.

15.6 Conclusions and Future Work

Creep Lifetime prediction is a challenging task primarily due to the stress breakdown phenomena and the lack of accurate quantification and consideration of cavitation damage. The modern approach on creep damage modelling is either strain or stress driven cavitation models. Over the past two decades, X-ray synchrotron tomography technique has emerged as an efficient tool to characterise creep cavities. This study

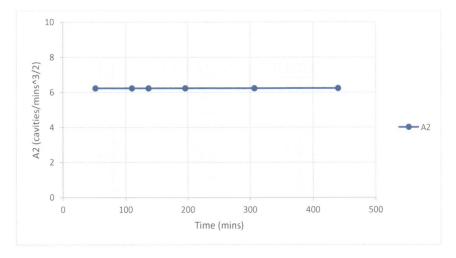

Fig. 15.7: Evolution of nucleation rate coefficient with time.

has reported the importance of creep cavitation data and how it can be used to aid creep lifetime prediction. In addition, this study has demonstrated how to model cavity nucleation, growth, and size distribution over the different creep time which will be used further to aid the development of creep cavitation prediction model based on extrapolation.

The following key points have been highlighted as future work:

- To develop damage criteria based on the cavitated area fraction along a grain boundary as well as cavitated volume fraction.
- To establish functional relationships between the damage criteria and creep exposure time.
- Using the developed relationship, extrapolate early-stage creep data to predict final rupture time.
- To develop an optimization software or program capable of solving for the parameters in the cavity size distribution function $N(R,t)$.
- To extend the application of this method of creep lifetime prediction to a wider range of materials and possibly to characterise other failure modes like fatigue.

References

[1] Nabarro FRN, de Villiers F (1995) Physics of Creep and Creep-Resistant Alloys. CRC Press, London
[2] Norton FH (1929) The Creep of Steel at High Temperatures. McGraw-Hill Book Company, Incorporated, New York

[3] Larson FR, Miller J (2022) A Time-Temperature Relationship for Rupture and Creep Stresses. Transactions of the American Society of Mechanical Engineers **74**(5):765–771

[4] Kachanov LM (1967) The Theory of Creep. National Lending Library for Science and Technology

[5] Dyson B (2000) Use of CDM in Materials Modeling and Component Creep Life Prediction. Journal of Pressure Vessel Technology **122**(3):281–296

[6] Abdallah Z, Gray V, Whittaker M, Perkins K (2014) A critical analysis of the conventionally employed creep lifing methods. Materials **7**(5):3371–3398

[7] Lee JS, Armaki HG, Maruyama K, Muraki T, Asahi H (2006) Causes of breakdown of creep strength in 9cr–1.8w–0.5mo–vnb steel. Materials Science and Engineering: A **428**(1):270–275

[8] Ennis PJ, Zielinska-Lipiec A, Wachter O, Czyrska-Filemonowicz A (1997) Microstructural stability and creep rupture strength of the martensitic steel p92 for advanced power plant. Acta Materialia **45**(12):4901–4907

[9] Yin Y, Faulkner RG, Morris PF, Clarke PD (2010) Modelling and experimental studies of alternative heat treatments in Steel 92 to optimise long term stress rupture properties. Energy Materials **3**(4):232–242

[10] Basirat M, Shrestha T, Potirniche GP, Charit I, Rink K (2012) A study of the creep behavior of modified 9cr–1mo steel using continuum-damage modeling. International Journal of Plasticity **37**:95–107

[11] Chen Y, Yan W, Hu P, Shan Y, Yang K (2011) CDM modeling of creep behavior of T/P91 steel under high stresses. Acta Metall Sin **47**(11):1372–1377

[12] Yang X, Xu Q, Lu Z (2013) The development and validation of the creep damage constitutive equations for p91 alloy. In: Proceedings of the 2013 World Congress in Computer Science and Computer Engineering and Application, CSREA Press, pp 121–127, 2013 World Congress in Computer Science and Computer Engineering and Application, Las Vegas, USA, 22nd June - 5th July 2013

[13] Xu Q, Yang X, Lu Z (2017) On the development of creep damage constitutive equations: a modified hyperbolic sine law for minimum creep strain rate and stress and creep fracture criteria based on cavity area fraction along grain boundaries. Materials at High Temperatures **34**(5-6):323–332

[14] Kassner ME, Hayes TA (2003) Creep cavitation in metals. International Journal of Plasticity **19**(10):1715–1748

[15] Michel B (2004) Formulation of a new intergranular creep damage model for austenitic stainless steels. Nuclear Engineering and Design **227**(2):161–174

[16] Riedel H (1987) Fracture at High Temperatures. Springer, Berlin Heidelberg

[17] Sklenička V, Kuchařová K, Svoboda M, Kloc L, Buršík J, Kroupa A (2003) Long-term creep behavior of 9–12% cr power plant steels. Materials Characterization **51**(1):35–48

[18] Balluffi R, Seigle L (1957) Growth of voids in metals during diffusion and creep. Acta Metallurgica **5**(8):449–454

[19] Lin J, Liu Y, Dean TA (2005) A review on damage mechanisms, models and calibration methods under various deformation conditions. International Journal of Damage Mechanics **14**(4):299–319

[20] Escalante C, Sierra E (2019) Fundamentals of transmission electron microscopy, the technique with the best resolution in the world. screen

[21] Du XW, Zhu J, Kim YW (2001) Microstructural characterization of creep cavitation in a fully-lamellar tial alloy. Intermetallics **9**(2):137–146

[22] Jiménez JA, Carsí M, Frommeyer G, Knippscher S, Wittig J, Ruano O (2005) The effect of microstructure on the creep behavior of the ti–46al–1mo–0.2si alloy. Intermetallics **13**(10):1021–1029

[23] Seo DY, Beddoes J, Zhao L, Botton GA (2002) The influence of aging on the microstructure and creep behaviour of a γ-ti–48% al–2% w intermetallic. Materials Science and Engineering: A **329-331**:810–820

[24] Sposito G, Ward C, Cawley P, Nagy PB, Scruby C (2010) A review of non-destructive techniques for the detection of creep damage in power plant steels. NDT & E International **43**(7):555–567

[25] Furtado HC, Le May I (1996) Metallography in life assessment of power plants. Materials Characterization **36**(4):175–184, international Metallography Conference MC95

[26] Masuyama F (2006) Creep degradation in welds of Mod. 9Cr-1Mo steel. International Journal of Pressure Vessels and Piping **83**(11):819–825, international Conference WELDS 2005

[27] Shibli IA, Holdsworth SR, Merckling G (eds) (2005) Creep & Fracture in High Temperature Components - Design & Life Assessment Issues. DEStech Publications, Inc.

[28] Stock SR (2008) Recent advances in X-ray microtomography applied to materials. International Materials Reviews **53**(3):129–181

[29] Martin CF, Josserond C, Salvo L, Blandin J, Cloetens P, Boller E (2000) Characterisation by X-ray micro-tomography of cavity coalescence during superplastic deformation. Scripta Materialia **42**(4):375–381

[30] Maire E, Withers PJ (2014) Quantitative X-ray tomography. International Materials Reviews **59**(1):1–43

[31] Yadav SD, Sonderegger B, Sartory B, Sommitsch C, Poletti C (2015) Characterisation and quantification of cavities in 9cr martensitic steel for power plants. Materials Science and Technology **31**(5):554–564

[32] Isaac A, Sket F, Reimers W, Camin B, Sauthoff G, Pyzalla AR (2008) In situ 3d quantification of the evolution of creep cavity size, shape, and spatial orientation using synchrotron X-ray tomography. Materials Science and Engineering: A **478**(1):108–118

[33] Gupta C, Toda H, Mayr P, Sommitsch C (2015) 3d creep cavitation characteristics and residual life assessment in high temperature steels: a critical review. Materials Science and Technology **31**(5):603–626

[34] Renversade L, Ruoff H, Maile K, Sket F, Borbély A (2014) Microtomographic assessment of damage in P91 and E911 steels after long-term creep. International Journal of Materials Research **105**(7):621–627

[35] He J, Sandström R (2016) Formation of creep cavities in austenitic stainless steels. Journal of Materials Science **51**(14):6674–6685

[36] Davanas K (2021) The critical role of Monkman–Grant ductility in creep cavity nucleation. Materials at High Temperatures **38**(1):1–6

[37] Sandström R, He J (2022) Prediction of creep ductility for austenitic stainless steels and copper. Materials at High Temperatures **39**(6):427–435

[38] Yu Y, Liu Z, Zhang C, Fan Z, Chen Z, Bao H, Chen H, Yang Z (2020) Correlation of creep fracture lifetime with microstructure evolution and cavity behaviors in g115 martensitic heat-resistant steel. Materials Science and Engineering: A **788**:139,468

[39] Huang L, Sauzay M, Cui Y, Bonnaille P (2021) Theoretical and experimental study of creep damage in alloy 800 at high temperature. Materials Science and Engineering: A **813**:140,953

[40] Zheng X, Xu Q, Lu Z, Wang X, Feng X (2020) The development of creep damage constitutive equations for high cr steel. Materials at High Temperatures **37**(2):129–138

[41] Zheng X (2021) The development and application of creep damage constitutive equations for high Cr steels over a wide range of stress. PhD thesis, University of Huddersfield, Zheng X.pdf (hud.ac.uk), accessed on 28th May 2023

[42] Fu G, Xu Q (2021) Further calibration of creep cavitation model for 316H steel. In: HIDA—8 International Conference, High Temperature Plant Cracking, Damage & Life Assessment, ETD Ltd, London

[43] Xu Q, Lu Z (2022) Recent progress on the modelling of the minimum creep strain rate and the creep cavitation damage and fracture. Materials at High Temperatures **39**(6):516–528

[44] Isaac A (2009) In-situ tomographic investigation of creep cavity evolution in brass. PhD thesis, Ruhr-Universität Bochum

[45] Sket F, Dzieciol K, Borbély A, Kaysser-Pyzalla AR, Maile K, Scheck R (2010) Microtomographic investigation of damage in e911 steel after long term creep. Materials Science and Engineering: A **528**(1):103–111, special Topic Section: Local and Near Surface Structure from Diffraction

Chapter 16
Self-heating Analysis with Respect to Holding Times of an Additive Manufactured Aluminium Alloy

Lukas Richter, Holger Sparr, Daniela Schob, Philipp Maasch, Robert Roszak, and Matthias Ziegenhorn

Abstract Materials exhibiting a rate-dependency in a mechanical loading regime enclose a variety of deformation mechanisms depending on their microstructure. This holds true for material classes from plastics to metals and is increasingly important for high-performance structural components. Material models covering viscoplastic deformation with hardening effects for metals have been widely studied in the last decades. The deformation mechanisms contribute to stored energy and dissipation and are reflected in the balance of energy. The current temperature measurement techniques give new opportunities to exploit an accurate temperature field to prove and validate material models. Especially, contact-free thermography with a small resolution range up to 1 mK is becoming more popular in mechanical testing set-ups. The paper examines a thermomechanical approach and an experimental concept for a material law verification and validation for self-heating in small temperature ranges. The focus lies on loading regimes incorporating holding times and the unloading path. An advanced thermographic measurement method is applied. It is pointed out that the thermomechanical approach is valuable and informative to assess the observed deformation processes and to describe the material behaviour with a thermodynamically valid parameter set.

16.1 Introduction

In the thermomechanical approach the processes of deformation as reason of mechanical loading and the temperature evolution as a consequence of heat generation

---

Lukas Richter · Holger Sparr · Daniela Schob · Philipp Maasch · Robert Roszak · Matthias Ziegenhorn
Brandenburg University of Technology Cottbus-Senftenberg, Universitätsplatz 1, 01968 Senftenberg, Germany,
e-mail: lukas.richter@b-tu.de, holger.sparr@b-tu.de, daniela.schob@b-tu.de, philipp.maasch@b-tu.de, robert.roszak@b-tu.de, matthias.ziegenhorn@b-tu.de

and heat fluxes are coupled. In the scope of this article the process is mainly focusing on self-heating phenomena, which is best illustrated by a balanced state in reference configuration. Without mechanical loading, a constant temperature distribution of the observed object(s) and no heat sources, the heat flux in the three-dimensional domain vanishes completely. Introducing then a mechanical load leads in general to an inhomogeneous temperature evolution.

The idealized case of thermoelasticity with negligible inelastic deformations is a reversible thermodynamic process, when adiabatic boundary conditions can be realized, which was already pointed out in the 19th century [1]. After mechanical unloading, the stored energy is released and the object returns to the homogeneous temperature field of the reference state. The adiabatic boundary condition is a sensible approximation for fast loading regimes and often serves as a limit case in numerical calculations. For most common thermal boundary conditions, a heat flux to the surroundings occurs and the thermodynamic irreversibility.

Therefore, any process has to be considered as irreversible, which is mathematically expressed by the inequality of the second law of thermodynamics often referred to as the CLAUSIUS-DUHEM inequality. This fundamental law in the context of deformable bodies leads to the framework of continuum thermomechanics, which is still discussed and successfully applied to a large variety of problems [2, 3]. The reflections on the CLAUSIUS-DUHEM inequality involved the idea of *admissible* processes and resulted in the Coleman-Noll procedure formulating restrictions on the constitutive equations and a reduced dissipation inequality [4]. In this sense, any material model must not violate the second law of thermodynamics or, in other words, needs to be thermodynamical consistent. The modelling approach often applies the concept of internal variables [5], which are related to specific deformation mechanisms.

These mechanisms comprise permanent modifications of the material's microstructure and generate permanent deformation. Taylor and Quinney [6] introduced a material specific ratio that only a fraction of plastic work is transferred into dissipation and therefore into heat. This observation was characterized by the term *stored energy of cold work* and was the central subject in the monograph of Bever et al in 1973 [7], which recorded a significant variance in the non-standardized testing set-ups.

Current infrared (IR) camera systems increased the accuracy of the measured temperature field on the surface dramatically, which lead to several substantial contributions w.r.t. the experimental methodology and the analytical material description [8–10]. The universal digital image correlation (DIC) was applied in several ways, which introduced very high experimental achievements in some applications [11–16].

The proposed set-up in this article reduces the experimental requirements to a minimum and is still able to access highly accurate temperature evolutions for complex loading regimes at room temperature. The influence of the strain rate is of major interest, which is reflected in the focus on material models for viscoplasticity. In the sense of continuum thermodynamics, the authors want to refer to the research groups headed by P. Rosakis [17], W. Egner [18], W. Oliferuk, [19] and M. Ristinmaa [20] and the references made therein. All these researchers pay special attention to thermomechanical consistency and the detailed evaluation of thermodynamical state variables to draw further conclusions to the material's microstructure regarding e.g. the yield initiation or strain hardening effects.

In particular, this paper deals with the numerical description of a selective laser melting (SLM) manufactured aluminium alloy. The nonlinear thermoviscoplastic material model described by Bröcker and Matzenmiller [21] is used for the material modelling. Motivated by the complex but clearly presented considerations, this model determines the components for the stored energy of cold work. Bröcker and Matzenmiller performed their investigations for aluminium with good agreement to the measured material behaviour based on a simple tension load without holding times. This paper introduces experiments with multiple defined holding times and two different strain rates. These complex experiments are observed and described by the thermomechanical approach.

The article summarizes the thermomechanical experiment in Sect. 16.2, which serves as motivation for the analytical and numerical approach. The numerical methods are outlined in Sects. 16.3 and 16.4. The results presented in Sect. 16.5 are discussed in detail in Sect. 16.6 and an outline for future research prospects is given in Sect. 16.7.

16.2 Thermomechanical Experiment

16.2.1 Experimental Set-Up

The thermomechanical experiments were performed by a servo-hydraulic testing machine with a maximum load cell of 25 kN from the company ZWICK&ROELL (Fig. 16.1 *left*). The machine has a temperature chamber, which is used as an additional thermal and convective shield of the system. On the right side of Fig. 16.1, the concept of the thermomechanical set-up is shown. In order to evaluate the displacement and temperature field in combination, an IR camera (No. 1), a mechanical testing system (No. 2) and an external radiator (No. 3) are necessary. The radiator produces a contrast in the thermogram which is needed for the later DIC for the strain evaluation. The IR camera ImageIR 8300 hs from INFRATEC GmbH was used to record the temperature field. The specifications of the camera are listed in Table 16.1. Based on the experimental set-up experiments for a SLM printed sample were realised. These specimens are made of the aluminium alloy AlSi7Mg0.6 and shown in Fig. 16.2 as detail d1. The illustrated geometry describes a dogbone sample according to DIN 50125 *shape E* [22]. The samples were provided by the Faculty of Mechanical Engineering at Wrocław University of Science and Technology. Detailed studies on the material structure were conducted and can be found in [23].

The material behaviour is to be characterised first. One approach is the consideration of displacement-controlled tests. The implementation of defined holding times can determine the rate dependency of the material behaviour. The holding time defines that part of the load path where the position of upper clamp is kept at a constant

Fig. 16.1: Thermomechanical experimental set-up, on the left side: testing machine an on the right side: experimental set-up (1 - IR camera, 2 - clamped specimen, 3 - radiation heat source).

Table 16.1: Technical Specifications of the ImageIR 8300 hs.

Property/Parameter	Unit	Specification
Spectral region	μm	1.5...5.5
Detector format	(px) × (px)	640 × 512
Temperature resolution at 30 °C measuring range	mK	≤ 20
Measuring range	°C	−40...1500
IR frame rate	Hz	125

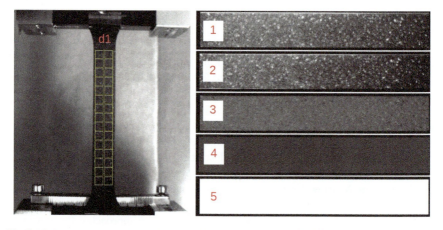

Fig. 16.2: Arrangement of the evaluation tool for temperature and displacement measurement.

level. If the stress decreases in this time, a rate-dependent material behaviour will be identified. These effect is known as stress relaxation. In case of constant stress during the holding time, the material behaviour is rate-independent [24].

The paper introduces complex experiments, which are defined by multiple holding times at defined displacement levels. In addition, these experiments include an unloading at the end. In the case of rate-dependent behaviour, relaxation effects become visible. Considering multiple holding times, different relaxation effects can be identified for one sample. The additional information on material behaviour is used to determine inelastic effects for material modelling in more detail. The phenomena of the holding times can provide information on the heat conduction effects of the system. Including an unloading path creates an additional identification possibility for the elastic deformation of the material description. If a chosen material model can describe these complex experiments, the determined parameters will show an enhanced physicality. According to the requirements mentioned above, an experiment with slow strain rates is performed. The value of the strain rate is motivated from [25] for the determination of the yield point. This slow experiment has a strain rate of $\dot{\varepsilon}_1 = 7.7 \cdot 10^{-5}$ s^{-1} and is furthermore referenced as $\dot{\varepsilon}_1$. The defined load path of $\dot{\varepsilon}_1$ is shown in Fig. 16.3.

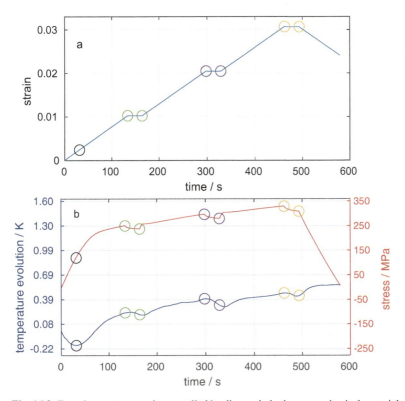

Fig. 16.3: Experiment $\dot{\varepsilon}_1$, a: strain controlled loading path; b: thermomechanical material response.

The slow strain rate $\dot{\varepsilon}_1$ and the thermal properties of aluminium create heat conduction effects that will have a significant impact on the thermal characteristics of the material. In order to determine the thermomechanical material behaviour without the influence of thermal conduction effects, a nearly adiabatic test is performed. This fast experiment has a strain rate of $\dot{\varepsilon}_2 = 7.7 \cdot 10^{-3} \text{s}^{-1}$ and is furthermore referenced as $\dot{\varepsilon}_2$.

In summary, experiment $\dot{\varepsilon}_2$ is used for near adiabatic thermomechanical analysis. Experiment $\dot{\varepsilon}_1$ is performed to describe the thermal effects of the system.

16.2.2 Temperature and Deformationfield Measurement by Digital Image Correlation

The displacement field and the temperature field can be determined by using the IR camera.The samples have to be prepared for DIC in order to measure the displacement field. Based on Fig. 16.2 *left* detail *d*1, the requirements for sample preparation can be described. In order to eliminate the reflection of the surrounding radiation almost completely the specimen is sprayed with black matt varnish. This procedure enables the evaluation of the temperature field. To create reference points for the DIC, the specimen is sprayed with random dot pattern. The spray contains metal particles which contrast with the black background in the thermal images. The applied radiator helps to increase the contrast. The yellow boxes in Fig. 16.2 *left* detail *d*1 show the number and the distribution of the tracker points set for the DIC.

In addition to the DIC, the emissivity correction is performed on the saved IR camera images, because the dot pattern disturbs the temperature measurement. The analyses are performed by a post-processing tool, whose function can be described as follows. Using the evaluation tool for the defined sub-surfaces allows the calculation of the strain tensor. This is achieved by applying an affine transformation algorithm on the in-plane displacement field [26]. It was followed by the emissivity correction (or computational pattern removal) and the mapping of the corrected temperature field evolution on the reference configuration. Finally, the results are evaluated with the necessary resolution in space and time using averaging strategies on the level of geometric objects. For more detailed explanations see [27].

Figure 16.2 *right* illustrates the results of the evaluation tool. No. 1 describes the original camera data. No. 2 shows the data with motion compensation and No. 3 the emissivity correction data. No. 3 demonstrates that the inhomogeneous emissivity distribution caused by the dot pattern is removed from the temperature data. No. 4 presents the strain calculation in the longitudinal direction of the specimen (tensile direction) and No. 5 the strain calculation in the transverse direction of the specimen. Using the FIJI software IMAGEJ, the measured temperature and strain fields were comfortably evaluated in time and space.

16.2.3 Experimental Results with Respect to Holding Time

The experimental results show the characteristics in the centre of the sample 16.2 *left d*1. Using the experimental set-up, Fig. 16.3 shows the defined load path and the thermomechanical material behaviour of experiment $\dot{\varepsilon}_1$. Figure 16.3 b) describes the mechanical material behaviour by a red curve and the temperature material behaviour by a blue curve. The temperature decreases to the black circle mark and increases then. The decrease results from the thermoelastic effect while the increase of the temperature shows the start of inelastic hardening phenomena. With reference to the red curve, the black circle mark indicates the yield point of the material. The beginning and the end of the holding times are shown by the green, purple and yellow marks. Relaxation processes can be seen in these areas. This suggests that the material behaviour is rate-dependent. For the description of the relaxation, the area between the green marks in the first holding time is considered in more detail.

At the beginning of the holding time (Fig. 16.4 first green mark), the stress decreases. In contrast, the temperature increases first and begins to decrease at 148 s. When the holding time ends (Fig. 16.4 second green mark), the stress increases

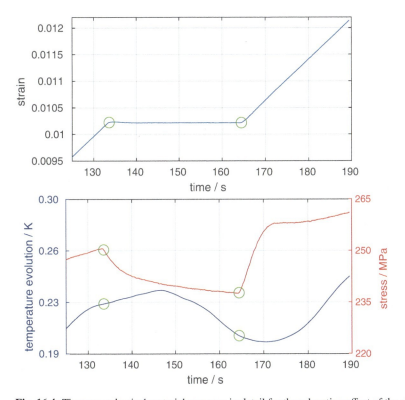

Fig. 16.4: Thermomechanical material response in detail for the relaxation effect of the slow load path.

immediately, while the temperature decreases at beginning of the load continuation and increases at 170 s.

It is possible that the phenomenon between the two green marks is based on the material effect (microstructure effects) or on thermal effects. In order to investigate this effect more, a near adiabatic experiment was motivated. The case of this adiabatic experiment shows the influence of the internal material behaviour. Fig. 16.5 represents the load path and the corresponding thermomechanical material response of the experiment $\dot{\varepsilon}_2$.

Compared to experiment $\dot{\varepsilon}_1$ (Fig. 16.3 b), the mechanical stresses (Fig. 16.5 b) and the resulting temperature changes are different to $\dot{\varepsilon}_1$. This confirms the rate dependence of the thermomechanical material behaviour. For the holding time between the purple marks (Fig. 16.5 a), the detailed view of the thermomechanical material behaviour demonstrates the following.

Similar to experiment $\dot{\varepsilon}_1$ (Fig. 16.4), the experiment $\dot{\varepsilon}_2$ proves the same effects during the relaxation process (Fig. 16.6). For this reason, it can be assumed that the phenomena shown are motivated by the material behaviour. Remarkably, these effects cause such a strong heating that it dominates the heat conduction and heats up the system.

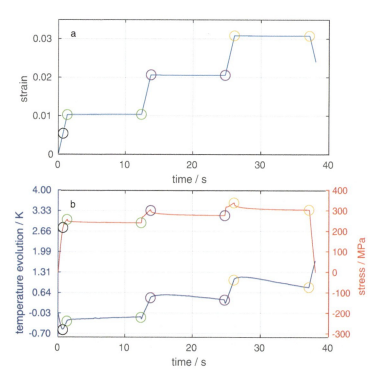

Fig. 16.5: Experiment $\dot{\varepsilon}_2$, a) strain controlled loading path and b) the thermomechanical material response.

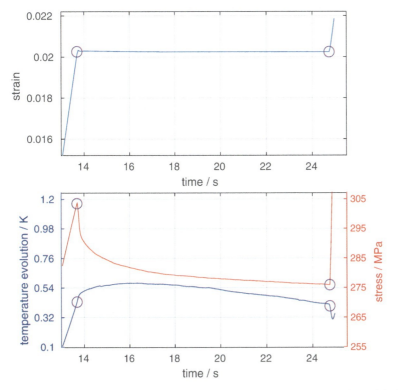

Fig. 16.6: Thermomechanical material response in detail for the relaxation effect of the fast load path.

16.3 Theoretical Framework

16.3.1 Energy Balance and Heat Conduction

For the numerical representation of the thermomechanical phenomena to be seen in Sect. 16.2, this subsection deals with the general derivation of the heat conduction equation and the mechanical dissipation. The principles are formulated by the material-independent equations of continuum mechanics.

If the fundamental laws of kinematics and energy balance are consistently applied, it leads to a coupling of the independent mechanical and thermal variables by the related partial differential equations. To evaluate the self-heating effect of a material, the energy and the entropy balance are refined by characteristic terms through the concept of internal variables, which is specifically laid out in Subsect. 16.3.2.

To begin the thermodynamic approach, the description refers to the displacements of the reference configuration using the displacement gradient tensor. Assuming small deformations, the norm of the displacement gradient tensor becomes much smaller than 1. Consequently, the deformations of the material particle can be described by the linearised deformation-displacement relation. The linearised approximation allows to formulate the balance equations of the momentary configuration at the undeformed body. As a result, the physical entities of the EULERIAN and the LAGRANGIAN description coincide.

The first law of thermodynamics describes the change in internal energy by the sum of the work done and the heat supplied. The energy balance is written as a local form by:

$$\dot{e} = -\frac{1}{\rho}\text{div}(\boldsymbol{q}) + r + \frac{1}{\rho}\boldsymbol{T}\cdot\dot{\boldsymbol{E}}, \tag{16.1}$$

where the boldface symbols characterize tensorial entities. $\boldsymbol{T}$ defines the CAUCHY stress tensor, $\boldsymbol{E}$ the linearised GREEN-LAGRANGE strain tensor and the variable $\dot{\boldsymbol{E}}$ specifies the strain rate tensor. The operation $\boldsymbol{T}\cdot\dot{\boldsymbol{E}}$ expresses the inner product of these two tensors resulting in a scalar value. The variable $\boldsymbol{q}$ describes the specific heat flux, $\dot{e}$ is the specific internal energy rate and r defines a radiation heat source.

The second law of thermodynamics describes the entropy development $\dot{s}$ in thermodynamic system as a reversible ($\dot{s} = 0$) or an irreversible ($\dot{s} > 0$) process. In the case of the thermodynamical system, the entropy can be seen as a measure of spontaneously occurring processes in the system. Specific processes can be the (pure) heat conduction or dissipation due to changes in the material structure, e. g. dislocation movement. In summary, entropy can be understood as a measure of the irreversibility of a system [24]. The local form of the resulting CLAUSIUS-DUHEM inequality is given by:

$$\dot{s} - \frac{r}{\theta} + \frac{1}{\rho\theta}\text{div}(\boldsymbol{q}) - \frac{1}{\rho\theta}\boldsymbol{q}\cdot\boldsymbol{g} \geq 0, \tag{16.2}$$

where $\boldsymbol{g}$ describes the temperature gradient, θ is the temperature and ρ is the mass density. Combining Eq. (16.1) and Eq. (16.2) results in one of the basic equations of thermomechanics, called the dissipation inequality

$$\Delta = \theta\dot{s} - \dot{e} + \frac{1}{\rho}\boldsymbol{T}\cdot\dot{\boldsymbol{E}} - \frac{1}{\rho\theta}\boldsymbol{q}\cdot\boldsymbol{g} \geq 0. \tag{16.3}$$

Equation (16.3) shows that the internal dissipation Δ depends on the internal energy e. The thermodynamic potential e is a function of the strain tensor $\boldsymbol{E}$ and the entropy s. For the later practical application with the requirement of a simple experimental set-up, the entropy s represents an indirectly measurable quantity. For this reason, the thermodynamic laws Eq. (16.1) and (16.2) can be transferred to another thermodynamic potential with the help of the LEGENDRE transformation [24]. The potential of the free energy ψ thereby offers a dependence on the strain tensor $\boldsymbol{E}$ and the temperature θ. These two quantities can be directly determined in most experimental configurations. The dissipation inequality as a function of the free energy ψ reads as:

$$\Delta = -s\dot{\theta} - \dot{\psi} + \frac{1}{\rho}\boldsymbol{T}\cdot\dot{\boldsymbol{E}} - \frac{1}{\rho\theta}\boldsymbol{q}\cdot\boldsymbol{g} \geq 0. \tag{16.4}$$

In order to fulfil the dissipation inequality Eq. (16.4), a simplification leads to the specification. This simplification suggests that the thermal strains E_{th} due to small temperature changes are negligible. The result of the beginning simplification of the linearised strain tensor is, that the strain tensor E can additively be split into an elastic component E_e and an inelastic component E_{ie}. It is common to understand the inelastic strain E_{ie} as an internal variable in the case of inelastic processes [28]. According to this, the thermodynamic state of the system depends on the difference between the small strain tensor E and the inelastic strain tensor E_{ie}

$$E = E_{th} + E_e + E_{ie} \quad \text{and} \quad E_e = E - E_{ie} \quad \text{with} \quad E_{th} = 0. \tag{16.5}$$

Considering Eq. (16.5), the free energy ψ results in:

$$\psi = \psi(E_e, \theta, g, a_1, \ldots, a_n) = \psi_e(E_e, \theta, g) + \psi_{ie}(\theta, g, a_1, \ldots, a_n), \tag{16.6}$$

where ψ_e describes the elastic part and ψ_{ie} the inelastic part. Eq. (16.6) introduces the variables $a_1, \ldots, a_n$. These variables are employed by the concept of internal variables and evolve as a result of inelastic material behaviour. Depending on the chosen material model, their number can differ. With the implementation of Eq. (16.5) and Eq. (16.6) in Eq. (16.4), it follows:

$$\Delta = \left(\frac{1}{\rho}T - \frac{\partial \psi}{\partial E_e}\right)\dot{E}_e + \left(-s - \frac{\partial \psi}{\partial \theta}\right)\dot{\theta} + \frac{1}{\rho}T \cdot \dot{E}_{ie} - \sum_{j=1}^{n}\left(\frac{\partial \psi}{\partial a_j}\dot{a}_j\right) - \frac{\partial \psi}{\partial g}\dot{g} - \frac{1}{\rho\theta}q \cdot g \geq 0. \tag{16.7}$$

In order to satisfy the inequality from Eq. (16.7), each summand of the equation can be evaluated individually [21]. The dicussion with respect to admissible thermodynamic processes leads to restrictions on the constitutive equations, which is known as the Coleman-Noll procedure. The procedure leads to:

$$T = \rho\frac{\partial \psi}{\partial E_e}, \quad s = -\frac{\partial \psi}{\partial \theta} \quad \text{and} \quad \frac{\partial \psi}{\partial g} = 0. \tag{16.8}$$

By substituting Eq. (16.8) into Eq. (16.7), the dissipation inequality is reduced to:

$$\Delta = \frac{1}{\rho}T \cdot \dot{E}_{ie} - \sum_{j=1}^{n}\frac{\partial \psi}{\partial a_j}\dot{a}_j - \frac{1}{\rho\theta}q \cdot g \geq 0, \tag{16.9}$$

In the case of an loaded thermomechanical system with adiabatic boundary conditions, heat transfer between the system and its environment is not possible. When the stress is in the range of elastic deformations E_e, the dissipation inequality becomes zero and the process is reversible. The process is irreversible, when the thermomechanical approach is formulated as physically consistent. Then inelastic deformations lead the internal variables of the system (E_{ie} and a_j) evolve and the dissipation inequality becomes greater than zero. The formulation of the corresponding evolutionary equations for the internal variables need to be thermodynamically consistent.

Under isothermal boundary conditions, heat transfer between the thermomechanical system and its surroundings is allowed. In this case q and g get a higher influence in the dissipation inequality. According to FOURIER, the heat flux vector q is negatively proportional to the temperature gradient g. In case of elastic deformations E_e, the inequality is greater than zero and the process is irreversible. As described for the adiabatic process, the statement for the inelastic deformations E_{ie} holds.

The dissipation inequality (Eq. (16.4)) motivates the analysis of the mechanically influenced part that applies to:

$$\delta_m = \frac{1}{\rho} T \cdot \dot{E}_{ie} - \sum_{j=1}^{n} \frac{\partial \psi}{\partial a_j} \dot{a}_j \geq 0, \quad (16.10)$$

with

$$p_{ie} = \frac{1}{\rho} T \cdot \dot{E}_{ie} \quad \text{and} \quad p_s = \sum_{j=1}^{n} \frac{\partial \psi}{\partial a_j} \dot{a}_j. \quad (16.11)$$

Equation (16.11) *left* defines the inelastic stress power p_{ie} which is not completely converted into heat. The energy fraction p_s stored in the material structure is described in Eq. (16.11) *right*. This stored part can be released under certain circumstances. An example of this is the change in the microstructure as a result of a dislocation pile-up at grain boundaries [21]. Using Eq. (16.11), a corresponding material characteristic can be identified. This is known as the *energy transformation ratio* (ETR) φ [29]. The ETR gives a general overview of the energy storage characteristics on the material under inelastic deformation and is defined as:

$$\varphi = \frac{w_s}{w_p} \quad (16.12)$$

with

$$w_p = \int_{t_0}^{t} p_e(\bar{t}) \, d\bar{t}; \quad w_s = \int_{t_0}^{t} p_s(\bar{t}) \, d\bar{t},$$

where w_p describes the plastic strain work and w_s the *stored energy of cold work*.

To calculate the temperature evolution on the surface of the sample, the heat conduction equation must be formulated in addition to the mechanical dissipation. The first law of thermodynamics (Eq. (16.1)) is used as the basis for the evaluation. Under consideration of the LEGENDRE transformation, Eq. (16.8) and the GIBBS relation [24], the general form of the heat conduction using the potential of the free energy equation is given by:

$$-\theta \frac{\partial^2 \psi_e}{\partial \theta^2} \dot{\theta} = -\theta \frac{\partial^2 \psi_e}{\partial \theta \partial E_e} \dot{E}_e + \frac{1}{\rho} T \dot{E}_{ie} - \theta \sum_{j=1}^{n} \left(\frac{\partial^2 \psi_{ie}}{\partial \theta \partial a_j} \dot{a}_j \right) - \frac{1}{\rho} \text{div}(q) + r. \quad (16.13)$$

16.3.2 Material Model

With the general energetic and thermal description of the thermomechanical modelling in Subsect. 16.3.1, the accurate description of the used material model follows in this chapter. Motivated by the experiment (Sect. 16.2), the chosen material model should represent a linear elastic behaviour followed by a non-linear viscoplastic characteristic. The material model of Bröcker and Matzenmiller [21] was chosen for this paper. Based on a rheological network, which is created by a clear structure and demonstrates the interaction of the different mechanical mechanisms. Before describing the material model in more detail, some simplifications should be made. The experiments that are shown Sect. 16.2 are uniaxial tensile tests. For this reason, the system is converted into an isotropic 1D formulation. In addition, the given displacements are in the technically relevant range of small deformations (Fig. 16.5 a) and Fig. 16.3 a). The total strain ε is split up into a thermal ε_{th}, an elastic ε_{el} and a visoplastic ε_{vp} part:

$$\varepsilon = \varepsilon_{th} + \varepsilon_{el} + \varepsilon_{vp}. \tag{16.14}$$

Motivated by the experiments (Fig. 16.5 b), small temperature changes are to be expected. For this reason, the thermal strains ε_{th} can be neglected. The elastic deformation is adequately covered by HOOKE's law. As a consequence of the rheological network, the kinematic and isotropic hardening are mathematically controlled by the viscoplastic strain ε_{vp}. In Fig. 16.7, the elements with dissipative and energy storage character are clearly separated from each other. All the elements coloured grey symbolise components with dissipative character and the rest of elements take into account the stored (free) energy [21]. The Bröcker and Matzenmiller model deals with a "small number" of parameters to be identified, which simplifies the solution of the system of ordinary differential equations. In order to create a full thermomechanically consistent material model, Bröcker and Matzenmiller motivate the flow function and the flow rule by simple calculations based on the rheological network.

The yield function f is defined by:

$$f = |\sigma - \xi| - (k_0 + \kappa), \tag{16.15}$$

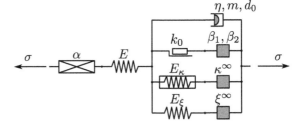

Fig. 16.7 Rheological model of the Bröcker Matzenmiller material model [21].

with the yield condition:

$$f = \begin{cases} f \leq 0 & \text{elastic domain} \\ f > 0 & \text{viscoplastic domain} \end{cases} \quad (16.16)$$

were ξ describes the backstress component that represents the kinematic hardening. The yield point is defined by k_0 and κ describes the isotropic hardening. The evolution equation of the true stress σ with respect to Eq. (16.14) is given by the stress rate $\dot{\sigma}$:

$$\dot{\sigma} = E\dot{\varepsilon}_{el} = E\left(\dot{\varepsilon} - \dot{\varepsilon}_{vp}\right) \quad , \quad (16.17)$$

where E is the Young's modulus.

In the case of the evolution in mechanical loads, which raises the yield function into the viscoplastic domain, irreversible effects are initiated [30]. These effects are characteristic of the hardening behaviour and will be described by defining internal variables. In the case of the Bröcker and Matzenmiller model, the evolution equations of the internal variables are given by:

$$\dot{\kappa} = E_\kappa \left(1 - \frac{\kappa}{\kappa^\infty}\right) \dot{\bar{\varepsilon}}_{vp}, \quad (16.18)$$

$$\dot{\xi} = E_\xi \left(\dot{\varepsilon}_{vp} - \frac{\xi}{\xi^\infty} \dot{\bar{\varepsilon}}_{vp}\right) \quad (16.19)$$

and

$$\dot{\varepsilon}_{vp} = \frac{1}{\eta} \left\langle \frac{f}{d_0} \right\rangle^m \text{sign}(\sigma - \xi) \quad (16.20)$$

In Eq. (16.18) E_κ describes the kinematic hardening modulus and κ^∞ defines the saturation value of isotropic hardening stress [21]. The kinematic hardening modulus E_ξ and the saturation value of kinematic hardening stress ξ^∞ are shown in Eq. (16.19). Eq. (16.20) introduces η for the strain rate scaling factor, m for the nonlinear rate dependency exponent and d_0 for regularising the exponential term as a dimensionless expression. After solving the shown system of differential equations, the mechanical stress is calculated by the expression in Eq. (16.17).

By using the example of the Subsect. 16.3.1, the system can be enhanced to the thermomechanical equations. As a consequence of the Coleman-Noll procedure in Eq. (16.8), the mechanical dissipation is described by:

$$\delta_m = \frac{1}{\rho}\left[f + k_0\left(1 - \beta_1 e^{-\beta_2 \bar{\varepsilon}_{vp}}\right) + \frac{\kappa^2}{\kappa^\infty} + \frac{\xi^2}{\xi^\infty}\right]\dot{\bar{\varepsilon}}_{vp} \geq 0, \quad (16.21)$$

where the parameter condition $0 \leq \beta_1 \leq 1$ and $\beta_2 \geq 0$ are defined by Bröcker and Matzenmiller. These introduced parameter can be used for scaling the stored energy and the resulting self-heating behaviour. For more details on the derivation of the mechanical dissipation see the Bröcker and Matzenmiller [21].

Based on the assumptions that have been made, the heat transfer equation according to the example in Eq. (16.13) follows to:

$$c_\varepsilon \dot{\theta} = -\frac{1}{\rho} E\alpha\theta\dot{\varepsilon}_{el} + \delta_m + \frac{1}{\rho} k \operatorname{div}(q) + r \quad , \tag{16.22}$$

where c_ε is the specific heat capacity and α is the thermal expansion coefficient [24].

Finally, each term of Eq. (16.22) should be explained. The first term $c_\varepsilon \dot{\theta}$ describes the material-dependent heat change. The second term

$$\frac{1}{\rho} E\alpha\theta\dot{\varepsilon}_{el}$$

defines the thermoelastic effect. As a result of this effect, the system is cooling in case of an elastic expansion and heating in case of an elastic compression. The third term δ_m symbolises the system's self-heating effect and is defined by Eq. (16.21). The fourth term

$$\frac{1}{\rho} k \operatorname{div}(q)$$

characterises the heat conduction effect and r defines an external heat source.

16.4 Modelling Methods

16.4.1 Parameter Identification

This section deals with the description of the parameter estimation by implementing the system of differential equations given in Subsect. 16.3.2 in MATLAB. For the parameter identification, the described material model was implemented in the programmable computing environment MATLAB. The calculation of the material behaviour is performed in the midpoint m_c of the sample (compare Fig. 16.8) with ideal adiabatic boundary conditions. Because of the adiabatic boundary conditions, only the experiment $\dot{\varepsilon}_2$ (Sec. 16.2.1) is qualified for a full thermomechanical parameter identification. To control the identification, the stress-strain characteristic is used for experiment $\dot{\varepsilon}_1$. In terms of the parameter estimation, the material model from Subsect. 16.3.2 can be classified in directly and indirectly determinable parameters.

Direct means that the parameters are directly established by measurements or are taken from data sheets. The Young's modulus E is determined through the experimental unloading curve (Subsect. 16.2.3). To evaluate mass density ρ, the sample weight is measured and the sample volume is calculated by the given dimensions. The coefficient of thermal expansion α, the heat capacity c_ε and the thermal conductivity k are specified in corresponding data sheets with respect to the applied standards. The regularising parameter d_0 is defined as 1 MPa.

Table 16.2: Optimisation results.

	Unit	Lower Limits	Upper Limits	Determined Value
Indirect Parameters				
k_0	MPa	200	220	211
m	–	1	5	4.485
η	s	$1 \cdot 10^6$	$1 \cdot 10^9$	$79.791 \cdot 10^6$
E_ξ	MPa	6500	7500	7000
ξ^∞	MPa	165	185	180
Direct Parameters				
E	MPa			45000
β_1	–			0.4
ρ	g cm^{-3}			2.66
α	K^{-1}			$2.2 \cdot 10^{-5}$
c_ε	Jkg^{-1}K^{-1}			910
k	Wm^{-1}K^{-1}			170

In addition, it is assumed a pure kinematic hardening to describe the hardening mechanisms. Therefore E_κ and κ^∞ become zero. The energy-motivated parameters β_1 and β_2 can be determined through the temperature characteristic. Assuming purely kinematic hardening, β_2 will be zero [21]. Table 16.2 shows the values of these directly determined parameters. All other parameters are subjected on an optimisation procedure. Before the optimisation starts, an initial sensitivity analysis for a limited parameter space offers an overview of the influence these parameters have on the thermoviscoplastic behaviour. The parameter space is limited by lower limits (LL) and upper limits (UL), which are shown in Table 16.2.

The sensitivity analysis is performed by Monte Carlo analysis. This simulation generates random parameter sets in the parameter space and evaluates their effect on the objective function. The advantage of the chosen simulation is that all parameters are varied and a global assessment related to the limits is achieved [31]. The most sensitive parameters of the Bröcker and Matzenmiller model are m and k_0. The parameters E_κ and κ^∞ have no influence on the system by assuming pure kinematic hardening.

For the optimisation the genetic algorithm, which is implemented in MATLAB, was used [31]. The optimisation takes place in the selected parameter space from the sensitivity analysis (Table 16.2). After finding a first parameter set, the parameters are limited further and optimisation is performed for the temperature and the stress-strain characteristics as a cost function. The determined values can be found in Table 16.2.

16.4.2 Concept for Thermomechanic FE Analysis

The implementation of the material model in MATLAB (Subsect. 16.4.1) created a fast possibility of the parameter identification. But this is only limited to the dissipative effects, which are motivated by the mechanical material behaviour. The simplified MATLAB solution is reduced to the solution of systems in which only small heat conduction effects take place. In order to describe the thermomechanical problem, the material model was transferred into a user subroutine for the FE software package ABAQUS (UMAT). In the case of the Bröcker and Matzenmiller model from Subsect. 16.3.2, the UMAT solves the mechanical evolution equations, defines the resulting energy and dissipation components and transfers them to the ABAQUS solver. The ABAQUS solver calculates the coupled problem with the representation of realistic boundary conditions. The implementation of the material model enables the numerical simulation of the slow test from Sect. 16.3.

To understand the simulation results, the model should be described in more detail in relation to the defined boundary conditions. As the material model Subsect. 16.3.2 is limited to a one-dimensional problem, the FE model is represented by 3-node quadratic displacement and linear temperature *truss* elements (*T2D3T*). In addition to the measurement length l_m (Fig. 16.8), the clamping range l_c was modelled to account for the heat exchange. Motivated by the experimental set-up (Subsect. 16.2.1), the radiator is introduced as an additional heat source. This heat source (DFLUX) is assumed to be constant for the measurement length l_m with $r = 0.012$ Jsm^{-2}. In addition, a heat transfer coefficient (FILM) was defined at the points A and B (Fig. 16.8) with $h = 10$ Js^{-1}m^{-2}K^{-1}. The initial temperature for all elements is constant with $T_{ini} = 20°$C. A mechanical load is realised by the displacement u of the node B.

16.5 Results

The parameters from Table 16.2 are integrated in the numerical model. The generated numerical results show the FE simulation. Based on the experimental evaluation (Subsect. 16.2.3), the results are shown in the point m_c (Fig. 16.8). The numerical results for experiment $\dot\varepsilon_1$ can be seen in Fig. 16.9.

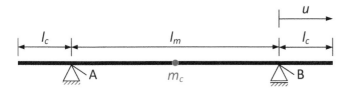

Fig. 16.8: Schematic of the FE model.

The simulated thermomechanical material response shown in Fig. 16.9 illustrates that the parameter set describes the slow experiment with good accuracy. A pronounced difference between experiment and simulation is in the area of the yield point k_0 and amounts to 14 MPa in the mechanical characteristic (Fig. 16.9 a) and 0.03 K in the temperature characteristic (Fig. 16.9 b). In addition, a larger difference can be seen at the end of the last holding time. This amounts to 10 MPa in the mechanical characteristic (Fig. 16.9 a) and 0.12 K in the temperature characteristic (Fig. 16.9 b).

The results for experiment $\dot{\varepsilon}_2$ are shown in Fig. 16.10. Similar to experiment $\dot{\varepsilon}_1$, a larger difference is seen at the end of the last holding time. In this case it amounts to 16 MPa in the mechanical characteristic (Fig. 16.10 a) and 0.04 K in the temperature characteristic (Fig. 16.10 b). The yield point k_0 is better reproduced for experiment $\dot{\varepsilon}_2$ than for experiment $\dot{\varepsilon}_1$. The simulation of experiment $\dot{\varepsilon}_2$ is more accurate.

To explain the differences in the temperature evolution between the two experiments, the ETR characteristics of the simulations can be considered. As already discussed in Subsect. 16.3.1, there are no corresponding experiments for these characteristics. Nevertheless, it illustrates the energy storage behaviour of the material. According to the literature, the plotting of the ETR over the plastic work w_p (Fig. 16.11) has been established for a better visualisation of energetic characteristic [29].

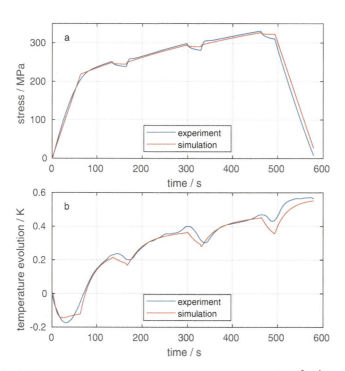

Fig. 16.9: Simulation results for the slow strain rate with $\dot{\varepsilon} = 7.7 \cdot 10^{-5}$ s^{-1}. a) stress evolution over time, b) temperature evolution over time.

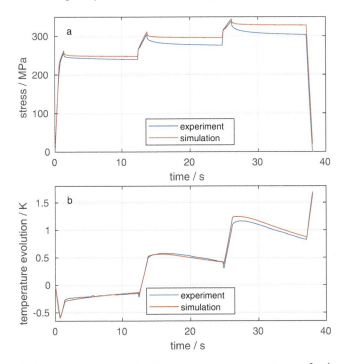

Fig. 16.10: Simulation results for the fast strain rate with $\dot{\varepsilon} = 7.7 \cdot 10^{-3} \text{ s}^{-1}$. a): stress evolution over time, b): temperature evolution over time.

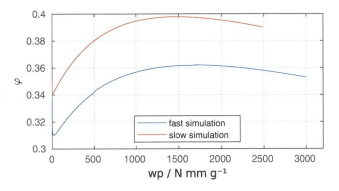

Fig. 16.11: Energy transformation ratio for both strain rates.

In accordance with the ETR definition (Eq. (16.12)), the lower ETR of the faster simulation results in less energy being stored in the material structure. The larger value of the plastic work w_p in the experiment with $\dot{\varepsilon}_2$ together with the always smaller ETR in comparison to the slower experiment $\dot{\varepsilon}_1$ generates more mechanical dissipation δ_m (Eq. (16.21)). The corresponding temperature evolution can only be discussed according to the actual heat conduction problem.

16.6 Discussion

The aim of this paper is to describe the material behaviour of an SLM manufactured aluminium alloy for complex load paths with a thermomechanical approach. For the applied tests several holding times and two different strain rates were defined. Corresponding to the selected speed differences, the image processing of the IR camera was able to determine values that could be evaluated accurately. Even effects such as the relaxation for the classification of the material can be represented as a full thermomechanical process. The high accuracy of the applied measuring system is supported by the outlined evaluation strategies. The IR camera is a helpful tool for the experimental determination of the coupled thermal and mechanical field components.

For the initial conditions, the initial temperature was assumed to be constant. A symmetric temperature profile is assumed for the boundary conditions. These hypotheses is in a first approximation confirmed by he IR measurement and can be refined in the future.

The results of the numerical simulations can reproduce the rate dependent material behaviour with good quality on different strain rates. The region of the yield strength shows a higher difference between the simulation and the experiment. This is a consequence of the strict decomposition of the total strain in the pure elastic and the elastic-viscoplastic domain by considering the yield function (Eq. (16.15)) and the yield condition (Eq. (16.16)). These approaches are constitutive for Perzyna [30] and Chaboche [32].

Another possibility of the thermomechanical representation of the relaxation processes (Fig. 16.10) is to introduce an additional internal variable. This would raise the degree of non-linearity in the material model. Such a principle example was considered and integrated by Kamlah and Haupt [29]. The internal variable could modify the viscoplastic strain or be added to the holding process as a deformation condition.

16.7 Conclusion and Outlook

This paper demonstrates that the Bröcker and Matzenmiller thermomechanical approach can simulate the complex material behaviour including the relaxation processes with high accuracy. Differences occur as soon as more complex phenomena such as relaxations and different velocities are to be represented. As one more option for a refined modelling, other material models than the chosen viscoplasticity model can be used for the evaluation. One option is the investigation of a unified viscoplastic model, *e.g.* the model proposed by Bodner and Lindenfeld [33]. A different approach splits the stress into the rate independent equilibrium stress and a rate dependent drag stress. This model type was suggested by Krempl [34] and was introduced in a pure mechanical framework by neglecting the temperature evolution. The thermomechanical approach can be implemented as described in the paper.

The discussed viscoplastic material model gives an idea of the complex processes of inelastic behaviour and creates an overview of dealing with parameter identification. The strategy with starting with a sensitivity analysis and then proceeding to an optimisation task offers a valuable approach. Such an approach can be used if enough experimental characteristics (like the thermal and the stress-strain characteristic) are available. In this paper, the thermomechanical analysis has shown that the optimisation tasks for a complete analysis of the system are limited to approximately adiabatic states.

Using the IR camera makes it possible to evaluate the temperature profile over the length of the sample. In the future, this system information can be integrated for the control of the thermal initial boundary conditions in the FE models. So, the heat conduction effects and external influences such as the radiator can generally be integrated at the sample length. In this sense, a precise and effective evaluation strategy needs to be formulated. This approach could be used to extend the scope of the diagrams and to define a more general objective function for the optimisation task on the foundation of a surface characterisation as shown in Fig. 16.12.

Acknowledgements The realisation of this work has been made possible by the valuable support and collaboration from Andrzej Pawlak and Irina Smolina at the Centre of Advanced Manufacturing Technologies (CAMT) of the Technical University of Wrocław. Expertise and dedication in providing the necessary specimens were essential in the successful completion of the study.

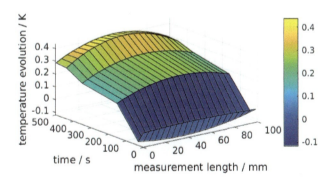

Fig. 16.12: Temperature evolution over time and over measurement length.

References

[1] Thomson W (1855) On the thermoelastic and thermomagnetic properties of matter, Part I. Quarterly Journal of Mathematics pp 57–77
[2] Truesdell C, Noll W (1992) The Non-Linear Field Theories of Mechanics, 3rd edn. Springer-Verlag, Berlin
[3] Gurtin ME, Fried E, Anand L (2013) The Mechanics and Thermodynamics of Continua. Cambridge University Press, Cambridge [et al.]
[4] Coleman BD, Noll W (1963) The thermodynamics of elastic materials with heat conduction and viscosity. Archive for Rational Mechanics and Analysis **13**(1):167–178
[5] Coleman BD, Gurtin ME (1967) Thermodynamics with internal state variables. The Journal of Chemical Physics **47**(2):597–613
[6] Taylor GI, Quinney H (1934) The latent energy remaining in a metal after cold work. Prog Mater Sci A **143**:307–326
[7] Bever M, Holt D, Titchener A (1973) The stored energy of cold work. Progress in Materials Science **17**:5 – 177
[8] Chrysochoos A, Maisonneuve O, Martin G, Caumon H, Chezeaux J (1989) Plastic and dissipated work and stored energy. Nuclear Engineering and Design **114**(3):323–333
[9] Muracciole J, Wattrisse B, Chrysochoos A (2008) Energy balance of a semicrystalline polymer during local plastic deformation. Strain **44**(6):468–474
[10] Chrysochoos A (2012) Infrared thermography applied to the analysis of material behavior: A brief overview. Quantitative InfraRed Thermography Journal **9**(2):193–208
[11] Maynadier A, Poncelet M, Lavernhe-Taillard K, Roux S (2012) One-shot measurement of thermal and kinematic fields: InfraRed Image Correlation (IRIC). Experimental Mechanics **52**(3):241–255
[12] Fedorova A, Bannikov M, Terekhina A, Plekhov O (2014) Heat dissipation energy under fatigue based on infrared data processing. Quantitative InfraRed Thermography Journal **11**(1):2–9
[13] Oliferuk W, Maj M, Zembrzycki K (2015) Determination of the energy storage rate distribution in the area of strain localization using infrared and visible imaging. Experimental Mechanics **55**(4):753–760
[14] Knysh P, Korkolis Y (2015) Determination of the fraction of plastic work converted into heat in metals. Mechanics of Materials **86**:71 – 80
[15] Cholewa N, Summers PT, Feih S, Mouritz AP, Lattimer BY, Case SW (2016) A technique for coupled thermomechanical response measurement using infrared thermography and digital image correlation (TDIC). Experimental Mechanics **56**(2):145–164
[16] Rose L, Menzel A (2020) Optimisation based material parameter identification using full field displacement and temperature measurements. Mechanics of Materials **145**:103,292

[17] Rosakis P, Rosakis A, Ravichandran G, Hodowany J (2000) A thermodynamic internal variable model for the partition of plastic work into heat and stored energy in metals. Journal of the Mechanics and Physics of Solids **48**(3):581–607
[18] Egner W, Egner H (2016) Thermo-mechanical coupling in constitutive modeling of dissipative materials. International Journal of Solids and Structures **91**:78 – 88
[19] Oliferuk W, Raniecki B (2018) Thermodynamic description of the plastic work partition into stored energy and heat during deformation of polycrystalline materials. European Journal of Mechanics - A/Solids **71**:326–334
[20] Håkansson P, Wallin M, Ristinmaa M (2008) Prediction of stored energy in polycrystalline materials during cyclic loading. International Journal of Solids and Structures **45**(6):1570 – 1586
[21] Bröcker C, Matzenmiller A (2013) An enhanced concept of rheological models to represent nonlinear thermoviscoplasticity and its energy storage behavior. Continuum Mechanics and Thermodynamics **25**(6):749–778
[22] Deutsches Institut für Normung e.V. (2022) DIN 50125: Prüfung metallischer Werkstoffe - Zugproben
[23] Smolina I, Gruber K, Pawlak A, Ziółkowski G, Grochowska E, Schob D, Kobiela K, Roszak R, Ziegenhorn M, Kurzynowski T (2022) Influence of the AlSi7Mg0.6 aluminium alloy powder reuse on the quality and mechanical properties of LPBF samples. Materials 2022 **15**:5019
[24] Haupt P (2002) Continuum Mechanics and Theory of Materials, 2nd edn. Springer-Verlag
[25] Deutsches Institut für Normung e.V. (2020) DIN EN ISO 6892-1: Metallische Werkstoffe - Zugversuch - Teil 1: Prüfverfahren bei Raumtemperatur
[26] Göttfert F, Sparr H, Ziegenhorn M, Dammass G, Krauß M (2022) Combining thermography and stress analysis for tensile testing in a single sensor. In: Mendioroz A, Avdelidis NP (eds) Thermosense: Thermal Infrared Applications XLIV, International Society for Optics and Photonics, SPIE, vol 12109, p 121090A
[27] Sparr H (2022) Thermomechanische Analyse von inelastischen Deformationsvorgängen bei komplexer Beanspruchung in Modellierung und Experiment. PhD thesis, Brandenburgische Technische Universität Cottbus Senftenberg
[28] Kratochvil J, Dillon OW (1969) Thermodynamics of elastic-plastic materials as a theory with internal state variables. Journal of Applied Physics **40**(8):3207–3218
[29] Kamlah M, Haupt P (1998) On the macroscopic description of stored energy and self heating during plastic deformation. International Journal of Plasticity **13**(10):893–911
[30] Perzyna P (1963) The constitutive equations for rate sensitive plastic materials. Quarterly of Applied Mathematics **20**:321–332
[31] The MathWorks Inc (2020) Simulink design optimization
[32] Chaboche JL (1993) Cyclic viscoplastic constitutive equations, Part I: A thermodynamically consistent formulation. Journal of Applied Mechanics **60**(4):813–821

[33] Bodner SR, Lindenfeld A (1995) Constitutive modelling of the stored energy of cold work under cyclic loading. European Journal of Mechanics - A/Solids **14**(3):333–348
[34] Krempl E (1987) Models of viscoplasticity some comments on equilibrium (back) stress and drag stress. Acta Mechanica **69**(1):25–42

Chapter 17
Creep Under High Temperature Thermal Cycling and Low Mechanical Loadings

Romana Schwing, Stefan Linn, Christian Kontermann, and Matthias Oechsner

Abstract In this paper the effect of an accelerated creep strain rate under thermal cycling conditions is introduced. The effect occurs in the industrial furnace sector, where thin walled structures are exposed to very high temperatures up to 80% of the melting temperature and low mechanical loadings which represent the dead weight of the components. The effect mentioned will be discussed extensively and a systematic examination of the effect itself and some influencing factors will be identified. Furthermore some explanatory approaches will be presented and discussed.

17.1 Introduction

Operators of industrial furnaces often report premature failure of thin walled components accompanied by large creep deformations [1]. The components show a significantly shorter lifetime than expected by calculations using linear damage accumulation modelling [2, 3]. In addition to very high temperatures, which are up to 80% of the melting temperature (K) of the deployed materials and the corrosive and oxidative ambient atmosphere, the complex loading conditions in such industrial furnaces often include temperature changes due to burner on-off-cycle, batch-wise operation or belt infeed.

Experiments under corresponding conditions of very high temperatures and low mechanical loadings showed that the creep strain of some typical metallic materials under anisothermal testing turned out to be significantly higher than the creep strain of an isothermal creep experiment even at maximum cycle temperature (Fig. 17.1). This effect of an accelerated creep strain rate is unexpected and further investigations

Romana Schwing · Stefan Linn · Christian Kontermann · Matthias Oechsner
Institut für Werkstoffkunde TU Darmstadt, Grafenstraße 2, 64283 Darmstadt, Germany,
e-mail: romana.schwing@tu-darmstadt.de, stefan.linn@tu-darmstadt.de, christian.kontermann@tu-darmstadt.de, matthias.oechsner@tu-darmstadt.de

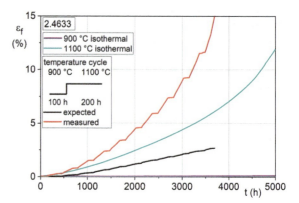

Fig. 17.1 Expected and measured creep strain of a temperature cycle creep test and isothermal creep tests at 900 °C and 1100 °C.

under anisothermal conditions were performed to generate a principal understanding of the phenomenon.

In this study the results of a systematically performed experimental campaign are presented. These results reveal relevant parameters that have a specific influence on the effect of accelerated creep strain rates under anisothermal temperature. Furthermore the relevance of microstructural properties and a formed metastable dislocation structure is discussed. In detail, microstructural processes are suspected to cause the phenomenon of *repriming*, which occurs after every temperature change. Finally, an approach for explaining these underlying creep mechanisms will be introduced briefly.

17.2 Experimental Methods

In the following the four examined alloys 2.4633, 2.4879, Centralloy 60 HT R and 1.4841 are introduced and the testing methods and temperature cycles are explained.

17.2.1 Materials

Creep tests were carried out on the above mentioned materials. All materials are common in industrial furnace applications.

The main focus of the investigation lies on alloy 2.4633, also known as Alloy 602 CA, NiCr25FeAlY or Nicrofer 6025 HT, which is an nickel-chrome-iron alloy with rather high chromium amounts of around 25% [4]. In combination with a comparatively high carbon content of around 0.2wt% this alloy is characterized by an austenitic matrix and the formation of $M_{23}C_6$ carbides [5, 6]. Those carbides are stable up to temperatures of around 1250 °C [7]. This is one of the reasons for the good temperature stability of this alloy [8–10]. The aluminum content in combination with the chromium is responsible for the formation of an aluminum oxide layer, which is also stable at temperatures above 1000 °C [11].

Furthermore two cast alloys were investigated, the alloys 2.4879 and Centralloy 60 HT R. Alloy 2.4879 shows an austenitic matrix with primary precipitated carbides. Those are formed with either tungsten or chromium [12]. The cast alloy Centralloy 60 HT R also shows an austenitic Ni-Cr-Fe-matrix with primary precipitated M_7C_3 or MC-carbides. This alloy is also an aluminum oxide former [13]. The fourth examined alloy is the steel 1.4841 with an austenitic matrix [14]. Figure 17.2 shows light micrographs of the initial microstructural state of all four alloys.

17.2.2 Creep Test Equipment

Creep tests for the investigation of the behavior under thermal cycling condition were carried out on single-specimen-testing machines with continuous strain measurement. For comparison, isothermal creep tests have been performed with comparable thermal and mechanical loadings as the anisothermal creep tests. The tested specimens were cylindrical round specimens with a diameter of 8.4 mm and a gauge length of 42 mm.

Two different kinds of thermal cycles are investigated. First low frequency thermal cycles are tested to emulate batch operation or belt infeed. These cycles are characterized by two holding phases in the range of 1.5 h up to 200 h at minimum and maximum cycle temperature. The temperature ramps have a moderate rate of temperature change of around 6.67 K/min in most cases which leads to a duration of 0.5 h for 200 K. The mechanical loading is constant in the range of 1-10 MPa (Fig. 17.3).

These creep tests, which represent low frequency thermal cycling, are carried out in convection furnaces. The strain is measured with axial extensometers, the temperature with three thermocouples of type S (Pt/PtRd).

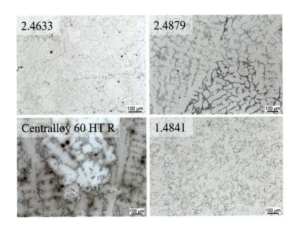

Fig. 17.2 Light micrographs of the four alloys in their initial state.

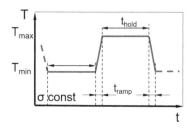

Fig. 17.3 Low frequency temperature cycle, schematic.

The second type of thermal cycles are high frequency cycles. These cycles are characterized by a triangular cycle without holding phases. The cycle duration of these tests is in the range of minutes. As with the low-frequency cycles, the mechanical loading is constant over the entire duration of the creep test (Fig. 17.4).

The high frequency thermal cycling is realized by induction heating. For the strain measurement also axial extensometers are used, but since the measuring linkage lies within the induction coil, only ceramic material is used. The temperature is measured and calibrated by a thermal imaging camera. To be able to measure absolute temperature values, the specimens are coated with a thermal paint with a defined constant emissivity. For temperature control a thermocouple of type S is used.

17.2.2.1 Strain Recording

The strain recording is carried out continuously. Just before the creep test starts, after heating up to the lower cycle temperature and a subsequent warm-up, the measuring system is reset. This way it is possible to determine the creep elongation after applying the load.

In case of anisothermal creep tests this procedure is performed after the lower cycle temperature is reached. In the ramp phases and the upper holding phase thermal expansion components of the difference between minimum and maximum cycle temperature of the specimen and measuring rods are included in the measured total strain. For this reason the lower envelope of the creep curves are shown in the following creep strain diagrams as for example in Fig. 17.5.

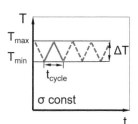

Fig. 17.4 High frequency temperature cycle, schematic.

17 Creep Under High Temperature Thermal Cycling and Low Mechanical Loadings

Fig. 17.5 Lower envelope of an anisothermal creep curve, example.

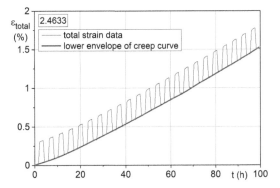

At minimum cycle temperature, no thermal strain components are included in the recorded strain, due to zeroing of the measuring system after the heating phase. Therefore, the lower envelope of the creep curve can be used for comparison purposes.

17.3 Observation of Accelerated Creep Under Anisothermal Testing Conditions

The effect of accelerated creep strain under thermal cycling condition has been reproduced by using different cycle types. Figure 17.6a) and b) each show two low frequency anisothermal creep curves and the corresponding isothermal creep curve at maximum cycle temperature. The strain curves of the temperature cycle tests lie above the isothermal creep curve. For the shorter holding phases of 1.5 h as well as for the longer holding phases of 3.5 h the repetition test confirms the original

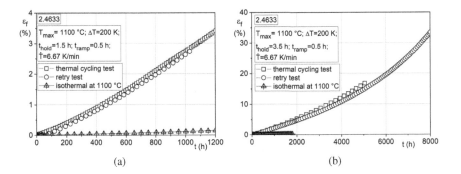

Fig. 17.6: Reproduction of accelerated creep rate with (a) short holding phases (b) longer holding phases.

observations. This effect is not a material specific phenomenon as illustrated in Fig. 17.7.

In this diagram isothermal and anisothermal creep curves from comparable testing conditions are plotted for different materials. The anisothermal creep tests were all carried out with a maximum cycle temperature of 1100 °C and a temperature range of 200 K. All thermal cycling tests had symmetrical holding phases at minimum and maximum temperature of 1.5 h and a ramp duration of 0.5 h. The isothermal creep tests were carried out at 1100 °C. As shown in the diagrams above, the material 2.4633 shows the effect of an accelerated creep rate under thermal cycling condition. The three other tested materials also show a similar behavior. The creep curve of the anisothermal condition always lies above the isothermal creep curve. This leads to the conclusion that the observed effect is not material specific but rather a general phenomenon, which occurs in face centered cubic materials under thermal cycling condition.

In Figs. 17.5 and 17.6 it can be seen that the effect of the higher creep rate under anisothermal conditions appears shortly after the start of the experiment. To provide evidence of this effect in more detail, Fig. 17.8 shows the first 30 h of a thermal cycling test in comparison to the isothermal creep test.

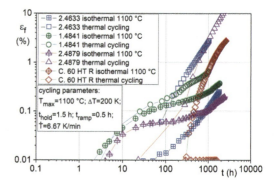

Fig. 17.7 Independence of the effect from the type of material.

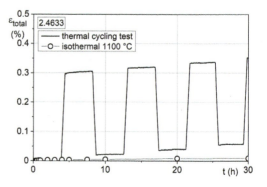

Fig. 17.8 First 30 h of a thermal cycling creep test in comparison to an isothermal creep test.

In this example it can be observed that the creep curve of the thermal cycling test lies already after the first cycle of around 8 h above the isothermal creep curve and with every further cycle the distance between the anisothermal and the isothermal creep curve increases. This signifies that the temperature changes in every cycle result in an additional creep increment on the total strain.

In summary, what can be said about the observed effect of accelerated creep strain under thermal cycling stress is that it is reproducible and occurs after a short time, and that it occurs in all the materials studied. This suggests that it is not an artifact of measurement, but a generic material behavior.

17.3.1 Anisothermal Creep Tests

As already mentioned above, the effect of an accelerated creep rate under thermal cycling condition can be observed in a somewhat unusual parameter range which finds its application in the industrial furnace industry. It seems to appear at very high temperatures of up to 80% of the melting temperature in Kelvin and very low mechanical loadings. For these conditions the Norton-plot shows that there is a linear relationship between creep rate and stress [15]. The creep rate is very low at these low mechanical loadings. At this parameter combination the dominant deformation mechanism is diffusion creep. Generally, diffusion can take place through the grain boundaries or the grain itself [16]. Material is transported from areas which are under pressure (transverse contraction at uniaxial loading) to areas which are under tension (parallel to the loading axis at uniaxial loading). This type of creep is a rather slow process.

These processes and the classification in the diffusion creep area holds up for the isothermal creep tests (see blue bubble in Fig. 17.9). The observed effect, that under these parameter conditions thermal cycling leads to an increased strain rate in comparison to the strain rate at isothermal maximum cycle temperature, seems not to be dominated only by diffusion creep, since the measured creep strain rates are too high (orange bubble in Fig. 17.9).

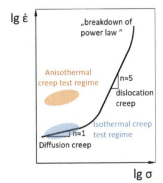

Fig. 17.9 Norton plot according to [15].

17.3.2 Influencing Factors

In the following some influencing factors on the effect of accelerated strain rate due to thermal cycling condition will be presented. It can be shown that grain size, mechanical loading, holding phases and the temperature changing rate have an impact on the effect.

17.3.2.1 Grain Size

The dependence of the effect on the grain size is shown by comparing different creep tests with identical cycle and testing parameters on different melts of one material. In Fig. 17.10, creep curves of three different batches of alloy 2.4633 are shown. Those differ in their average grain size. The melt with the biggest grains has an average grain size of 63 µm, the medium grain size is 44 µm and the melt with the smallest grain size has an average value of 33 µm.

For each melt an anisothermal creep test with low frequency cycle and a maximum cycle temperature of 1100 °C, a temperature range of 200 K and symmetrical holding times of 1.5 hours were carried out. Furthermore, isothermal creep tests at maximum cycle temperature were performed on the same three melts.

It is evident that the melts with the smaller grain sizes show higher creep strains. In all three cases the anisothermal creep tests show higher creep strains than the corresponding isothermal creep tests.

In Fig. 17.11 it is demonstrated that the relationship between achieved creep strain of the anisothermal and isothermal creep test is not equal for the three melts. By calculating the quotient between the measurement from the corresponding experiments, it can be shown that the effect of the accelerated creep strain under anisothermal conditions is more pronounced in melts with larger grain sizes.

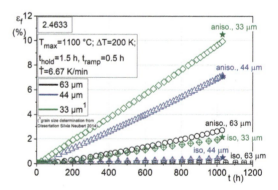

Fig. 17.10 Dependence of the effect of accelerated creep due to thermal cycling condition on the grain size.

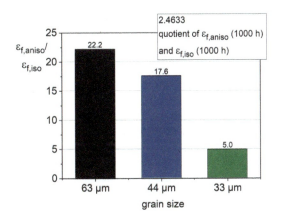

Fig. 17.11 Quotient of achieved creep strain of the anisothermal creep test after 1000 hours and achieved creep strain of the corresponding isothermal creep test after 1000 h.

17.3.2.2 Stress

As a further influencing factor the applied stress was investigated. For this, isothermal and anisothermal creep tests with similar temperature cycle parameters were carried out with different applied stresses. The result of those creep tests on the alloys Centralloy 60 HT R and 1.4841 are shown in Fig. 17.12. The cast alloy Centralloy 60 HT R shows for both stress levels a higher creep strain in the anisothermal case in comparison to the isothermal creep curve.

The alloy 1.4841 on the other hand shows a change in the arrangement of the isothermal and anisothermal creep curves with increasing stress. For the creep tests with the lower stress level (black curves), the anisothermal creep curve lies above the isothermal creep curve but at a higher stress level the alloy shows the expected behavior (blue curves). The isothermal creep curve at maximum cycle temperature lies above the anisothermal creep curve.

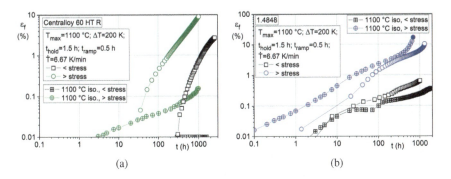

Fig. 17.12: Anisothermal and isothermal creep tests with different stresses on the alloy (a) Centralloy 60 HT R and (b) 1.4841.

To examine this behavior more in depth, creep tests on alloy 2.4633 were carried out on three different stress levels as well. For the lowest stress level and the medium stress level the anisothermal creep curve lies above the corresponding isothermal curve, but as it can be seen in Fig. 17.13a the distance between the anisothermal and isothermal creep curve decreases with increasing stress. For the highest stress level, this relationship has changed. The isothermal creep curve shows a much higher creep rate in comparison to the anisothermal one. This observation is also evident in the application of the quotient of the creep strain of the anisothermal creep strain after 400 h with the isothermal creep strain after 400 h (Figure 17.13b). Hence it can be concluded that the effect of the accelerated creep strain under anisothermal testing condition only appears at low stresses and it is more pronounced the lower the stress is.

17.3.2.3 Holding Phases

Another influencing factor on the accelerated creep rate under thermal cycling condition might be the duration of the holding phases. To examine this, anisothermal creep tests with different holding phases and besides this same cycle parameters were carried out.

Figure 17.14 shows low frequency creep tests with holding phases of 1.5 h and 3.5 h at minimum and maximum cycle temperature. For both materials, 2.4879 and 2.4633, it can be observed that the creep tests with the shorter holding phases show higher creep strains after a comparable time period. This is interesting, since after a comparable run time, the tests with longer holding times spend proportionally more time at the maximum temperature than the tests with shorter holding times (due to similar ramp durations). On the other hand a specimen in a test with longer holding

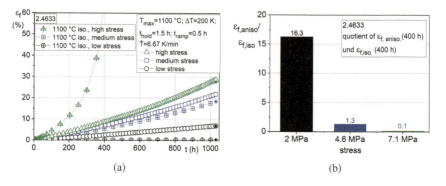

Fig. 17.13: (a) creep tests on 2.4633 on different stress levels (b) quotient of achieved creep strain of the anisothermal creep test after 400 h and achieved creep strain of the corresponding isothermal creep test after 400 h.

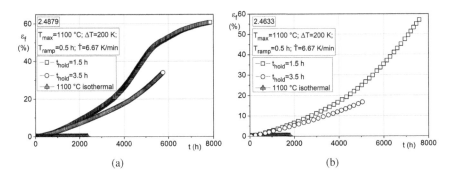

Fig. 17.14: Creep tests with holding phases of 1.5 h and 3.5 h at minimum and maximum cycle temperature at (a) 2.4879 and (b) 2.4633.

times experiences fewer cycles in a given loading time than the sample in a test with shorter holding times.

This indicates that the amount of cycles has a stronger impact on the effect of accelerated creep rate under anisothermal conditions than the amount of time spent at maximum temperature.

Another example of the influence of holding times in combination with the influence of grain size can be observed in Fig. 17.15.

In Fig. 17.15a two anisothermal and one isothermal creep test are shown. One anisothermal creep test was carried out with holding phases of 1.5 h at minimum and maximum cycle temperature. A second anisothermal creep test has holding phases of 100 h at minimum cycles temperature and 200 h at maximum cycle temperature. Both creep curves lie above the isothermal creep curve at 1100 °C.

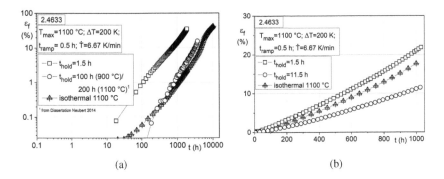

Fig. 17.15: Creep tests at two different melts of 2.4633 with (a) holding phases of 1.5 h at minimum and maximum cycle temperature and 100 h at minimum and 200 h at maximum cycle temperature and (b) of 1.5 and 11.5 h at minimum and maximum cycle temperature.

In Fig. 17.15b two different creep curves on a different melt are shown. The creep curve with holding phases of 1.5 h lies still above the isothermal creep curve, but the anisothermal creep curve with holding phases of 11.5 h lies below the isothermal creep curve and therefore shows the expected behavior.

These results suggest that a grain size-dependent threshold of the holding time exists, at which the creep strain under anisothermal conditions lies above the strain of the isothermal creep test at maximum cycle temperature.

17.3.2.4 Temperature Changing Rate

More creep tests were carried out to examine the influence of cycle type or rate of temperature change as can be seen from Fig. 17.16.

The comparison of low and high frequency anisothermal creep tests shows in both diagrams, Fig. 17.16a and b, that the creep strain of the high frequency tests lie above the creep strain of the low frequency tests with comparable parameters. In Fig. 17.16a the high frequency cycle was carried out with a temperature range of 50 K whereas the temperature range of the high frequency test in Fig. 17.16b was 200 K. By comparison of both diagrams it can be assumed that the higher temperature changing rate in Fig. 17.16b leads to higher creep strain.

17.3.3 Observations on the Creep Behavior Within a Cycle

The creep curves shown so far are the lower envelope of the recorded total strain curve. To improve the understanding of the effect of accelerated creep strain rate

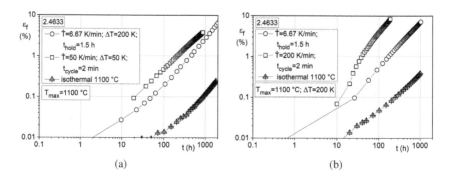

Fig. 17.16: Comparison of a low frequency temperature cycle creep test with a holding phase of 1.5 h with a high frequency temperature cycle creep test; a) with a temperature change rate of 50 K/min, b) with a temperature change rate of 200 K/min.

17 Creep Under High Temperature Thermal Cycling and Low Mechanical Loadings 301

under anisothermal conditions, the individual cycle parts should now be examined more intensively.

For this purpose creep tests were carried out on a servo-hydraulic fatigue testing machine with a lateral extensometer. With this test setup it is possible to determine the thermal strain parts in the upper holding phase and the temperature ramps and subtract them from the total strain. Figure 17.17 schematically shows the temperature cycle with constant stress of the tested specimen for the investigation of the individual cycle sections.

The results of this test with the modified setup are shown in the diagrams of figure 17.18. In the diagram on the left hand side the reset creep strain of six different cycles is shown. The temperature cycle consists of holding phases at 900 °C and 1100 °C for each 300 s and temperature ramps in between of 300 s.

The creep strain in the four phases (holding phase at 900 °C, temperature ramp 900 °C to 1100 °C, holding phase at 1100 °C and temperature ramp 1100 °C to 900 °C) applied over the cycle duration shows a characteristic course. In the first phase, the holding phase at 900 °C, a small strain increase is noticeable, which can be identified as a kind of "primary creep" in figure 17.18b, since the strain rate decreases in the first phase.

In the second holding phase at 1100 °C the creep strain shows the same behavior. However, the strain increase is 10 times higher than in the 900 °C phase. During the temperature rise a clear increase in strain with a progressive course can be seen. In the temperature drop phase the strain increase is extremely low.

The diagram of the average creep rate in Fig. 17.18b shows that *repriming* occurs in both holding phases independent of the kind of temperature ramp before the hold time. This means that after every temperature change a new primary creep period seems to occur, which apparently leads to the accelerated creep strain development. This *repriming* could be an explanatory approach for the influence of temperature cycles on the creep behavior under anisothermal conditions.

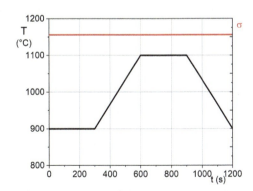

Fig. 17.17 Schematic diagram of stress and temperature for the investigation of the individual cycle sections.

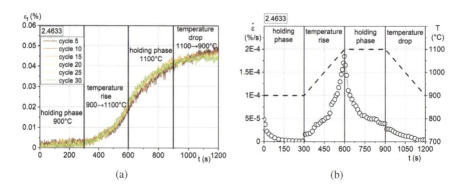

Fig. 17.18: (a) average zeroed creep strain of six cycles (b) average creep strain rate of the creep test.

17.4 Discussion

In the examinations of the creep behavior under anisothermal conditions of materials, which are used in industrial furnace construction it could be shown that the creep response to tests at very high temperatures of around 80% of the melting temperature and low mechanical loadings shows an unexpected behavior. Creep curves of anisothermal tests lie above the isothermal creep curve at maximum cycle temperature and not as expected between the isothermal creep curves at minimum and maximum cycle temperature.

It could be shown, that the effect of the accelerated creep rate under anisothermal conditions is independent of the material, occurs only at low mechanical stresses, shows a dependence on the grain size and on different temperatures of the cycle and it is already measurable after short times. These observations lead to the conclusion, that the observed effect is no material specific effect, but rather a generic material behavior. Different explanatory approaches for the effect described will be introduced in the following.

17.4.1 Possible causes of accelerated creep under thermal cycling

In the following, three explanations (thermally induced stresses, oxidation and microstructural processes) are analyzed.

17.4.1.1 Thermally Induced Stresses

Thermally induced stresses are the consequence of temperature gradients over the specimen cross section. Cause of thermally induced stresses are a combination of the thickness of the specimen or the component, very high temperature changing rates and the materials limited thermal conductivity. FEM calculations to examine the expected thermally induced stress were carried out.

During the temperature rise from 900 °C to 1100 °C in 30 minutes a temperature difference between specimen surface and bulk of <0.1 °C arises (Fig. 17.19a). Around 300 s after the ambient temperature reaches its maximum, the difference between specimen inside and outside vanishes.

The corresponding thermally induced stress is plotted in Fig. 17.19b. As can be seen from the plot, a maximum thermally induced stress of 0.15 MPa arises due to the temperature differences. This stress is considered to be negligible.

Another observation also suggests that thermally induced stresses are not the cause of the accelerated creep strain development under thermal cycling. An anisothermal creep test at a specimen with a smaller cross-sectional area showed even higher creep strains than the thicker sample (Fig. 17.20). In case of thermally induced stresses as the main cause of the accelerated creep strain development under anisothermal conditions the sample with the smaller diameter should show a less pronounced effect than the thicker sample.

17.4.1.2 Oxidation

Another possible cause of the accelerated creep strain under thermal cycling are oxidative effects at these high temperatures. However, since the phenomenon of accelerated can already be observed within very initial cycles, and after short times, where no significant oxide layer is present, oxidation should not be considered to be a main cause for the presence of the phenomenon.

Furthermore, microscopic examinations revealed no significant differences between isothermal and anisothermal creep tests with comparably long terms (Fig. 17.21).

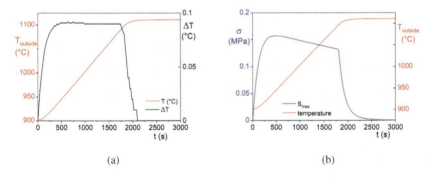

Fig. 17.19: (a) calculated expected temperature difference for the temperature rise of 200 K in 30 min (b) corresponding expected thermally induced stress, elastic calculation.

Fig. 17.20 Comparison of two creep curves at comparable cycle temperatures and loads with different specimen diameters.

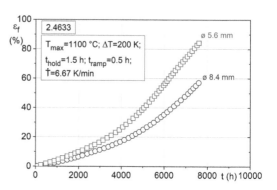

The light micrographs of the material 2.4633 show in both cases, isothermal and anisothermal creep test, a carbide-depleted area under the specimen surface. Moreover, areas of inner oxidation can be identified. Beneath the surface carbides dissolve and pores are left behind. The isothermal sample tends to show a slightly wider depletion zone than the anisothermally tested sample. Beyond that, no further significant differences are recognizable. This also applies to the other materials and samples examined.

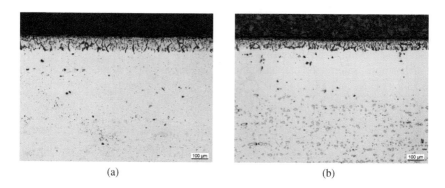

Fig. 17.21: Lightmicroscopy images, 2.4633 unetched, (a) isothermal creep test at 1100 °C for 1030 h (b) anisothermal creep test: T_{max}=1100 °C; $\Delta T = 200$ K, $\dot{T} = 6.67$ K/min, t_{hold}= 1.5 h, t_{ramp}= 0.5 h for 1032 h.

17.4.2 *Microstructural Processes Under Anisothermal Creep Testing*

As already mentioned in Subsect. 17.3.1 the observed strain rates are too high for pure diffusional creep. This leads to the conclusion that dislocations are the carriers of deformation and therefore microstructural processes could be the cause of accelerated creep under thermal cycling.

Due to the observed *repriming*, firstly the microstructural processes in the primary creep regime of an isothermally tested specimen are discussed.

With the application of the load, the dislocation density increases and new dislocations are formed. Afterwards these dislocations begin to move and climb and start to form areas with dense, parallel dislocation walls and areas without structure. With increasing duration, a uniform dislocation structure and globular subgrains form (Fig. 17.22).

There are different parameters to describe the substructure. In addition to the dislocation density, which can be divided into dislocation density in areas with and without subgrain structure (ρ_S and ρ_R), the subgrain diameter φ and the misorientation angle Θ can be used to describe the substructure. All of these parameters tend towards saturation when the minimum creep strain rate is reached (Fig. 17.23).

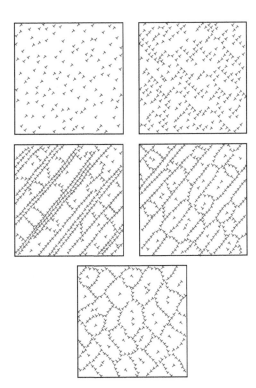

Fig. 17.22 Formation of a substructure in the isothermal case according to [17].

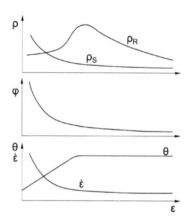

Fig. 17.23 Development of the mobile dislocation densities (ρ_R in areas without substructure; ρ_S in areas with existing substructure), the subgrain diameter φ and the misorientation angle Θ in the area of primary creep plotted against strain according to [18].

To examine the anisothermal case even more in depth, a special creep test was carried out. First, the temperature was kept isothermal at 1100 °C until the test reached the minimum creep rate, respectively reached the secondary creep zone. By this stage the ordered substructure should have developed.

In the next step thermal cycles are started. The resulting creep curve is shown in Fig. 17.24.

From this plot it can be seen that the creep strain rate increases sharply with the start of the temperature cycles. In a more detailed plot in Fig. 17.25 the transition region between the isothermal and anisothermal loading shows, that already after the first temperature cycle, the creep strain rate lies above the previous isothermal creep strain rate. The creep strain rate in Fig. 17.25b in the first 1100 °C holding phase after the first temperature cycle is higher by a factor of 100 than at the end of the isothermal phase.

These results suggest that, when the minimum strain rate is reached at correspondingly low stresses, no stable substructure is built or it is disrupted by the temperature change. The formation of a substructure in the primary creep region in the isothermal case, which is known from literature [17], together with the occurrence of *repriming* and the deductions from the experiment just shown, allow the conclusion that thermal

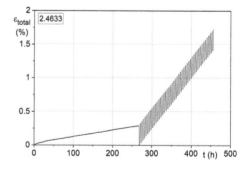

Fig. 17.24 Creep curve of a test which was started isothermally and after reaching the minimum creep strain rate temperature cycles were started.

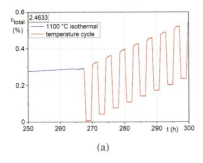
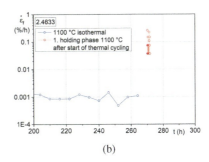

Fig. 17.25: (a) creep curve in the region between isothermal and anisothermal loading (b) creep strain rate at the end of the isothermal range and in the first 1100 °C holding phase after the first temperature change.

cycling does prevent the formation of a stable substructure, or that the formation process is restarted with each new holding phase.

17.5 Summary

In this paper the effect of an accelerated creep strain development under temperature cycling condition was discussed. The appearance of the effect was analyzed and influencing factors could be identified. In this examinations it could be shown that the effect is stronger with large grain sizes. Another influencing factor is the mechanical stress. The effect appears only at low mechanical stresses and is more pronounced the lower the stress. Shorter holding phases in the temperature cycle lead also to a more pronounced appearance of the effect. Comparisons of low- and high-frequency anisothermal tests have shown that the high-frequency creep tests with higher rates of temperature change have higher creep strain rates. In further examinations a so called *repriming* in every holding phase was observed.

Thermally induced stresses or oxidative effects could be ruled out as possible reasons for this effect. The investigations led to the conclusion that this phenomenon is based on microstructural processes. During the temperature cycles it is not possible to form a stable substructure, causing a repriming mechanism to occur which leads to increased creep strain rates during anisothermal loading.

References

[1] Karthik NK, Schmitz N, Pfeifer H, Schwing R, Linn S, Kontermann C, Oechsner M (2019) Einfluss von Temperaturwechselbeanspruchung auf das Verfor-

mungsverhalten von Ofenkomponenten und deren Lebensdauer. BHM Berg- und Hüttenmännische Monatshefte **164**(9):364–371

[2] Saravanan P, Sahoo G, Srikanth S, Ravi K (2011) Failure analysis of radiant tube burners in continuous annealing line (CAL) of an integrated steel plant. Journal of Failure Analysis and Prevention **11**(3):286–292

[3] Tari V, Najafizadeh A, Aghaei MH, Mazloumi MA (2009) Failure analysis of ethylene cracking tube. Journal of Failure Analysis and Prevention **9**(4):316–322

[4] VDM Metals GmbH (2007) VDM (R) Alloy 602 CA Nicrofer 6025 HT: Werkstoffdatenblatt Nr. 4137

[5] Neubert S (2014) Hochtemperaturverhalten von nickel-chrom-aluminium-legierungen für den industriellen ofen- und apparatebau. Dissertation, TU Darmstadt, Aachen

[6] Schwing R, Linn S, Kontermann C, Neubert S, Oechsner M (2021) Isothermal and anisothermal creep behavior of the nickel base alloy 602 ca. Materialwissenschaft und Werkstofftechnik **52**(2):231–247

[7] El-Magd E, Gebhard J, Stuhrmann J (2007) Effect of temperature on the creep behavior of a ni–cr–fe–al alloy: a comparison of the experimental data and a model. Journal of Materials Science **42**(14):5666–5670

[8] Brill U (1994) Eigenschaften und Einsatzgebiete der neuen warmfesten Legierung Nicrofer 6025 HT. STAHL (3):32–35

[9] Brill U (1995) Praktische Erfahrungen mit dem neuen Werkstoff Nicrofer 6025 HT im Ofen- und Wärmebehandlungsanlagenbau. STAHL (6):37–40

[10] Brill U (1999) Neue Ergebnisse mit dem Werkstoff Nicrofer 6025 HT im Ofen- und Wärmebehandlungsanlagenbau. STAHL (3):54–56

[11] Brill U (2005) Advanced nickel-based and nickel-iron-based superalloys for civil engineering applications. Acta Metallurgica Sinica **18**(4):453–462

[12] Asteman H, Hartnagel W, Jakobi D (2013) The influence of Al content on the high temperature oxidation properties of state-of-the-art cast Ni-base alloys. Oxidation of Metals **80**(1-2):3–12

[13] Schmidt + Clemens GmbH und Co KG (2015) Centralloy 60 HT R material data sheet: Designation G-NiCr25Fe10Al

[14] Schmid A, Mori G, Hönig S, Strobl S, Haubner R (2019) High temperature corrosion of steel X15CrNiSi25-21 (1.4841) in a mixed gas atmosphere – sample preparation and damage mechanism. Practical Metallography **56**(7):457–468

[15] Bürgel R, Maier HJ, Niendorf T (2011) Handbuch Hochtemperatur-Werkstofftechnik. Vieweg+Teubner, Wiesbaden

[16] Blum W, Eisenlohr P (2009) Dislocation mechanics of creep. Materials Science and Engineering: A **510-511**:7–13

[17] Takeuchi S, Argon AS (1976) Review steady-state creep of single-phase crystalline matter at high temperature. Journal of Materials Science **11**:1542–1566

[18] Orlová A, Pahutová M, Čadek J (1972) Dislocation structure and applied, effective and internal stress in high-temperature creep of alpha iron. Philosophical Magazine **25**(4):865–877

Chapter 18
The Development and Application of Optimisation Technique for the Calibrating of Creep Cavitation Model Based on Cavity Histogram

Qiang Xu, Bilal Rafiq, Xuming Zheng, and Zhongyu Lu

Abstract It is generally accepted that the creep rupture is primarily determined by the creep cavitation at grain boundaries, hence it is vital important to develop an accurate cavitation model. Most of the existing creep cavity models used some simplifications such as using average diameter of cavities, assumed nucleation, et al. Recently, the concept of calibrating the creep cavity models using 3D cavity histogram without any aforementioned simplifications was conceived and practical trial and error method was devised and used. Whilst the use of such trial and error method had produced results, arguably very accurate too, but its use heavily relies on the user's insight knowledge of the characteristics of the cavity density distribution function and intervention. Here, we present the development and application of optimisation techniques for the calibration of creep cavitation model using Excel Solver, via the minimising the difference of the predicted cavity distribution density (number of cavities per volume) over cavity size and the experimental measured one. Its application produces an updated creep cavitation model without any suspicions doubt of its mathematical accuracy. We anticipate this optimisation implementation via Excel Solver will be widely used in future.

18.1 Introduction

Creep causes various microstructural changes and property deteriorations over time; and cavitation at grain boundary is, arguably, the most important one for structural integrity. However, this has not been adequately appreciated by research communities and high temperature industries [1–3]. Furthermore, the phenomenological approach

---

Qiang Xu · Bilal Rafiq · Xuming Zheng · Zhongyu Lu
Department of Technology and Engineering, School of Computing and Engineering, University of Huddersfield, Huddersfield, HD1 3DH, UK,
e-mail: Q.Xu2@hud.ac.uk, U2181632@unimail.hud.ac.uk, Xuming.Zheng@hud.ac.uk, j.lu@hud.ac.uk

based continuum creep damage mechanics do suffer a few fundamental criticisms [1–10] such as:

1. The phenomenological approach does not actually model the creep cavity damage, only its effect was considered;
2. hence, the coupling of creep cavitation damage and creep deformation is not mechanism based; and
3. the inability to meet the deformation consistency condition in the multi-axial version of creep damage constitutive equations generalized by the creep damage equivalent stress approach. Some pioneer and tentative works were carried out and can be seen in [9–18], but not limited by these.

The challenges involved with creep damage mechanics and creep damage models have been analysed in the literature review such as those shown initially in [1] and formally in [2, 3], and it was summarised in [3] as:

1. Characterizing and quantifying creep cavitation and developing damage criterion for parent metal and weld, respectively; experimental work (uniaxial and multi-axial interrupted creep test) to be carried out or gathered under low stress; cavitation to be quantified, ideally using X-ray micro-tomography. A new damage criterion shall be developed.
2. Quantifying the microstructural evolutions and their effects on the creep deformation.
3. Developing and applying the novel creep formulation suitable for a wider range of stress and incorporating the damage criterion developed in 1.
4. Generalizing uniaxial version into a three-dimensional creep damage model.

Bearing the above thoughts in mind, some preliminary research in order to overcome the above problems [11–18]. In 2013, on the sight of the published 3D synchrotron creep cavity data for high Cr steels [19], a research project was set up to model the cavitation and creep cavitation lifetime [20–22]. In pursuing the above research, we did utilize and calibrated Riedel's generic creep cavitation model [23] and developed creep rupture model based on area fraction along the grain boundary concept. Its success has encouraged research to investigate other alloys and to use other 3-dimensional cavitation data such as those produced by small angle neutron deflection [24, 25] and publications [26–28].

Among these research, one of the core activities is how to calibrate the creep cavitation model based on histogram. Initially, due to the priority of research, we decided to be practical to find the numerical answers without having to resort to optimization techniques though we did have some experience [29, 30]. During the course of research, we have progressively devised and used the trial and error based method with and/or without semi-optimisation. Now, we have adopted the Solver of Microsoft Excel for full optimisation. The original trial and error methods were reported in [31], and this paper will report the further progress on the semi-optimisation and full optimisation. The cavity data of E911 [32] was chosen and used in this paper for illustration. This paper contributes to the method and knowledge of how to calibrate the creep cavitation model based on histogram.

18.2 Background Theories and Knowledge

18.2.1 Cavitation Model Theory

The cavity nucleation and cavity growth models are proposed as [23]:

$$J^* = A_2 t^\gamma, \qquad (18.1)$$

$$\dot{r} = A_1 r^{-\beta} t^{-\alpha} \qquad (18.2)$$

where J^* is the cavity nucleation rate, r and $\dot{r}$ is the cavity radius and its growth rate respectively, and t is time, while A_1, A_2, α, β, and γ are material cavitation constants. It is implicitly assumed that they are not changing during the creep process and they might be dependent stress. Some preliminary explanation of their physical meaning and their significance can be found from literature, such as in [23, 32].

Generic cavity size distribution function was derived by Riedel [23] as:

$$N(R,t) = \frac{A_2}{A_1} R^\beta t^{\alpha+\gamma} \left(1 - \frac{1-\alpha}{1+\beta} \frac{R^{\beta+1}}{A_1 t^{1-\alpha}}\right)^{(\alpha+\gamma)/(1-\alpha)} \qquad (18.3)$$

The R is cavity radius at the specific time t. There are five cavitation constants here, namely, $\alpha, \beta, \gamma, A_1$ and A_2, in total. Beware of the definition of R in the $N-R$ space, and it differs from r normally.

18.2.2 Current Calibration Methods

Mathematically, Eqs. (18.1) will uniquely decide Eq. (18.3), and vice versa, so they are equivalent. Hence Xu concluded [3] that:

1. We can determine the values for the cavitation model based on the data in histogram alone, as long as there are five data points or more, there is no need of any other experimental data such as the direct measurement of the cavity growth rate and/or the change of cavity number over time.
2. The standard approach would be resorting to the optimization techniques.
3. Without resorting optimization, a set of values can be obtained by solving 5 independent equations, simultaneously. There are standard procedures for such task.
4. Even more practical, the trial and error method can be used to construct a theoretical histogram, and through comparison of the predicted histogram and experimental data, and this can be very easily achieved by programming with Excel.
5. The known typical values for any variable can be used as a good starting point, which will reduce the order of difficulty and complexity.
6. If the characteristic values for α and β have been obtained and the value for γ has been suggested for a specific test, then the inner code of the calibration is reduced

to find the solution for A_1 and A_2, given set values of α, β and γ. The predicted $r-t$ and $N-t$ at t_f was used to construct the $N(R, t_f)$.
7. The outside loop can be performed for various values for α, β and γ, the sensitivity of the values of α, β, γ can be explored afterwards.

The essence of item 6 and 7 is to reduce the number of variables to only A_1 and A_2 and find their values for a set of guessed values of α, β and γ. The typical values were suggested in literature as $\alpha = 1, \beta = 2, \gamma = 1$.

Further mathematical equations if $\alpha = 1$ for calibration are given bellow:

$$N(R, t_f) = \frac{A_2}{A_1} R^\beta t_f^{1+\gamma} \exp\left(-\frac{1+\gamma}{1+\beta} \frac{R^{\beta+1}}{A_1}\right) \tag{18.4}$$

$$\frac{1}{3} r^3 = A_1 \ln t + C, \tag{18.5}$$

$$\frac{1}{3} r_i^3 = A_1 \ln t_i + C, \tag{18.6}$$

$$\frac{1}{3} r_j^3 = A_1 \ln t_j + C, \tag{18.7}$$

$$A_1 = \frac{R_j^3 - R_i^3}{3(\ln t_j - \ln t_i)}, \tag{18.8}$$

$$A_1 = \frac{-2R_1^3 + 2R_2^3}{3\ln(N(R_1, t_f)R_2^2 / N(R_2, t_f)R_1^2)} \tag{18.9}$$

Substitute A_1 into Eq. (18.4) and using the point 2 (R_2), on the histogram, gives:

$$A_2 = \frac{N(R_2, t_f) A_1}{R_2^2 t_f^2 \exp(-2R_2^3 / 3A_1)} \tag{18.10}$$

Substitute A_1 into Eq. (18.4) and using the point 1 (R_1), gives:

$$A_2 = \frac{N(R_2, t_f) A_1}{R_1^2 t_f^2 \exp(-2R_1^3 / 3A_1)} \tag{18.11}$$

With a set of assumed α, β and γ, the forward method starts with a set of guessed A_1 and C (18.6) and a data point to calculate A_2 according to (18.10). The predicted curve can be plotted and the goodness of fitting can be visually checked with a series of set of A_1 and C; then iterate over a different set of assumed α, β and γ to find the best solution. The check of goodness of the fitting is visually assessed.

The backward method used two data points to directly calculate the A_1 and A_2 directly using (18.9) and (18.10) or (18.11). The predicted curve can be plotted and its goodness of fitting is visually checked, initially. Later on, the semi-optimisation method was devised by introducing the root-mean-square error (RMSE) to measure the goodness of a specific fitting and the trial and error exercise will stop when it is deemed that the minimum RSME has been achieved.

18.3 Optimisation with Excel Solver

Solver in Microsoft Excel offers the optimisation function. It shares the same fundamentals in optimisation:

1. the objective function,
2. the variables to be optimised,
3. the constraints and limits, and
4. the starting values.

In Excel, there are designated areas for them: the objective cell, decision variables cells, constraint and limit window, respectively, while the starting values are inserted into the decision variables cells. A Solver Parameters window and Solver Result window is shown by Fig. 18.1. The third author found it is very user-friendly as there is no complicated coding and the need for controlling the accuracy of the optimisation et al which other optimisation techniques.

18.4 Cavitation Data

The E911 high Cr high temperature steel was chosen for illustration of the optimisation method. The 3D cavities after creep rupture was measured and histogram was produced by [24]. The data points of the cavity histogram were digitized by Zheng [33] and it is shown by Fig. 18.2.

18.5 Results

The result obtained by backward method is included for completeness and shown in Fig. 18.3. The semi-optimised result is shown in Fig. 18.4. The full optimised result and the comparison is shown by Fig. 18.5. The different sets of results and the accuracy are listed in Table 18.1.

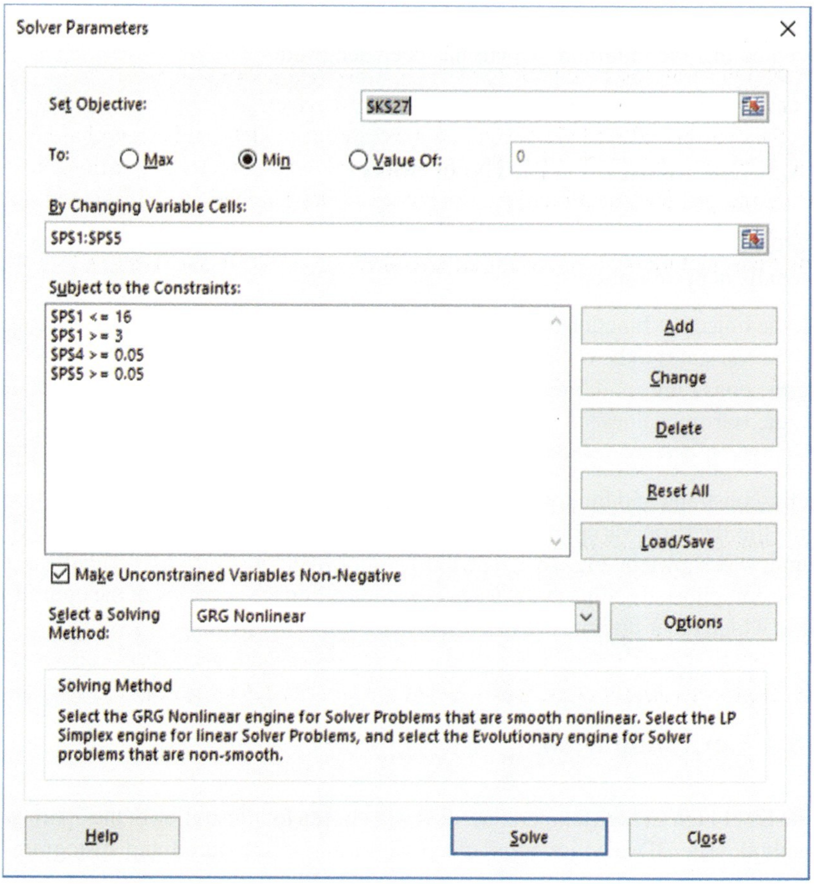

Fig. 18.1: The illustration of Solver Parameters window and Solver Results: a) Solver Parameter window, b) Solver Results window.

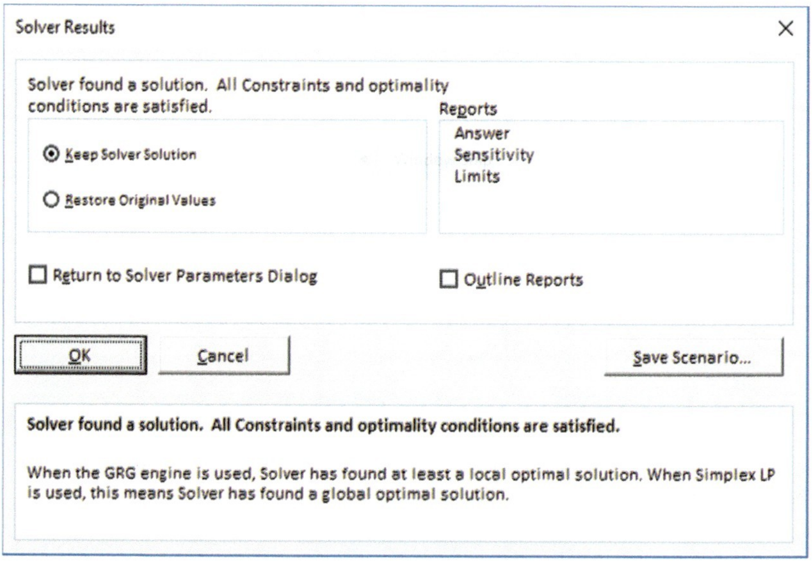

18 Optimisation Technique for the Calibrating of Creep Cavitation Model   315

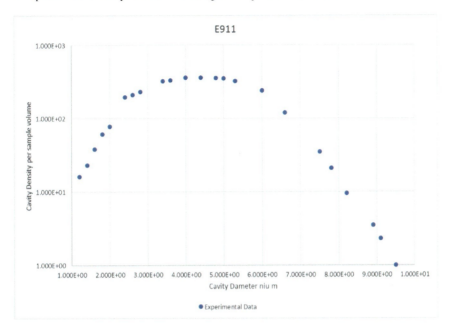

Fig. 18.2: Experimental data points of creep cavity histogram for E911, originally from [32], digitised by [24].

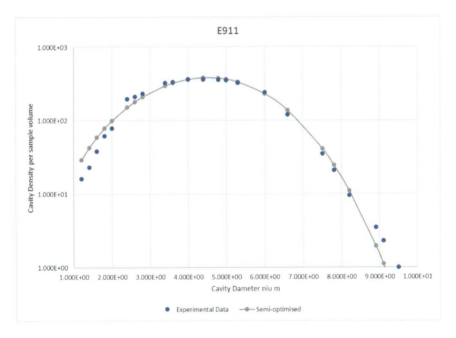

Fig. 18.3: Result of backward method from [24].

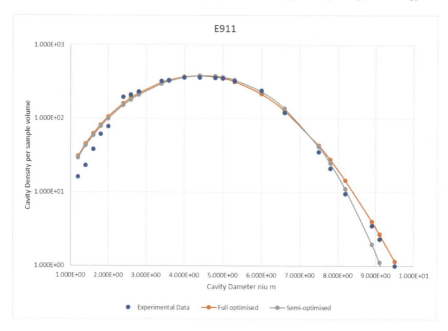

Fig. 18.4: Results with semi-optimisation and full-optimisation.

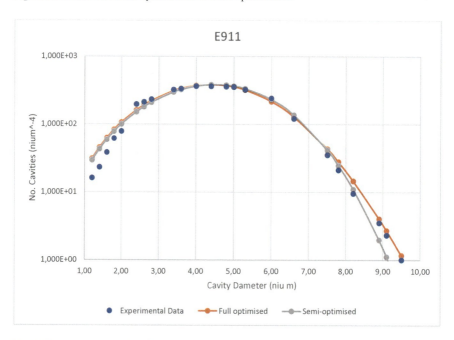

Fig. 18.5: Results with full-optimisation and semi-optimisation.

Table 18.1: The cavitation constants and the fitting accuracy (RMSE).

Material parameter	Backward Method	Semi-optimised	Full-optimised
A_1, $\mu m^3/\ln h$	8.76	13.67	15.05
A_2 (No. cavities $\mu m^{-3} h^{-1}$)	9.47E-07	1.00E-06	7.277E-03
α	1.0	1.0	1.072
β	2.0	2.472	2.498
γ	1.0	1.0	0.09811
Accuracy: RMSE	35.9255	18.0219	16.2952

18.6 Discussion and Conclusion

As far as the authors are aware of the literature, this is the first time that the method for calibrating a generic cavitation model based on cavity histogram was optimised. The following conclusions can be drawn:

1. The application of the backward method can produce a reasonable solution, see Fig. 18.3, and its success depends on the judicious choosing two representative points and the user-guided iteration in order to achieve the best fit.
2. The semi-optimisation was achieved by introducing the MSRE in the above iteration, and better results are obtained, see Fig. 18.4. The accuracy has been improved significantly as the MSRE has reduced from 35.9255 for un-optimised to 18.0219 for the semi-optimised, see Table 18.1.
3. The full-optimised via Solver in Excel has produced the best results, see Fig. 18.5, and with the lowest RMSE of 16.2952.
4. The semi-optimised method can produce very low MSRE (18.0219) which is very close to that of full-optimised (16.2952), the difference between these two methods shown in histograms is very small.
5. However, there is a noticeable difference in the values of the cavity constants, this is due to the different sensitivities of histogram on the cavitation parameters.

The recommended future work should investigate the sensitivity of the histogram on the cavitation parameters.

Acknowledgements Dr Qiang Xu is grateful for the Santander Postgraduate Mobility scheme which partially enabled Mrs Xueming Zheng to visit and present her research results to Prof Wang, Wuhan University, China, in 2017. Mr Bilal Rafiq is grateful for the School OF Computing and Engineering Department, the University of Huddersfield, and the supervisor for providing a MSc project in which he conducted the research work included in this paper.

References

1. Xu Q, Lu J, Wang X (2013) Damage modelling: the current state and the latest progress on the developing of creep damage constitutive equations for high Cr steels. In: Shibli A (ed) 6th International 'HIDA-6' Conference: Life/Defect Assessment & Failures in High Temperature Plant, ETD Ltd, UK, 6th International High-temperature Defect Assessment (HIDA) Conference : Life/Defect Assessment & Failures in High Temperature Plant, HIDA-6 Conference 02-04-12-2013
2. Xu Q, Yang X, Lu Z (2017) Damage modelling: the current state and the latest progress on the development of creep damage constitutive equations for high Cr steels. Materials at High Temperatures **34**(3):229–237
3. Xu Q, Yang X, Lu Z (2017) On the development of creep damage constitutive equations: a modified hyperbolic sine law for minimum creep strain rate and stress and creep fracture criteria based on cavity area fraction along grain boundaries. Materials at High Temperatures **34**(5-6):323–332
4. Xu Q (2004) The development of validation methodology of multi-axial creep damage constitutive equations and its application to 0.5Cr0.5Mo0.25V ferritic steel at 590°C. Nuclear Engineering and Design **228**(1):97–106, SMiRT 16. Selected and Updated Papers from the 16th International Conference on Structural Mechanics in Reactor Technology, Washington DC
5. Xu Q (2000) Development of constitutive equations for creep damage behaviour under multi-axial states of stress. In: Berveiller M (ed) Advances in Mechanical Behaviour, Plasticity and Damage: Proceedings of EUROMAT 2000, vol 2, Elsevier, pp 1375–1382
6. Xu Q (2000) Development of constitutive equations for creep damage behaviour under multi-axial states of stress. In: Structures and Materials, vol 6, Elsevier, pp 435–445, sixth International Conference on Damage and Fracture Mechanics: Computer Aided Assessment and Control, Damage and Fracture Mechanics 2000
7. Xu Q (2003) The development of phenomenological multi-axial isotropic creep damage constitutive equations. In: Mechanics and Material Engineering for Science and Experiments, pp 457–460, proceedings of the International Symposium of Young Scholars on Mechanics and Material Engineering for Science and Experiments; Changsha/Zhangjiajie
8. Xu Q (2001) Creep damage constitutive equations for multi-axial states of stress for 0.5cr0.5mo0.25v ferritic steel at 590°c. Theoretical and Applied Fracture Mechanics **36**(2):99–107
9. Xu Q, Hayhurst DR (2003) The evaluation of high-stress creep ductility for 316 stainless steel at 550 °c by extrapolation of constitutive equations derived for lower stress levels. International Journal of Pressure Vessels and Piping **80**(10):689–694
10. Xu Q, Barrans S (2003) The Development of Multi-axial Creep Damage Constitutive Equations for 0.5Cr0.5Mo0.25V Ferritic Steel at 590°C. JSME International Journal Series A **46**(1):51–59

[11] Xu Q, Wright M, Xu Q (2003) The development and validation of multi-axial creep damage constitutive equations for P91. In: Proceedings of 2011 17th International Conference on Automation and Computing, ICAC 2011, pp 177–182
[12] An L, Xu Q, Xu DL, Lu ZY (2012) Review on the current state of developing of advanced creep damage constitutive equations for high chromium alloy. In: Machinery, Materials Science and Engineering Applications, MMSE2012, Trans Tech Publications Ltd, Advanced Materials Research, vol 510, pp 776–780
[13] Xu QH, Xu Q, Pan YX, Short M (2012) Current state of developing creep damage constitutive equation for 0.5cr0.5mo0.25v ferritic steel. In: Machinery, Materials Science and Engineering Applications, MMSE2012, Trans Tech Publications Ltd, Advanced Materials Research, vol 510, pp 812–816
[14] An L, Xu Q, Xu D, Lu Z (2012) Preliminary analysing of experimental data for the development of high Cr alloy creep damage constitutive equations. In: 18th International Conference on Automation and Computing (ICAC), pp 207–211
[15] Xu QH, Xu Q, Pan YX, Short M (2013) Review of creep cavitation and rupture of low cr alloy and its weldment. In: Advances in Material Science, Mechanical Engineering and Manufacturing, Trans Tech Publications Ltd, Advanced Materials Research, vol 744, pp 407–411
[16] An L, Xu Q, Lu ZY, Xu DL (2013) Analyzing the characteristics of the cavity nucleation, growth and coalescence mechanism of 9Cr-1Mo-VNb Steel (P91) steel. In: Advances in Material Science, Mechanical Engineering and Manufacturing, Trans Tech Publications Ltd, Advanced Materials Research, vol 744, pp 412–416
[17] Liu DZ, Xu Q, Lu ZY, Xu DL, Tan F (2013) The validation of computational fe software for creep damage mechanics. In: Advances in Material Science, Mechanical Engineering and Manufacturing, Trans Tech Publications Ltd, Advanced Materials Research, vol 744, pp 205–210
[18] Yang X, Xu Q, Lu ZY, Barrans S (2014) Preliminary review of the influence of cavitation behavior in creep damage constitutive equations. In: Machinery, Materials Science and Engineering Applications 2014, Trans Tech Publications Ltd, Advanced Materials Research, vol 940, pp 46–51
[19] Gupta C, Toda H, Schlacher C, Adachi Y, Mayr P, Sommitsch C, Uesugi K, Suzuki Y, Takeuchi A, Kobayashi M (2013) Study of creep cavitation behavior in tempered martensitic steel using synchrotron micro-tomography and serial sectioning techniques. Materials Science and Engineering: A **564**:525–538
[20] Yang X (2018) The development of creep damage constitutive equations for high chromium steel based on the mechanism of cavitation damage. PhD thesis, The University of Huddersfield
[21] Xu Q, Yang X, Lu Z (2017) On the development of creep damage constitutive equations: a modified hyperbolic sine law for minimum creep strain rate and stress and creep fracture criteria based on cavity area fraction along grain boundaries. Materials at High Temperatures **34**(5-6):323–332
[22] Yang X, Lu Z, Xu Q (2015) The interpretation of experimental observation date for the development of mechanisms based creep damage constitutive equations

for high chromium steel. In: 21st International Conference on Automation and Computing (ICAC), pp 1–6
[23] Riedel H (1987) Fracture at High Temperatures. Springer, Berlin Heidelberg
[24] Zheng X, Xu Q, Lu Z, Wang X, Feng X (2020) The development of creep damage constitutive equations for high cr steel. Materials at High Temperatures **37**(2):129–138
[25] Fu GL (2020) Creep cavitation damage modeling – stress level on 316 alloy. Msc thesis, The University of Huddersfield, UK
[26] Zheng X, Xu Q, Lu Z, Wang X, Feng X (2020) The development of creep damage constitutive equations for high Cr steel. Materials at High Temperatures **37**(2):129–138
[27] Xu Q, Lu Z (2019) Modeling of creep deformation and creep fracture. In: Héctor JS, Avila JA, Chen C (eds) Strength of Materials, IntechOpen, Rijeka, chap 7
[28] Xu Q, Lu Z (2022) Recent progress on the modelling of the minimum creep strain rate and the creep cavitation damage and fracture. Materials at High Temperatures **39**(6):516–528
[29] Pola A, Soltani M, Xu Q (2014) The effect of initial estimated points on objective functions for optimization. In: Proceedings - 28th European Conference on Modelling and Simulation, ECMS 2014, pp 304–308
[30] Igual JZ, Xu Q (2015) Determination of the material constants of creep damage constitutive equations using Matlab optimization procedure. In: 21st International Conference on Automation and Computing (ICAC), pp 1–6
[31] Zheng X, Yang X, Lu Z, Xu Q (2021) The method for the determination of creep cavitation model based on cavity histogram. Materials at High Temperatures **38**(5):383–390
[32] Sket F, Dzieciol K, Borbély A, Kaysser-Pyzalla AR, Maile K, Scheck R (2010) Microtomographic investigation of damage in E911 steel after long term creep. Materials Science and Engineering: A **528**(1):103–111, special Topic Section: Local and Near Surface Structure from Diffraction
[33] Zheng XM (2021) The development and application of creep damage constitutive equations for high Cr steels over a wide range of stress. Phd thesis, The University of Huddersfield, Huddersfield, UK

Chapter 19
A Temperature-Dependent Viscoelastic Approach to the Constitutive Behavior of Semi-Crystalline Thermoplastics at Finite Deformations

Le Zhang, Bo Yin, Robert Fleischhauer, and Michael Kaliske

Abstract The contribution at hand aims at the formulation of a promising constitutive model for solids exhibiting thermo-viscoelastic characteristics. Temperature dependency and nonlinear creep properties are included into this material formulation. In general, a phenomenological constitutive formulation considering isotropic thermo-viscoelasticity at finite strains is introduced based upon a multiplicative split of the deformation gradient. The evolution equations for the inelastic deformation gradient are introduced in a thermo-dynamically consistent manner. In particular, the present approach focuses on an inelastic incompressibility condition and the principle of maximum of dissipation. The derivation starts from a well-defined HELMHOLTZ energy function, which also includes a volumetric thermal deformation. For simplicity, isotropic thermal conductivity behavior is taken into account. The set of constitutive equations is consistently linearized and incorporated into a NEWTON-type solver. The physical applicability of the present formulation is validated by a promising numerical study, which has also demonstrated favourable numerical stability and robustness.

19.1 Introduction

A special class of polymers with entangled but un-crosslinked macromolecules are thermoplastics. They show significant creep phenomena, when subjected to static mechanical loads, due to the special characteristics of their micro-structure. The

---

Le Zhang · Robert Fleischhauer · Michael Kaliske
Institute for Structural Analysis, Technische Universität Dresden, 01062 Dresden, Germany,
e-mail: le.zhang@mailbox.tu-dresden.de, robert.fleischhauer@tu-dresden.de, michael.kaliske@tu-dresden.de

Bo Yin
Ansys Germany GmbH, 99423 Weimar, Germany,
e-mail: liam.yin@ansys.com

contribution at hand aims at a thermo-dynamically consistent constitutive modeling of the material behavior of semi-crystalline thermoplastics at finite deformations. The material property, that is especially focused on, is characterized by an elastic and viscous deformation, in order to model the creep behavior of thermoplastics.

The concept of the split of the deformation gradient into volumetric and isochoric parts is applied. The isochoric part is further split into elastic and viscous contributions and the volumetric part is considered to account for thermal and elastic deformations. Based on these multiplicative kinematics, the isochoric elastic right CAUCHY-GREEN deformation tensor is introduced such that it is not influenced by change of temperature. The determinant of the volumetric part of the deformation gradient is used to account for the thermal expansion and stress-inducing volumetric elastic deformations. This kinematic approach is based on the work of [1].

The specific heat capacity of thermoplastics is incorporated into the HELMHOLTZ energy and is assumed to be a material constant. The heat flux vector is assumed to follow FOURIER's law and is a function of the thermal conductivity coefficient for the appropriate thermoplastics, compare [2]. A suitable specific formulation of the HELMHOLTZ energy is introduced, based on [3], consisting of volumetric, isochoric, thermal and latent parts. The energy formulation is used to derive the first PIOLA-KIRCHHOFF stress as well as the external power, which is used to define the change of entropy inside the thermoplastic material. Furthermore, the energy is used to specify the dissipative behavior of the material for considering a change of mechanical into thermal energetic parts, compare [4].

The viscous part of the deformation gradient is driven by its thermodynamic consistent evolution equation. This evolution equation is based on a constitutive viscous flow potential with respect to the viscous intermediate configuration and the respective MANDEL stress. The second internal variable, the hardening strain is driven by the latent part of the HELMHOLTZ energy and its thermodynamic consistent evolution equation. The presented contributions are based on the developments in [3].

All constitutive descriptions and developments are incorporated into a two-field global finite element solver, considering the balance laws of non-linear thermo-inelasticity at finite deformations [2], with respect to the reference configuration. The NEWTON-type solver is based on the consistently linearized field equations for the displacement and the temperature field, which form the global unknown fields. The implicit function theorem is applied to consider the change of these global unknowns, due to a change of the local internal variables, namely the viscous part of the deformation gradient and the hardening strains. All computational developments and constitutive descriptions are successfully validated by a representative numerical study, where the experimental creep data for polyoxymethylene (POM) at 60°C, taken from [5], is used to identify the introduced material constants.

The framework of this paper is outlined as follows. In Sect. 19.2, the preliminaries of the finite thermo-viscoelasticity are introduced, which pay particular attentions to the basic kinematics and the multiplicative decomposition of the total deformation gradient. In Sect. 19.3, a brief overview of the constitutive framework is summarized, including the HELMHOLTZ free energy and the nonlinear creep law. In the sequel, a representative numerical example is studied in Sect. 19.4, consisting of both stress-controlled and strain-controlled loading conditions. Sect. 19.5 closes the paper by summarizing the present work and proposing future perspectives.

19.2 Preliminaries of the Finite Thermo-Viscoelasticity

This section depicts the fundamental theoretical background of the constitutive framework of the isotropic thermo-viscoelasticity. To classify the deformation process, let $\mathcal{B}_0$ be the solid body in the reference configuration as a subset of the EUCLIDEAN space $\mathcal{B}_0 \subset \mathbb{R}^3$ at $t_0 \in \mathcal{T} \mid \mathcal{T} \subset \mathbb{R}_+$. For each material point P_{t_0} of $\mathcal{B}_0$, its position vector is $X \in \mathcal{B}_0$. At time $t \in \mathcal{T}$, the current configuration is denoted as $\mathcal{B}_t$ and the corresponding material point P_t has a position vector $x \in \mathcal{B}_t$. The mapping $\varphi_t : \mathcal{B}_0 \times \mathcal{T} \to \mathbb{R}^3$ denotes the motion of the solid domain at the time interval $\mathcal{T}$. The motion φ_t is a non-linear and bijective mapping, reading $\varphi_t : X \mapsto x = \varphi_t(X)$. The deformation gradient F is now defined as

$$F(X) := \operatorname{Grad}(\varphi_t(X)) = \frac{\partial \varphi_t(X)}{\partial X} = g_i \otimes G^i \quad (19.1)$$

having a determinant $J(X) := \det(F)$. The basis vectors G_i and g_i are defined with respect to the reference and the current configuration, respectively.

The absolute temperature for a point of $\mathcal{B}_t$ is denoted as $\theta \geq 0\,\mathrm{K}$. Furthermore, it is common to denote the reference temperature for a material point of $\mathcal{B}_0$ as θ_0 and define the change of temperature ϑ as

$$\vartheta := \theta - \theta_0. \quad (19.2)$$

As a convention, the temperature is transferred into units of Kelvin [K] instead of Celsius [°C] or Fahrenheit [°F], which naturally enables that temperature parts of the HELMHOLTZ energy can be introduced by a logarithmic form due to the strict positiveness.

For a finite thermo-viscoelastic formulation, the deformation gradient F is split into the thermal and viscoelastic parts

$$F = F^\vartheta F^{ve}. \quad (19.3)$$

The thermal part of the deformation is constitutively assumed to be of purely volumetric nature, so that an increase of temperature ensures an increase of the volume of the solid body and vice versa. Here, the volumetric thermal expansion is modeled by

$$F^\vartheta = \left(J^\vartheta\right)^{\frac{1}{3}} \mathbf{1}, \quad \text{where} \quad J^\vartheta = \exp(3\alpha_t \vartheta), \quad (19.4)$$

where $\mathbf{1}$ is the identity tensor and α_t denotes the isotropic thermal expansion coefficient. Based on [4], herein an exponential form of the thermal expansion is introduced.

The non-thermal part

$$F^{ev} = (J^{ev})^{\frac{1}{3}} \bar{F}^{ev}$$

can also be further split based on a volume preserving formulation, namely

$$\bar{F}^{ev} = \bar{F}^e \bar{F}^v \qquad (19.5)$$

for the elastic and viscous parts; see e.g. [1, 6–10]. Therefore, the determinant of deformation gradient J satisfies $J = J^\vartheta J^{ev}$.

The viscous part of the deformation $\bar{F}^v$ describes the irreversible and inelastic part of the total isochoric deformation. This portion can evolve e.g. due to the creep law, leading to micro-structural rearrangements of the material and, thus, to dissipation and temperature changes. Once $\bar{F}^v$ evolves, the incompressibility condition

$$\det\left(\bar{F}^v\right) = J^v = 1 \ \forall \ \bar{F}^v \qquad (19.6)$$

has to be ensured for most of the materials, especially for metals, see [11]. The time derivative of inelastic deformation gradient is defined by

$$\dot{\bar{F}}^v = \bar{L}^v \bar{F}^v , \qquad (19.7)$$

where $\bar{L}^v$ is the rate of deformation. The elastic part is defined by

$$\bar{F}^e = \bar{F}^{ev} \bar{F}^{v^{-1}} . \qquad (19.8)$$

Using Eq. (19.8), the deformation measure

$$\bar{C}^e = \bar{F}^{e^T} g \bar{F}^e = \bar{F}^{v^{-T}} \bar{F}^T g \bar{F} \bar{F}^{v^{-1}} \qquad (19.9)$$

is introduced as a function of the current metric tensor g, which represents a key kinematic quantity for defining specific constitutive equations.

Furthermore, the heat flow q_n out of the surface $\partial \mathcal{B}_t$ of the current configuration can be expressed as $q_n =: \boldsymbol{q} \cdot \boldsymbol{n}$, where $\boldsymbol{q}$ denotes the spatial heat flux through a point $\boldsymbol{x} \in \partial \mathcal{B}_t$ and $\boldsymbol{n}$ is the current outward normal at the observed point, see Fig. 19.1. The spatial heat flux vector $\boldsymbol{q}$ at point $\boldsymbol{x} \in \mathcal{B}_t$, describing the heat conduction inside of the solid body, is assumed to follow FOURIER's law

$$\boldsymbol{q} = -\frac{k}{J} \text{grad}(\vartheta) , \qquad (19.10)$$

where k is a material constant. Eq. (19.10) sufficiently describes the spatial heat conduction phenomena by the current state of the temperature gradient $\text{grad}(\vartheta)$. Additionally, the conductive dissipation

$$\mathcal{D}_{con} := -\frac{1}{\theta} \boldsymbol{q} \cdot \text{grad}(\vartheta)$$

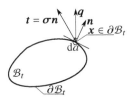

Fig. 19.1 Schematic depiction of the boundary $\partial \mathcal{B}_t$ of the current configuration and its normal n at point x as well as tractions t and heat flux q

is also fulfilled.

19.3 Constitutive Formulation of Finite Thermo-Viscoelasticity

19.3.1 Helmholtz Energy

In order to define the isotropic thermo-viscoelastic material in a systematic manner, the HELMHOLTZ energy ψ is particularly defined herein, reading

$$\psi = \psi_{vol}^e + \psi_{iso}^e + \psi^\vartheta , \tag{19.11}$$

where ψ_{vol}^e denotes the volumetric part from both the elastic deformation and the temperature changes. ψ_{iso}^e represents the isochoric energy density. Additionally, a pure thermal contribution ψ^ϑ captures the change of free energy due to any change of temperature. The volumetric and isochoric part of Eq. (19.11) take the definition of [12, 13], reading

$$\rho_0 \psi_{vol}^e (J, J^\vartheta) = \frac{K}{2} \cdot \ln(J^{ev})^2, \quad \text{and} \quad \rho_0 \psi_{iso}^e (\bar{C}^e) = \frac{\mu}{2}(\bar{I}_1 - 3), \tag{19.12}$$

respectively, noting, $\bar{I}_1 = \mathrm{tr}\,(\bar{C}^e) = \delta_{ij}\bar{C}_{ij}^e$. The thermal part $\rho_0\psi^\vartheta = \rho_0\psi^\vartheta(\vartheta)$ is defined as

$$\rho_0 \psi^\vartheta(\vartheta) = -\rho_0 c \left(\theta \ln\left(\frac{\theta}{\theta_0}\right) - \vartheta \right) + C(\vartheta) . \tag{19.13}$$

The corrector function $C(\vartheta)$ is required in order to ensure the assumed constant heat capacity at constant deformation. The specific heat capacity c is defined by

$$\rho_0 c := -\theta \frac{\partial^2 \rho_0 \psi}{\partial \vartheta \partial \vartheta} = -\theta \frac{\partial^2 \rho_0 \psi_{vol}^e(J, J^\vartheta)}{\partial \vartheta \partial \vartheta} - \theta \frac{\partial^2 \rho_0 \psi^\vartheta(\vartheta)}{\partial \vartheta \partial \vartheta} \tag{19.14}$$

and determines the temperature-dependent part of the HELMHOLTZ energy by

$$\frac{\partial^2 \rho_0 \psi^\vartheta(\vartheta)}{\partial \vartheta \partial \vartheta} = -\frac{\partial^2 \rho_0 \psi_{vol}^e(J, J^\vartheta)}{\partial \vartheta \partial \vartheta} - \frac{\rho_0 c}{\theta} . \tag{19.15}$$

An integration over the temperature domain together with the use of the constraints $\partial_\theta \psi^\vartheta(\vartheta=0) := 0$ and $\partial_{\theta\theta}^2 \psi^\vartheta(\vartheta=0) := 0$ as well as considering $\theta = \theta_0 \; \forall \; \vartheta = 0$ would yield the specific formulation of ψ^ϑ and $C(\vartheta)$, respectively.

Following [14], the Cauchy stress tensor σ and the KIRCHHOFF stress tensor τ are given by

$$\sigma = 2\rho \frac{\partial \psi}{\partial g} = \frac{2}{J} \frac{\partial \rho_0 \psi}{\partial g} = \frac{1}{J} \tau, \quad (19.16)$$

where ρ is the current density and g the current metric tensor. According to Eq. (19.11), the CAUCHY stresses are additively split into volumetric and isochoric parts, reading

$$\sigma_{\text{vol}} = \frac{1}{J} \left[\kappa \ln(J^{ev}) g^{-1} \right], \quad \text{and} \quad \sigma_{\text{iso}} = 2\mu J \frac{\partial \bar{I}_1}{\partial g}. \quad (19.17)$$

The external power

$$\rho w_{\text{ext}} := \rho \theta \left(2 \frac{\partial^2 \psi}{\partial g \partial \vartheta} \right) : d = \theta \frac{\partial \sigma}{\partial \vartheta} : d = \theta \frac{\partial \sigma}{\partial \vartheta} : \text{sym} l, \quad (19.18)$$

where $l = \dot{F} F^{-1}$ is the spatial velocity gradient and d its symmetric part, influences the change of entropy per time. After the specification of the HELMHOLTZ energy, ρw_{ext} takes the form

$$\rho w_{\text{ext}} = \frac{\kappa \theta}{J} \left[\frac{1}{J^{ev}} \frac{\partial J^{ev}}{\partial \vartheta} \right] g^{-1} : \text{sym} l \quad (19.19)$$

and represents the amount of energy that changes the temperature at x, due to an arbitrarily applied deformation rate. This part of the entropy change is a function of the volumetric energy, since temperature changes only affect the changes in volumetric deformation of most materials.

The internal power is expressed as

$$\rho w_{\text{int}} := \rho \left(\frac{\partial \psi}{\partial \bar{F}^v} - \theta \frac{\partial^2 \psi}{\partial \bar{F}^v \partial \theta} \right) : \dot{\bar{F}}^v, \quad (19.20)$$

which models any change of temperature at $x \in \mathcal{B}_t$ due to the evolution of internal variables in an irreversible manner, whenever $\dot{\bar{F}}^v \neq 0$. Eq. (19.19) captures the reversible change of temperature at any deformation rate, while Eq. (19.20) can be interpreted as an underlying ground state temperature change response. If $\dot{\bar{F}}^v = 0$ and $\dot{F} \neq 0$, an entropic cooling at tension and heating at compression is present at $x \in \mathcal{B}_t$ and, if $\dot{\bar{F}}^v \neq 0$ and $\dot{F} \neq 0$, the temperature is increased as time elapses.

19.3.2 Creep Law

Having Eq. (19.7) at hand, the viscous evolution operator $\bar{L}^v$ related to the viscous intermediate configuration, needs to be defined. The standard arguments for defining work conjugates at the intermediate configuration are the positiveness of the internal dissipation, see e.g. [15]. If $\rho_0 \psi^e_{iso}(\bar{C}^e)$ represents the change of energy stored due to viscoelastic loading, the internal dissipation can be defined by

$$\mathcal{D}_{int} := \mathcal{P} - \rho_0 \dot{\psi}^e_{iso}(\bar{C}^e) \geq 0, \qquad (19.21)$$

where no temperature change is assumed while viscous evolution. The stress power $\mathcal{P} := \bar{\Sigma} : \bar{L}$, as a function of the MANDEL stress

$$\bar{\Sigma} = 2\bar{C}^e \frac{\partial \rho_0 \psi^e_{iso}(\bar{C}^e)}{\partial \bar{C}^e} \qquad (19.22)$$

with respect to the intermediate configuration, is introduced, see [1]. The total rate of deformation $\bar{L} = \bar{F}^e \bar{l} \bar{F}^{e^{-1}}$ related to the intermediate configuration can be split into

$$\bar{L} = \bar{L}^e + \bar{L}^v, \qquad (19.23)$$

where $\bar{L}^e = \bar{F}^{e^{-1}} \dot{\bar{F}}^e$, compared to [6]. A further evaluation of Eq. (19.21) yields

$$\rho_0 \dot{\psi}^e_{iso}(\bar{C}^e) = \frac{\partial \rho_0 \psi^e_{iso}(\bar{C}^e)}{\partial \bar{C}^e} : \dot{\bar{C}}^e = \left[2\bar{C}^e \frac{\partial \rho_0 \psi^e_{iso}(\bar{C}^e)}{\partial \bar{C}^e} \right] : \bar{L}^e, \qquad (19.24)$$

for the inelastic and isochoric part of the HELMHOLTZ energy. Inserting Eq. (19.23) and Eq. (19.24) into Eq. (19.21) and applying the standard arguments for the strict positiveness of the internal dissipation leads to the constitutive description of $\bar{\Sigma}$ (compare Eq. (19.22)), reading

$$\mathcal{D}_{int} := \left[\bar{\Sigma} - 2\bar{C}^e \frac{\partial \rho_0 \psi^e_{iso}(\bar{C}^e)}{\partial \bar{C}^e} \right] : \bar{L} + \left[2\bar{C}^e \frac{\partial \rho_0 \psi^e_{iso}(\bar{C}^e)}{\partial \bar{C}^e} \right] : \bar{L}^v \geq 0. \qquad (19.25)$$

The following constitutive definition for $\bar{L}^v$ is introduced, reading

$$\bar{L}^v := \dot{\gamma}^v N = \dot{\gamma}^v \frac{\bar{\Sigma}}{\|\bar{\Sigma}\|}. \qquad (19.26)$$

Compare [16], the creep flow is defined as

$$\dot{\gamma}^v = \dot{\gamma}_0 \left[\exp\left(\left(\frac{\|\bar{\Sigma}\|}{s_0} \right)^n \right) - 1 \right], \qquad (19.27)$$

where $\dot{\gamma}_0$ denotes the pre-exponential shear strain rate factor. n and s_0 are another two material parameters. Here, both of them are given as constant coefficients. It ensures the inequality of internal dissipation, see Eq. (19.25). Recalling Eq. (19.7), the time integration is given by

$$\dot{\bar{F}}^v = [\dot{\gamma}^v N] \bar{F}^v .\tag{19.28}$$

Within a standard time discretization, the implicit update algorithm

$$\bar{F}^v_{n+1} = \exp\left[\Delta \gamma^v_{n+1} N_{n+1}\right] \bar{F}^v_n ,\tag{19.29}$$

is applied for a time increment $\{t_n, t_{n+1}\} \in \mathbb{R}$ as an approximation.

19.3.3 Governing Equations

This section briefly summarizes the spatial formulation of the driving partial differential equations (PDE) for evolving the global nodal unknowns, i.e. temperature, displacements, velocities and accelerations. The local forms, where local means the validity at any $x \in \mathcal{B}_t$, are given by

$$\rho \ddot{u} = \rho b + \text{div}(\sigma) ,\tag{19.30}$$

$$\rho c \dot{\theta} = -\text{div}(q) + \rho \left(\theta \cdot 2 \frac{\partial^2 \psi}{\partial g \partial \theta}\right) : d - \rho \left(\frac{\partial \psi}{\partial \bar{F}^v} - \theta \frac{\partial^2 \psi}{\partial \bar{F}^v \partial \theta}\right) : \dot{\bar{F}}_v \tag{19.31}$$

$$= -\text{div}(q) + \rho w_{\text{ext}} - \rho w_{\text{int}} .\tag{19.32}$$

The focus of the contribution at hand is a consistent solution of the path dependent problem by use of a NEWTON-type solver considering initial condition at t_0 for all $X \in \mathcal{B}_0$ and boundary conditions for all $x \in \mathcal{B}_t$ and $X \in \mathcal{B}_0$. The boundary of B is divided into $\partial B = \{\partial B_u , \partial B_\theta\} = \{\partial B_\sigma , \partial B_q\}$, compare Fig. 19.1, with prescribed displacements, temperatures or tractions and heat flows for each of the configurations.

19.4 Numerical Study

In this section, a numerical example is conducted in comparison with the related experimental investigation, in order to intuitively demonstrate the capability of the present model. Based on [5], a flat specimen of polyoxymethylene (POM) with stress-controlled load and strain-controlled load is modeled, respectively. The material constants are shown in Table 19.1 The specimen geometry is depicted in Fig. 19.2 (a) and the finite element discretization is shown in Fig. 19.2 (b).

Table 19.1: Material parameters.

κ	μ	α_t	k_t	$\rho_0 c$	θ_0	$\dot{\gamma}_0$	s_0	n
2000 MPa	110 MPa	$2e^{-3}$	0.3 N/s·K	2 N/(mm²·K)	333 K	$7.5e^{-9}$	12.5 MPa	2.37

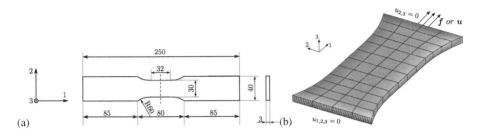

Fig. 19.2: (a) geometry of the flat specimen with all measures in mm; (b) setup for finite element model.

19.4.1 Stress-Controlled loading

In the stress-controlled loading simulation, a set of nominal stresses $\sigma = \{27.5, 30, 33, 35\}$ MPa is considered, which can be mimicked alternatively by applying a force f to yield a same level of stress in the cross-section of the central region in the specimen, see Fig. 19.2 (b). The longitudinal strain $\varepsilon_1 = \ln(l/l_0)$ and lateral strain $\varepsilon_2 = \ln(\omega/\omega_0)$ in a logarithmic formulation are two essential parameters for analyzing the experimental outcomes. Hence, in the numerical evaluation, a similar behavior is obtained for an effective comparison.

In comparison with experimental outcomes from [5], the simulation results by the present model based on a stress-controlled loading are validated in Fig. 19.3 with respect to all the stress states $\{27.5, 30, 33, 35\}$ MPa. It can be evidently seen that the numerical prediction shows a good agreement to the experimental investigation for all setups. Furthermore, the model also captures the characteristic that the strain growth rate increases obviously with the larger nominal stress application. Therefore, the given creep law in Eq. (19.27) is demonstrated to have a good performance in fitting the overall experimental results.

Furthermore, the deformation process with the nominal stress $\sigma = 33$ MPa is also shown in Fig. 19.4 for a straightforward visualization of the creep behavior. Considering the original length 80 mm, the creep deformation is obviously observed as time increases, e.g., it reaches more than 100 mm at $t = 1.78e^4$ s. In addition, the temperature evolution along the time is investigated. Figure 19.5 depicts the temperature change ϑ in the central point of the specimen along the time increase.

Moreover, another interesting comparison about temperature evolution is shown in Fig. 19.6, which describes the temperature distribution for all nominal stress applications but at the same specimen elongation state.

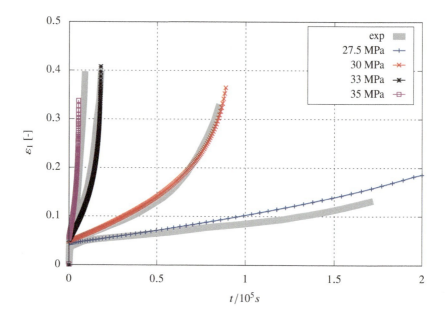

Fig. 19.3: Model results under stress-controlled loading.

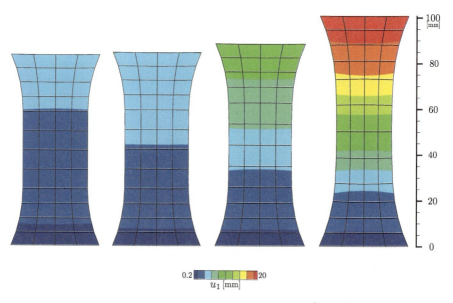

Fig. 19.4: Creep deformation of the numerical model at $t = [10, 4.3\mathrm{e}^3, 1.12\mathrm{e}^4, 1.78\mathrm{e}^4]$ s with $\sigma = 33$ MPa.

19 A Temperature-Dependent Viscoelastic Approach to the Constitutive Behavior

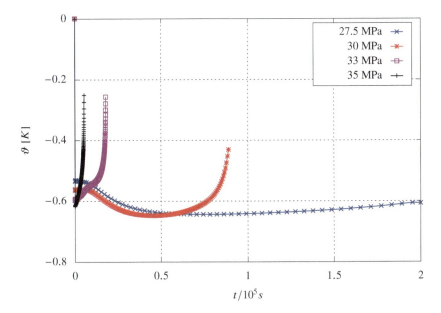

Fig. 19.5: Temperature course in the central point of specimen under stress-controlled loading.

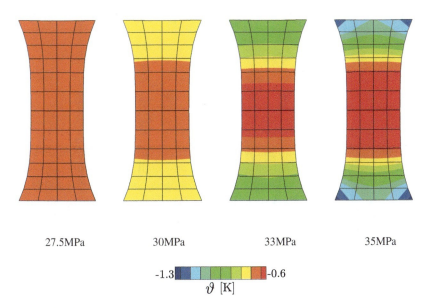

Fig. 19.6: Temperature distribution of the flat specimen with the stain state $\varepsilon_1 = 0.185$ under stress-controlled loading.

19.4.2 Strain-Controlled loading

The strain-controlled tension test is experimentally conducted by SCHLEGEL & BEINER. For the detail of experimental setup, one is referred to [5, 17]. As shown in Fig. 19.2, the same material constants, the same specimen geometry and finite element model are used with the only difference being the load application. The applied strain rate is determined from the free length of the specimen between the clamps. Herein, a machine speed of $v = 0.8$ mm/min for a global strain rate $\dot{\varepsilon} = 0.01$ min^{-1} and $v = 8$ mm/min for $\dot{\varepsilon} = 0.1$ min^{-1} approximation are used. With respect to the experiment, results in the following are presented with the nominal strain definition $\varepsilon_1 = l/l_0 - 1$ for the longitudinal strain and $\varepsilon_2 = w/w_0 - 1$ for the lateral strain. Nominal stress σ is again given by measured force divided by the initial size of the middle cross-section of the specimen.

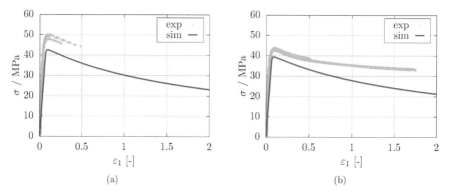

Fig. 19.7: Longitudinal results for tension tests with strain rates: (a) $\dot{\varepsilon} = 0.1$min^{-1}; (b) $\dot{\varepsilon} = 0.01$min^{-1}.

Results from the tension tests with two different strain rates are illustrated in Fig. 19.7. Experiments show an initially stiff response until a nominal stress maximum σ_{crit} is reached. In the sequel, significant softening is observed until rupture. The simulation results match the experimental results during the initially stiff phase. Nevertheless, the peak stresses are unfortunately not appropriately predicted, which yields an underestimation of the peak stress values. The possible reasons behind it can be from the material constants fitting, as well as the measurement deviations. But the good side is that the softening trends for both cases are also similar to the experimental results. Therefore, further efforts, from both the experimental measurements and model fitting aspects, are certainly required to overcome such imperfect predictions.

19.5 Summary

The developments presented in this contribution successfully introduces a stable and robust modeling approach to simulate a class of thermo-viscoelastic materials at finite deformations. This is achieved by a consistent numerical framework, which is introduced with respect to the linearization of the global and local driving evolution equations. The constitutive descriptions, such as the presented HELMHOLTZ energy or the creep law, are examples. Herein, the applicability of the aforementioned formulations is demonstrated mainly for polymeric materials. Nevertheless, it can be easily exchanged to other materials, e.g., when thermo-plastic materials are required to be modeled. Further physical validation with respect to temperature changes and larger deformations of other classes of inelastic materials or composites is certainly possible. Additionally, other multi-physical phenomena can also be properly addressed by considering further multiplicative splits of the deformation gradient.

References

[1] Miehe C (1992) Kanonische Modelle multiplikativer Elasto-Plastizität. Thermodynamische Formulierung und numerische Implementation. Habilitation thesis, Universität Hannover
[2] Fleischhauer R, Platen J, Kato J, Terada K, Kaliske M (2022 (under review)) A finite anisotropic thermo-elasto-plastic modeling approach to additive manufactured specimens. Engeneering Computations
[3] Zerbe P, Schneider B, Moosbrugger E, Kaliske M (2017) A viscoelastic-viscoplastic-damage model for creep and recovery of a semicrystalline thermoplastic. International Journal of Solids and Structures **110**:340–350
[4] Miehe C (1988) Zur numerischen Behandlung thermomechanischer Prozesse. Phd thesis, Universität Hannover
[5] Zerbe P (2017) Constitutive modeling of a semicrystalline thermoplastic under cyclic creep loading - Konstitutive Modellierung eines teilkristallinen Thermoplast. Phd thesis, Technische Universität Dresden
[6] Lee EH (1969) Elastic-Plastic Deformation at Finite Strains. Journal of Applied Mechanics **36**:1–6
[7] Moran B, Ortiz M, Shih CF (1990) Formulation of implicit finite element methods for multiplicative finite deformation plasticity. International Journal for Numerical Methods in Engineering **29**:483–514
[8] Lubarda VA (2004) Constitutive theories based on the multiplicative decomposition of deformation gradient: Thermoelasticity, elastoplasticity, and biomechanics. Applied Mechanics Reviews **57**:95–108
[9] Yin B, Kaliske M (2020) Fracture simulation of viscoelastic polymers by the phase-field method. Computational Mechanics **65**:293–309
[10] Yin B (2022) Phase-field fracture description on elastic and inelastic materials at finite strains. Phd thesis, Technische Universität Dresden

[11] Simo JC (1998) Numerical analysis and simulation of plasticity. Handbook of Numerical Analysis **6**:183–499
[12] Kaliske M (2000) A formulation of elasticity and viscoelasticity for fibre reinforced material at small and finite strains. Computer Methods in Applied Mechanics and Engineering **185**:225–243
[13] Fleischhauer R, Thomas T, Kato J, Terada K, Kaliske M (2020) Finite thermoelastic decoupled two-scale analysis. International Journal for Numerical Methods in Engineering **121**:355–392
[14] Hughes TJR, Marsden JE (1983) Mathematical Foundations of Elasticity. Dover Publications, Inc., New York
[15] Truesdell C, Noll W (2004) The non-linear field theories of mechanics. Springer Berlin, Heidelberg
[16] Boyce MC, Parks DM, Argon AS (1988) Large inelastic deformation of glassy polymers. Part I: rate dependent constitutive model. Mechanics of Materials **7**:15–33
[17] Schlegel R, Beiner M (2015) Zyklische Versuche und Zugsversuche an POM mit optischer Dehnungsmessung. Report, Fraunhofer-Institut für Mikrostruktur von Werkstoffen und Systemen IWMS, Halle (Saale)

Printed by Printforce, the Netherlands